NUMERICAL METHODS FOR ENGINEERS AND SCIENTISTS

A SERIES OF PROGRAMMES ON MATHEMATICS
FOR
ENGINEERS AND SCIENTISTS

CONSULTANT EDITOR

A.C. BAJPAI

PROFESSOR OF MATHEMATICAL EDUCATION AND DIRECTOR OF CAMET

Titles already available in this series:

FORTRAN AND ALGOL	A.C. Bajpai, H.W. Pakes, R.J. Clarke, J.M. Doubleday and T.J. Stevens
MATHEMATICS FOR ENGINEERS AND SCIENTISTS – *Volume 1*	A.C. Bajpai, I.M. Calus, J.A. Fairley
MATHEMATICS FOR ENGINEERS AND SCIENTISTS – *Volume 2*	A.C. Bajpai, I.M. Calus, J.A. Fairley, D. Walker
NUMERICAL METHODS FOR ENGINEERS AND SCIENTISTS	A.C. Bajpai, I.M. Calus, J.A. Fairley
STATISTICAL METHODS FOR ENGINEERS AND SCIENTISTS	A.C. Bajpai, I.M. Calus, J.A. Fairley

CAMET

(CENTRE FOR ADVANCEMENT OF MATHEMATICAL EDUCATION IN TECHNOLOGY)

DEPARTMENT OF ENGINEERING MATHEMATICS

LOUGHBOROUGH UNIVERSITY OF TECHNOLOGY

NUMERICAL METHODS FOR ENGINEERS AND SCIENTISTS

A STUDENTS' COURSE BOOK

A. C. BAJPAI

I. M. CALUS

J. A. FAIRLEY

LOUGHBOROUGH UNIVERSITY OF TECHNOLOGY

JOHN WILEY & SONS

Chichester · New York · Brisbane · Toronto

Printed and bound in Great Britain

PREFACE

This is a volume of programmes on numerical methods which is part of a
course written for undergraduate science and engineering students in
universities, polytechnics and other colleges in all parts of the world.
Numerical methods are now included in the syllabuses for all such students
and this book covers most of the work that these students are likely to
require.   The emphasis is on the practical side of the subject and the
more theoretical aspects have been omitted.   Numerical methods are, of
course, closely linked to the use of the computer and several references
will be found as to the suitability of various methods for programming
on to a computer.   As different programming languages are in use, the
various techniques discussed have not, with one exception, been translated
into computer programs, but, wherever appropriate, flow diagrams have been
incorporated into the text.   References have, however, been given to
other books in which typical computer programs can be found.   A list of
these references appears on page 377.

When one is using numerical methods as a tool, the majority of the
calculations would be done on a computer, as, except in a few simple cases,
the amount of arithmetic involved is far too complicated to do any other
way.   However, when learning the subject there is nothing, in general, to
be gained by taking very complicated examples or by carrying working
through to a very large number of significant figures.   Doing relatively
simple examples manually does give the student an appreciation of what is
involved and so the actual numerical working of many examples has been
included.

If the reader has access to a calculating aid, such as a pocket
calculator, it will be found very helpful.   However, if not, it will
often mean that the working can only be done to fewer significant figures
or decimal places than is indicated.   It is suggested, therefore, that,
irrespective of the number of decimal places asked for, working is only
done to such a number as can conveniently be obtained.   Any points that
consequently might not be obvious can still be followed from the working
in the text.

The volume comprises three Units in which are grouped programmes on allied
topics.   Before reading a programme, the student should be familiar with
the items listed under the heading of Pre-requisites at the beginning of
each Unit.   The programmed method of presentation has been used throughout
and has many advantages.   The development of the subject proceeds in
carefully sequenced steps, the student working through these at his own
pace.   The active participation of the student is required in many places
where he or she is asked to answer a question or to solve, either
partially or completely, a problem.   The answers to these are always
given so that the student can check his attempt and thus obtain a
continuous assessment of his understanding of the subject.   Explanation
of the material covered is given in greater detail than is often to be
found in conventional style textbooks, especially at those points where
difficulties are most likely to occur.

In places where units are involved, the S.I. system has been used.   The
standard practice of using italic letters for quantities, e.g., C for
capacitance, has not, however, been followed as italic lettering is used
for the answer frames.   Where natural logarithms occur, the notation
ln is used.

Where references are made to frames in the same programme, only the frame numbers are given. Where page numbers are quoted as well, this indicates that references are to frames in another programme.

Several references are made to our volumes on Mathematics for Engineers and Scientists. The details of these books are:

Mathematics for Engineers and Scientists, Volume 1.
A.C. Bajpai, I.M. Calus, J.A. Fairley   Wiley, 1973.

Mathematics for Engineers and Scientists, Volume 2.
A.C. Bajpai, I.M. Calus, J.A. Fairley, D. Walker   Wiley, 1973.

We have also referred to

Fortran and Algol
A.C. Bajpai, H.W. Pakes, R.J. Clarke, J.M. Doubleday, T.J. Stevens
                                                         Wiley, 1972.

In spite of careful checking by the authors, it is possible that the occasional error has crept through. They would appreciate receiving information about any such mistakes which might be discovered.

A debt of gratitude to the following is acknowledged with pleasure:

Loughborough University of Technology for supporting this venture.

Staff and students of the university and other institutions for their help in various ways.

Mrs. Barbara Bell for preparing the camera-ready copy from which the book has been printed.

Taylor & Francis Ltd for their help and cooperation.

The *University of London* and the *Council of Engineering Institutions* for permission to use questions from their past examination papers. These are denoted by L.U. and C.E.I. respectively.

John Wiley and Sons Ltd for their help and cooperation.

INSTRUCTIONS

Each programme is divided up into a number of FRAMES
which are to be worked *in the order given*.   You will
be required to participate in many of these frames and
in such cases the answers are provided in ANSWER
FRAMES, designated by the letter A following the
frame number.   Steps in the working are given where
this is considered helpful.   The answer frame is
separated from the main frame by a line of asterisks:
********.   Keep the answers covered until you have
written your own response.   If your answer is wrong,
go back and try to see why.   Do not proceed to the
next frame until you have corrected any mistakes in
your attempt and are satisfied that you understand
the contents up to this point.

Suggestion to the Reader

It is strongly recommended that you make use of a pocket
calculator to help you with the arithmetic involved in
the examples.

CONTENTS

UNIT 1 - Equations and Matrices

UNIT 2 - Finite Differences and their Applications

LEAST SQUARES

FINITE DIFFERENCES

INTERPOLATION

NUMERICAL DIFFERENTIATION

NUMERICAL INTEGRATION

UNIT 3 - Differential Equations

FIRST ORDER ORDINARY DIFFERENTIAL EQUATIONS

UNIT 1
EQUATIONS and MATRICES

This Unit comprises four programmes:

    (a)   Basic Ideas, Errors and Evaluation of Formulae

    (b)   Solution of non-Linear Equations

    (c)   Simultaneous Linear Equations

    (d)   Matrices

Before reading these programmes, it is necessary that you are familiar with the following

Prerequisites

For (a):   Differentiation, including the definition of a derivative in terms of a limit.

For (b):   The binomial theorem, differentiation and Taylor's series for the main programme.

            Maxima and Minima for APPENDIX A.

            Partial differentiation and differentials for APPENDIX B.

            Notation of determinants and algebra of complex numbers for APPENDIX C.

            Taylor's series in two dimensions for APPENDIX D.

For (c):   Evaluation of Determinants, matrix notation for linear simultaneous equations, including the augmented matrix, for the main programme.

            Partial differentiation and differentials, the properties of inequalities of absolute values for the APPENDIX.

For (d):   The algebra of matrices. The meaning of eigenvalues and eigenvectors and the analytical method of their determination.

Basic Ideas, Errors and Evaluation of Formulae

Why Numerical Methods?

So far in your mathematics course, you have probably concentrated mainly on analytical techniques. Thus it is likely that you know how to find, for example,

$$\frac{d}{dx} \sin^2 3x \quad \text{and} \quad \int_0^{\frac{1}{2}\pi} x \cos 2x \, dx,$$

and also how to solve an equation such as

$$\frac{d^2 y}{dx^2} - 2 \frac{dy}{dx} - 3y = e^x$$

Again, you have probably met determinant and matrix methods for solving a set of simultaneous linear equations. So your reaction on encountering a book such as this may very well be – Why Numerical Methods? – or, perhaps, the even more fundamental question – What are Numerical Methods?

When studying, for example, integration, you learn many techniques for integrating a variety of functions. Some of these techniques are integration by substitution, integration by partial fractions, integration by parts, etc. But whatever methods you learn, there are still many functions that you just cannot integrate. Two examples of such functions are e^{-x^2} and $\sin \sqrt{x}$. Again, when dealing with differential equations, anything slightly different from one of the few standard types of equation can lead to a situation whose solution is extremely difficult or even impossible by standard techniques. So you will see that we are sometimes very restricted in what we can do by purely analytical methods. However, don't get the impression that all your troubles will be over when you have finished this book. Some of them will be – for example, you will know how to find $\int_0^{0.2} e^{-x^2} dx$ but you will be no nearer to finding the indefinite integral $\int e^{-x^2} dx$.

Turning for a moment to the solution of simultaneous linear equations, the use of Cramer's rule or of the formula $\frac{1}{\det A}$ adj A for A^{-1} does not present much trouble if, say, you have to solve three equations in three unknowns. However, if you have to solve fifty equations in fifty unknowns, such as can occur when dealing with space frames which are used in roof trusses, bridge trusses, pylons, etc., you are going to require some help with the arithmetic and for that help you will probably turn to a digital computer. This piece of equipment will almost certainly not use either of the two methods quoted above as the evaluation of determinants is a very time consuming process on a computer and 'time', where a computer is concerned, is simply another way of spelling 'money'. In such a situation, a numerical approach is adopted which, incidentally, is also purely mechanical in its operation.

Whilst on the subject of equations, there are many simple looking single algebraic equations that you would find very troublesome or impossible to solve analytically - for example, how would you set about solving the equation $e^x = 10 - x$?   Such an equation can, however, be solved numerically, to any required degree of accuracy, with very little difficulty.

Some problems which cannot be solved analytically do at least have an analytical look about them in the first place, for example, the equation $e^x = 10 - x$ and the integral $\int_0^2 e^{-x^2}\, dx$.   However you may quite well meet a problem that is not even formulated in analytical terms.   For example, suppose an experiment has been performed and a series of values of, say, the temperature θ of a body measured against a series of values of the time t.   Thus the values of θ may be measured at intervals of one minute.   Having performed the experiment, you may then be asked "What was the temperature after $5\frac{1}{2}$ minutes?" or "At what rate was the temperature changing after 10 minutes?"   You cannot use analytical means to answer questions such as these as the formula for θ in terms of t is not known.   Again a numerical method is necessary to determine the answers to such questions.

The following are some more examples of practical problems that require numerical methods for their solution:

The equation $\left\{\dfrac{\sin \alpha}{\alpha}\right\}^2 = \dfrac{1}{2}$ occurs in Fraunhofer diffraction. What value of α satisfies this equation?

The motion of a planetary gear system in a certain automatic transmission involves the equation $\sin \omega t - e^{-\alpha t} = 0$.   What is the smallest positive value of t for given values of α and ω?

A certain column buckles when kL has the least positive value that satisfies the equation $\tan kL - kL = 0$.   What is this value of kL?

The integral $\displaystyle\int_0^{x_0} \dfrac{x^4 e^x}{(e^x - 1)^2}\, dx$ occurs when obtaining the heat capacity of a solid by a method based on the vibrational frequencies of the crystal.   What is the value of this integral for a given x_0?

The integral $\displaystyle\int_{\lambda_1}^{\lambda_2} \dfrac{1}{\lambda^5 (e^{hc/\lambda kT} - 1)}\, d\lambda$ occurs in finding the fraction of total energy that is visible radiation of a black body.   What is the value of this integral for given values of λ_1 and λ_2?

Once a numerical method has been found for a particular type of problem, it can also be used for similar problems that do have analytical solutions. For example, the same numerical technique can be used for solving $x^2 - 2x - 14 = 0$ as for $e^x = 10 - x$. So your question now might be — If I can solve a greater variety of problems by numerical techniques than I can with analytical methods, why bother with the analytical techniques? This question has two answers:

(i) If an analytical technique exists, it is usually easier and more exact than the corresponding numerical method - for example, the easiest way of obtaining the solutions of $x^2 - 2x - 14 = 0$ is still by the use of the quadratic formula. But even this requires some method of evaluating $\sqrt{60}$.

(ii) Analysis forms the basis of many of the numerical techniques.

The conclusion we should draw is that both analytical methods and numerical methods have, in their own rights, their places in mathematics and that in any particular problem where there is a choice as to the method of solution, then the method chosen should be that which leads to the best combination of simplicity, speed and accuracy. As analytical methods have been considered in 'Mathematics for Engineers and Scientists, Vols. 1 and 2', by the present authors, this book will concentrate on numerical methods.

Aids to Calculation

The majority of problems which are tackled by numerical methods involve a considerable amount of arithmetic. Some help with this arithmetic is obviously desirable and in most cases essential. Some of the aids available you will have already met - for example, mathematical tables and slide rules. These are useful tools when the number of calculations to be performed is limited and relatively few significant figures are required in the answers. Furthermore they are restricted to multiplication and division, being of no help for addition and subtraction. The number of significant figures obtainable with these aids can be increased by the use of more comprehensive tables, quoting quantities to more significant figures (e.g., six-figure instead of four-figure logs), and larger slide rules.

Proceeding up the scale, so to speak, we next come to the mechanical desk calculator which is basically an adding and subtracting machine. However, as multiplication can be performed by a series of additions and division by a series of subtractions, many desk machines were modified so that they could carry out these operations either automatically or semi-automatically. The number of significant figures to which they could work was usually greater than is the case with either tables or a slide rule. Originally this type of machine was driven manually but later electrically driven models were introduced.

The mechanical type of desk machine has now been almost entirely superseded by electronic types and some of these are even programmable thus turning them into mini-computers. So rapid have been the recent

advances in this type of machine that electronic pocket calculators
are now widely used.

FRAME 9

Useful as desk and pocket calculators no doubt are, they pale into
insignificance when compared with a full-size digital computer. Not
only is such a machine extremely fast but it can be given a whole set of
instructions and left to get on with the job, whereas a desk machine
requires constant attention. It is really the advent of the computer
that has brought numerical methods into their own. Before computers
were available there just was not the means of doing vast amounts of
arithmetic at a reasonable speed. So although the numerical methods
were there, their use was considerably restricted.

It may happen that there is more than one method available for solving a
problem numerically. In that case the tendency these days is to use
that method which is best suited to the computer.

FRAME 10

When applying numerical methods in actual practice, it is the more
complicated type of problem (e.g., fifty simultaneous linear equations in
fifty unknowns) that is liable to occur. However the methods used in
the solution of such a problem are basically the same as those which can
be used in much simpler cases (e.g., three simultaneous linear equations
in three unknowns). As there is nothing to be gained when learning the
actual methods by having very complicated problems, the examples used in
these programmes to illustrate the methods will therefore be kept
relatively simple.

If you have access to a desk machine or a pocket calculator you will
find it a great help when working through the examples in this book.
If, however, you haven't, you will still be able to do most of the
working with the aid of a set of mathematical tables. This will mean
that in some places you will only be able to work to a fewer number of
significant figures with consequent loss of accuracy. In some examples
you will find this loss of accuracy very marked. Even so, you will
still be able to appreciate the techniques involved.

You may also have knowledge of a computing language. If so, then
having learnt the technique of a numerical method, it will be a good
idea for you to write a computer program for that method, and, if
possible, get it run on a computer, using suitable data.

FRAME 11

Accuracy and Errors - Types of Error

Whenever calculations are performed there are many possible sources of
error and errors will obviously affect the accuracy of the solution of a
particular problem. Errors can be introduced in three ways:

  (i)   Mistakes made by the person carrying out the calculations,
 (ii)   The use of inaccurate formulae,
(iii)   The use of inaccurate data, including the effects of round-off.

In the next few frames, we will have a look at each of these in turn.

Theoretically, all errors made under the heading (i) shouldn't be there at all. But, as a certain gentleman once remarked: "To err is human ...." and the operator is, of course, human. However, in some cases he may be forgiven - if, for example, he is using a machine that has developed a fault which he cannot detect. Even so, such a fault is still going to affect the accuracy of his result.

Mistakes commonly made by a human operator occur when copying and when doing mental arithmetic. Two common copying errors are:

(a) the reversal of two digits, e.g. writing down 236 721 instead of 263 721, and

(b) the repetition of the wrong digit, e.g. 233 721 instead of 223 721.

It is obviously best to avoid such mistakes as these but as the chances are that you will still make some, it is advisable to take steps to, firstly, reduce the number that you make and, secondly, try to detect any that you do make as soon as possible.

To reduce the probability of making mistakes, it is very advisable to keep your computational work neat and tidy - and also legible. Furthermore try and arrange your work so that you have to copy numbers as few times as possible. To assist in detecting mistakes you should arrange to check your working wherever possible - not by repetition but by some independent process. How this can be done in certain cases will be indicated later.

In some types of work, mistakes are automatically taken care of. This does not mean, of course, that you shouldn't take care not to make them as they can still cause you to waste time.

In the case of (ii), an inaccurate formula may arise due to chopping off an infinite series after a finite number of terms. For example, this is done when Simpson's rule is obtained by the use of Taylor series. f(a - h) and f(a + h) are expanded in powers of h but all terms involving powers greater than the second are dropped. An error introduced in this way is known as a TRUNCATION ERROR.

A truncation error such as that described in the last frame leads to an approximate formula being used instead of an exact one. Differentiation gives us another example where a true formula may be replaced by an approximate one. As you know, if $y = f(x)$ then the value of $\frac{dy}{dx}$ at the point $x = a$ is given by the formula

$$\frac{dy}{dx} = \lim_{h \to 0} \frac{f(a + h) - f(a)}{h}$$

Assuming h is small, $\frac{dy}{dx}$ is given approximately by the formula

$$\frac{f(a + h) - f(a)}{h} \qquad (14.1)$$

the accuracy increasing as h is decreased.   To illustrate this, if

$$y = \frac{e^x}{x} \, , \quad \frac{dy}{dx} = \frac{x-1}{x^2} \, e^x \quad \text{and so, when} \quad x = 2, \quad \frac{dy}{dx} = 1 \cdot 8473.$$

Now use the formula (14.1) to find approximate values for $\frac{dy}{dx}$ when
a = 2 and h is, in turn, 0·2, 0·1, 0·05, 0·01.
****

14A

2·038, 1·941, 1·892, 1·85.

FRAME 15

As we have already observed from our work on calculus, the result
obtained always becomes better as h is decreased.   At least it does
theoretically.   Practically there are certain other snags which may
upset the apple cart, as will be seen later.   You will also find later
that many numerical methods involve the choice of an h (Simpson's rule
is another example) and that when this is the case, the smaller it is,
the better.   However, there are other points to be noticed in
connection with the results obtained in 14A, and we now come to errors
introduced due to the use of inaccurate data [(iii) in FRAME 11].

When calculating the value of $\frac{dy}{dx}$ as given by the formula (14.1), you
had to evaluate

$$\frac{\dfrac{e^{2 \cdot 2}}{2 \cdot 2} - \dfrac{e^2}{2}}{0 \cdot 2}$$

for the case when h = 0·2.   What aids did you use in evaluating this
expression?
****

15A

Almost certainly you used exponential tables for $e^{2 \cdot 2}$ and e^2.   The
division by 2·2 you may have carried out on a desk or pocket calculator,
by logs, on a slide rule or without any such aid.

Division by 2 and 0·2 and also the subtraction you probably did
mentally.

FRAME 16

Accuracy and Errors — Round-off

Taking first the values of $e^{2 \cdot 2}$ and e^2 from (as we did) exponential
tables to four places of decimals, the figures 9·0250 and 7·3891 are
obtained.   It is, of course, extremely unlikely that these are the
exact values of $e^{2 \cdot 2}$ and e^2.   They are almost certainly subject to
ROUNDING ERRORS.   These occur whenever a number is quoted correct to so
many decimal places or significant figures, the quoted figure being thus
not quite the true value.   Various questions then arise such as:-   What
effect do such errors have on the result of the calculation?   Are any
other numbers in our original expression subject to error in this way?
If so, what effect will this have on the result?   Are any more round-
off errors likely to occur during the course of the calculation?

In our particular case, the answer to the second of these questions is 'No' - assuming we are considering the derivative at x = 2 exactly and not very close to it. Also we are at liberty to choose an exact value for h. Other round-off errors are certainly likely to occur, for example after a division is performed, or if other mathematical tables are used.

To answer the first of these questions, it is necessary to consider the effect of round-off errors when numbers are added, subtracted, multiplied or divided. But before doing this, we will remind you of the rules for rounding off.

When rounding one digit,

if that digit lies in the range 0 - 4, the previous digit is unchanged,

if that digit lies in the range 6 - 9, the previous digit is increased by 1,

if that digit is 5, the previous digit is unchanged or is increased by 1 according as it is even or odd.

Thus, for example,

$$
\begin{array}{lll}
7 \cdot 4727 & \text{becomes} & 7 \cdot 473 \\
76\,340 & \text{becomes} & 76\,300 \\
15 \cdot 235 & \text{becomes} & 15 \cdot 24 \\
15 \cdot 245 & \text{becomes} & 15 \cdot 24
\end{array}
$$

When rounding two digits,

if those digits lie in the range 00 - 49, the previous digit is unchanged,

if those digits lie in the range 51 - 99, the previous digit is increased by 1,

if those digits are 50, the previous digit is unchanged or is increased by 1 according as it is even or odd.

Thus, for example,

$$
\begin{array}{lll}
35 \cdot 671 & \text{becomes} & 35 \cdot 7 \\
430\,050 & \text{becomes} & 430\,000 \\
2 \cdot 732\,35 & \text{becomes} & 2 \cdot 732
\end{array}
$$

Similar rules apply when rounding more than two digits.

You will see from this that, if two or more digits are to be rounded, they should be done simultaneously, not one at a time.

What will 7·52, 6·345, and 182 377 become when rounded one digit and what will 73·6547, 180·273 and 541 500 499 become when rounded three digits?

****

17A

7·5, 6·34, 182 380; 73·7, 180, 541 500 000

The ERROR, ε, in any quantity is given by the expression

exact value - approximate value

(You will find that some authors adopt the alternative definition, ε = approximate value - exact value.   The use of two definitions is unfortunate, but if you use one consistently, you shouldn't get into difficulties.)

Thus the errors introduced in each of the seven examples quoted in the last frame, assuming the first figure in each case to be exact, are, respectively,

$$-0 \cdot 0003, \ 40, \ -0 \cdot 005, \ 0 \cdot 005: \ \ -0 \cdot 029, \ 50, \ 0 \cdot 000 \ 35$$

What are the corresponding errors introduced in the six examples you did in the last frame?

****

18A

$0 \cdot 02, \quad 0 \cdot 005, \quad -3: \quad -0 \cdot 0453, \quad 0 \cdot 273, \quad 499$

Accuracy and Errors - Effects of Errors on Calculations

Now find the errors introduced when (a) $18 \cdot 496$,   (b) $18 \cdot 493$, (c) $17 \cdot 208$ are rounded by one digit.   Then find (d) $18 \cdot 496 + 18 \cdot 493$, (e) $18 \cdot 496 + 17 \cdot 208$, (f) $18 \cdot 496 - 18 \cdot 493$, (g) $18 \cdot 496 - 17 \cdot 208$ and the errors introduced when (a), (b), (c) are first rounded by one digit and then the sums and differences corresponding to (d), (e), (f) and (g) are formed.

****

19A

(a)	$18 \cdot 496$	becomes	$18 \cdot 50$	introducing error	$-0 \cdot 004$
(b)	$18 \cdot 493$	"	$18 \cdot 49$	"	"	$0 \cdot 003$
(c)	$17 \cdot 208$	"	$17 \cdot 21$	"	"	$-0 \cdot 002$
(d)	$36 \cdot 989$	(e)	$35 \cdot 704$	(f) $0 \cdot 003$	(g) $1 \cdot 288$

$18 \cdot 50 + 18 \cdot 49 = 36 \cdot 99$ introducing error $-0 \cdot 001$
$18 \cdot 50 + 17 \cdot 21 = 35 \cdot 71$ "   " $-0 \cdot 006*$
$18 \cdot 50 - 18 \cdot 49 = 0 \cdot 01$ "   " $-0 \cdot 007*$
$18 \cdot 50 - 17 \cdot 21 = 1 \cdot 29$ "   " $-0 \cdot 002$

You will find the reason for the stars in the next frame.

If you examine the answers you obtained in the last frame you will notice that the error introduced when two numbers are rounded and then added is the sum of the separate errors introduced on rounding, and that when they are subtracted the total error is the difference of the two individual errors.

Now the individual errors may be positive or negative so that when they are either added or subtracted they may reinforce each other.   You will notice that this happened in the two cases starred in 19A.   Thus, if you had only been given the numbers $18 \cdot 50$, $18 \cdot 49$ and $17 \cdot 21$ with the information that they were subject to round-off errors, it would have been impossible for you to tell whether individual errors would reinforce each other or not upon addition or subtraction.   And this, of course, is the situation you normally encounter.   Furthermore you would not know the actual magnitudes of the individual errors.

The only thing you can do in such a situation is to estimate the magnitude of the maximum error that can be introduced.   This is the extreme case when each of the two numbers to be added or subtracted possesses an error of maximum magnitude and the signs of the individual errors are such that their effect is additive.

The magnitude of the maximum error that each of 18·50, 18·49, and 17·21 can contain is 0·005.   For example, 18·495 and 18·505 are the lowest and highest numbers that give 18·50 when rounded one digit.   The magnitude of the maximum error that can be introduced when any two of these three figures is added or subtracted is therefore 0·01.

More generally, if ε_1 and ε_2 are the errors contained in two numbers, then the magnitude of the maximum error in either their sum or their difference is

$$|\varepsilon_1| + |\varepsilon_2|$$

What do you think will be the magnitude of the maximum possible error if n numbers are added, the individual errors being

$$\varepsilon_1 , \varepsilon_2 , \varepsilon_3 , \ldots\ldots\ldots\ldots \varepsilon_n?$$

****

21A

$$|\varepsilon_1| + |\varepsilon_2| + |\varepsilon_3| + \ldots\ldots\ldots\ldots\ldots + |\varepsilon_n|$$

The way in which errors can very quickly accumulate can be illustrated by a simple difference table, shown on the following page.

The numbers in the left hand column are the values of $x^3/10$ for x = 0, 1, 2, ... 10.   Each number in any of the other columns is formed by subtracting two adjacent numbers in the column immediately to its

left.   Thus the number enclosed in △ is obtained by subtracting the

number in ◯ from the number in ▭ , even if the result is negative.

All entries in the columns to the right of the last one are zero, but are not shown.

Now form the new difference table obtained when each entry in the left hand column of the table is rounded to the nearest whole number. Continue your table to include 7 columns of differences.   Notice how it is not long before the differences start behaving very wildly, i.e. do not show any similarity to their original values.

0.0				
	0.1			
0.1		0.6		
	0.7		0.6	
0.8		1.2		0
	1.9		0.6	
2.7		1.8		0
	3.7		0.6	
6.4		2.4		0
	6.1		0.6	
12.5		3.0		0
	(9.1)		0.6	
21.6		△3.6		0
	[12.7]		0.6	
34.3		4.2		0
	16.9		0.6	
51.2		4.8		0
	21.7		0.6	
72.9		5.4		
	27.1			
100.0				

****

0						
	0					
0		*1*				
	1		*0*			
1		*1*		*0*		
	2		*0*		*2*	
3		*1*		*2*		*-5*
	3		*2*		*-3*	*6*
6		*3*		*-1*		*1*
	6		*1*		*-2*	*9*
12		*4*		*-3*		*10*
	10		*-2*		*8*	*-26*
22		*2*		*5*		*-16*
	12		*3*		*-8*	*27*
34		*5*		*-3*		*11*
	17		*0*		*3*	
51		*5*		*0*		
	22		*0*			
73		*5*				
	27					
100						

Each entry in a column (except in the first column) is formed by taking
the difference of two adjacent entries in the preceding column.   As you
go to the right in the table, you are thus finding differences of
differences and, as you can see, any small errors that exist in the
original tabulated values are propagated very quickly.   Worse still,

11

they rapidly become magnified with the effect that errors in the entries
in the right hand columns are so large that the entries in these columns
are completely meaningless.

Returning now to the answers in 19A, we can compare the errors in the
results with the actual values themselves.

Thus:

the error introduced in	36·989	is	−0·001
" " " "	35·704	"	−0·006
" " " "	0·003	"	−0·007
" " " "	1·288	"	−0·002

The first of these is approximately 1 in 37 000 as far as magnitude is
concerned, and, unless very great accuracy were required, would normally
be acceptable.   On the other hand the third error is equivalent to 7
in 3 and it is extremely unlikely that this would be acceptable.   Note
that this error has crept in as a result of subtracting two very nearly
equal numbers.   This is thus a process that should obviously be avoided
and will be referred to again later.

If a number x is subject to an error ε then the ratio $\dfrac{\varepsilon}{x}$ is the
RELATIVE ERROR.   (Here again, some authors adopt a different
definition.)   Thus the relative errors in each of the four examples
quoted above are −0·000 027, −0·000 17, −2·3 and −0·0016.   It is this
quantity which should be taken into account when considering how
meaningful a result is, a small relative error being more acceptable
than a large one.   Unfortunately, no exact figures can be laid down as
to acceptability of errors or otherwise.   For example, most people
would not worry if the statement that Loughborough is 175 km from
London is an odd kilometre or two out, but an auditor would certainly
expect a figure of £175 in book-keeping to be exact.

So far we have not considered the effect of errors where multiplication
or division is concerned.   To see the effect on, say, division, find
$\dfrac{1}{x}$, $\dfrac{1}{z}$, $\dfrac{1}{x + y}$ and $\dfrac{1}{x - y}$ if $x = 1·50$, $y = 1·48$, $z = 0·02$.   Then
assume that these figures are all subject to a maximum error of 0·005
either way and find the greatest and least values of $\dfrac{1}{x}$, $\dfrac{1}{z}$, $\dfrac{1}{x + y}$ and
$\dfrac{1}{x - y}$ by taking suitable combinations of up and down errors.

****

$\dfrac{1}{x} = \dfrac{1}{1·50} \simeq 0·6667$

Greatest value will obtain when $x = 1·495$.   *It will be* $\dfrac{1}{1·495} \simeq 0·6687$.

Least value when $x = 1·505$.   *It will be* $\dfrac{1}{1·505} \simeq 0·6646$.

25A   (continued)

$\dfrac{1}{z} = \dfrac{1}{0\cdot02} = 50$

Greatest value when $z = 0\cdot015.$ *It will be* $\dfrac{1}{0\cdot015} \simeq 66\cdot7$

Least value when $z = 0\cdot025.$ *It will be* $\dfrac{1}{0\cdot025} = 40$

$\dfrac{1}{x + y} = \dfrac{1}{2\cdot98} \simeq 0\cdot3356$

Greatest value will obtain when $x = 1\cdot495$ *and* $y = 1\cdot475.$ *It will be*
$\dfrac{1}{2\cdot97} \simeq 0\cdot3367.$

Least value when $x = 1\cdot505$ *and* $y = 1\cdot485.$ *It will be* $\dfrac{1}{2\cdot99} \simeq 0\cdot3344.$

$\dfrac{1}{x - y} = \dfrac{1}{0\cdot02} = 50$

Greatest value when $x = 1\cdot495$ *and* $y = 1\cdot485.$ *It will be* $\dfrac{1}{0\cdot01} = 100.$

Least value when $x = 1\cdot505$ *and* $y = 1\cdot475.$ *It will be* $\dfrac{1}{0\cdot03} \simeq 33\cdot3.$

FRAME 26

These answers illustrate far more vividly than words how errors can affect division.   You will see that the effect of an error or a combination of errors is far more severe when there are few significant figures.   In a similar way you could easily demonstrate to yourself the effect of errors in the numerator and also in a product.   Furthermore, you could go to cases where there are errors of different magnitudes in the various terms.

From the various results we have obtained you will realise that it is the number of significant figures that can be given in an answer that is more meaningful than the number of decimal places.   You will also realise that it is difficult to give any precise simple rule about the number of accurate figures in an answer.   As was pointed out in FRAME 20, most data are likely to be inexact.

Now return to the examples you worked out in FRAME 14.   If you look at your working for the various values of h, you will realise that it, i.e. your working, is becoming more inaccurate as h becomes smaller. Although, from the mathematical theory, the result is becoming _more_ accurate, the limitations of the numerical work, as done there, are counterbalancing this.   If we want a more accurate result it is necessary to resort to tables giving e^x to a greater number of significant figures - or else obtain these values in some other way. The point is that if the same number of significant figures are used for the various values of e^x, the number of significant figures in $f(a + h) - f(a)$ decreases as h becomes smaller.

13

We have already obtained formulae giving the errors involved when two numbers are added or subtracted. To obtain a similar formula when two numbers are divided, we proceed as follows.

If x_1 is the correct value of the numerator and X_1 the quoted, approximate value, then $x_1 - X_1 = \varepsilon_1$, the error. Similarly, for the denominator, we can write $x_2 - X_2 = \varepsilon_2$. The true quotient is x_1/x_2 and the calculated quotient (which is, of course, subject to error) is X_1/X_2. The error in the quotient is then

$$\frac{x_1}{x_2} - \frac{X_1}{X_2} = \frac{X_1 + \varepsilon_1}{X_2 + \varepsilon_2} - \frac{X_1}{X_2}$$

$$= \frac{\varepsilon_1 X_2 - \varepsilon_2 X_1}{(X_2 + \varepsilon_2)X_2}$$

$$= \frac{\varepsilon_1 X_2 - \varepsilon_2 X_1}{X_2{}^2 \left[1 + \dfrac{\varepsilon_2}{X_2}\right]}$$

If it is assumed that ε_2 is small in comparison with X_2, as you hope it would be, so that ε_2/X_2 is small in comparison with 1, then the error is approximately

$$\frac{\varepsilon_1 X_2 - \varepsilon_2 X_1}{X_2{}^2} = \left[\frac{\varepsilon_1}{X_1} - \frac{\varepsilon_2}{X_2}\right]\frac{X_1}{X_2}$$

If, as they might well be, ε_1 and ε_2 are of opposite signs, the two terms in the brackets are additive, assuming that X_1 and X_2 are both positive.

Now, once again, in any particular case, you are unlikely to know the actual errors ε_1 and ε_2. All you can then do is to estimate the maximum possible error in the quotient assuming ε_1 and ε_2 have their greatest possible values and are opposite in sign. In this case, the magnitude of the maximum possible error is given by

$$\left[\frac{|\varepsilon_1|}{X_1} + \frac{|\varepsilon_2|}{X_2}\right]\frac{X_1}{X_2}$$

To illustrate the result just obtained, let us take the quotient $25 \cdot 4/12 \cdot 37$. When divided this gives $2 \cdot 053$. To what extent can this be relied upon?

Now the magnitude of the maximum possible error in $25 \cdot 4$ is $0 \cdot 05$ and that in $12 \cdot 37$ is $0 \cdot 005$. Hence an estimate of the maximum possible error in the quotient is

$$\left[\frac{0 \cdot 05}{25 \cdot 4} + \frac{0 \cdot 005}{12 \cdot 37}\right]\frac{25 \cdot 4}{12 \cdot 37} \simeq (0 \cdot 002 + 0 \cdot 0004)2 \simeq 0 \cdot 005$$

The true quotient can therefore be expected to lie between $2 \cdot 048$ and

2·058.   The answer can therefore only be quoted as being 2, i.e., only given as correct to one significant figure.   If we try to quote even to two significant figures, it is uncertain whether it is 2·0 or 2·1.

Note that when performing the calculation $\left[\dfrac{0 \cdot 05}{25 \cdot 4} + \dfrac{0 \cdot 005}{12 \cdot 37}\right]\dfrac{25 \cdot 4}{12 \cdot 37}$, great accuracy is not necessary as you are only finding an estimate of the error.   As it is to be combined with your quotient, you only need the same number of decimal places as you have taken in your quotient.   As a matter of interest, the greatest possible value of the quotient is 25·45/12·365 = 2·058, and the least possible value 25·35/12·375 = 2·048.

Now use the method of FRAME 27 to find how many figures you can be sure of in the quotient 2·634/17·12.
****

28A

$$2 \cdot 634/17 \cdot 12 \ = \ 0 \cdot 153 \ 86$$

$$\left[\frac{0 \cdot 0005}{2 \cdot 634} + \frac{0 \cdot 005}{17 \cdot 12}\right]\frac{2 \cdot 634}{17 \cdot 12} \ \simeq \ 0 \cdot 000 \ 07$$

The quotient can be expected to lie between 0·153 79 and 0·153 93. Hence you can be sure of three significant figures, i.e. 0·154.

FRAME 29

Now see if you can find a formula giving you the error in a product. Assume the factors are exactly x_1 and x_2 but are given as X_1 and X_2, where, as before, $x_1 = X_1 + \varepsilon_1$ and $x_2 = X_2 + \varepsilon_2$.

****

29A

$$
\begin{aligned}
Error \ &= \ x_1 x_2 - X_1 X_2 \\
&= \ (X_1 + \varepsilon_1)(X_2 + \varepsilon_2) - X_1 X_2 \\
&= \ \varepsilon_1 X_2 + \varepsilon_2 X_1 + \varepsilon_1 \varepsilon_2 \\
&\simeq \ \varepsilon_1 X_2 + \varepsilon_2 X_1 \ , \quad \textit{assuming } \varepsilon_1, \varepsilon_2 \textit{ small in comparison with the} \\
&\qquad\qquad\qquad\qquad\qquad\qquad\qquad\quad \textit{other terms.}
\end{aligned}
$$

FRAME 30

Once again, in practice, all that will be possible will be an estimation of the maximum error in $X_1 X_2$ if X_1 and X_2 are subject to errors of maximum magnitude.   This time, if it is assumed that X_1 and X_2 are both positive, the greatest error will occur when ε_1 and ε_2 are of the same sign and its magnitude will be $|\varepsilon_1| X_1 + |\varepsilon_2| X_2$, approximately.

As an example, the approximate maximum error in 5·43 × 27·2 is 0·005 × 27·2 + 0·05 × 5·43 ≃ 0·4.   The product 147·7 can therefore be expected to lie between 147·3 and 148·1.   You cannot therefore really give more than 2 significant figures, i.e., 150.

The formulae for the maximum relative error in the case of either division or multiplication can easily be deduced. It is only necessary to divide the actual maximum error by the quotient or product respectively, i.e. by $\dfrac{x_1}{x_2}$ or $x_1 x_2$.

Thus for division, the relative error is $\left(\dfrac{|\varepsilon_1|}{X_1} + \dfrac{|\varepsilon_2|}{X_2} \right) \dfrac{X_1}{X_2} \Big/ \dfrac{x_1}{x_2}$, and for multiplication, it is $(|\varepsilon_1|X_2 + |\varepsilon_2|X_1)/(x_1 x_2)$. There is, however, one snag: x_1 and x_2 are not known and so neither is x_1/x_2 or $x_1 x_2$. The only figures that are known are X_1 and X_2 and hence X_1/X_2 and $X_1 X_2$.

As the formulae for relative error are only approximate anyway, it is sometimes assumed that x_1/x_2 and $x_1 x_2$ are not all that different from X_1/X_2 and $X_1 X_2$. If this is done

$$\left(\frac{|\varepsilon_1|}{X_1} + \frac{|\varepsilon_2|}{X_2} \right) \frac{X_1}{X_2} \Big/ \frac{x_1}{x_2} \quad \text{becomes} \quad \left(\frac{|\varepsilon_1|}{X_1} + \frac{|\varepsilon_2|}{X_2} \right) \frac{X_1}{X_2} \Big/ \frac{X_1}{X_2}$$

$$= \frac{|\varepsilon_1|}{X_1} + \frac{|\varepsilon_2|}{X_2}$$

and $(|\varepsilon_1|X_2 + |\varepsilon_2|X_1)/(x_1 x_2)$ becomes $(|\varepsilon_1|X_2 + |\varepsilon_2|X_1)/(X_1 X_2)$

$$= \frac{|\varepsilon_1|}{X_1} + \frac{|\varepsilon_2|}{X_2}$$

The relative error in both cases thus becomes

$$\frac{|\varepsilon_1|}{X_1} + \frac{|\varepsilon_2|}{X_2}$$

As you will realise this is not always an entirely satisfactory process as the true value of a quotient or product can be considerably different from the calculated value.

The various results can easily be combined if a more complicated calculation is carried out, for example

$$\frac{15 \cdot 36 + 27 \cdot 1 - 1 \cdot 672}{2 \cdot 36 \times 1 \cdot 043}$$

The maximum possible error in the numerator is of magnitude

$$0 \cdot 005 + 0 \cdot 05 + 0 \cdot 0005 = 0 \cdot 0555$$

The approximate maximum error in the denominator is

$$0 \cdot 005 \times 1 \cdot 043 + 0 \cdot 0005 \times 2 \cdot 36 \simeq 0 \cdot 0064$$

and the fraction as quoted is $\dfrac{40 \cdot 79}{2 \cdot 461}$ approximately.

This gives 16·57 subject to approximate error of magnitude

$$\left(\frac{0 \cdot 0555}{40 \cdot 79} + \frac{0 \cdot 0064}{2 \cdot 461} \right) \frac{40 \cdot 79}{2 \cdot 461} \approx 0 \cdot 07$$

The result therefore lies between approximately 16·50 and 16·64.
What will be the approximate values between which

$$\frac{1 \cdot 362(7 \cdot 54 - 13 \cdot 2)}{47}$$

can be expected to lie, assuming the maximum possible error in each of
the figures?

****

As figures are given, result $= - \dfrac{1 \cdot 362 \times 5 \cdot 66}{47}$

$$\approx - \frac{7 \cdot 709}{47}$$

$$\approx -0 \cdot 1640$$

Maximum possible error in difference $= 0 \cdot 005 + 0 \cdot 05 = 0 \cdot 055$

Approximate maximum error in numerator is

$$0 \cdot 0005 \times 5 \cdot 66 + 0 \cdot 055 \times 1 \cdot 362 \approx 0 \cdot 078$$

Approximate error in quotient $= \left(\dfrac{0 \cdot 078}{7 \cdot 709} + \dfrac{0 \cdot 5}{47} \right) \dfrac{7 \cdot 709}{47}$

$$\approx 0 \cdot 0034$$

$\therefore$ *Result lies between, approximately,* $-0 \cdot 1606$ *and* $-0 \cdot 1674$.

Evaluation of Formulae

In the examples that have been done in the previous section several
calculations have occurred involving fractions.   These have just been
worked through as they stood, no account having been taken as to whether
one procedure is better than another.   For example, in the calculation
$\dfrac{(13 \cdot 2 - 7 \cdot 54)1 \cdot 362}{47}$ which you have just done, three arithmetic
operations were necessary, a subtraction, a multiplication and a
division.   But what is the best order in which to perform these
calculations?   Remember that, apart from arithmetical errors, a
frequent source of error is in copying numbers.   Thus, the fewer
numbers you have to copy, the better.

Now, you no doubt have had plenty of experience with log tables and
possibly also with slide rules and pocket calculators.   It is less
likely that you have had so much experience with a desk calculator.
When you are proposing to make a calculation, whatever aid you may be
using, it is wise to think out before you start the best sequence of
operations so that, as much as possible, you avoid both noting down
numbers and making transfers unless you have facilities for doing this
latter operation automatically.

Apart from the direct evaluation of fractions, you will also sometimes find it necessary to evaluate formulae. When this happens, it will help if you first of all examine the formula to see if it is in the best form for computational purposes. It may be that some rearrangement is possible that will make the formula more suitable for calculation. We propose to consider just two cases of this. One such case is in the evaluation of a polynomial, e.g., in finding the value of

$$10{\cdot}03x^5 + 7{\cdot}42x^4 + 6{\cdot}53x^3 - 7{\cdot}67x^2 - 18{\cdot}32x + 143{\cdot}21$$

when $x = 2{\cdot}36$. Of course this can be evaluated as it stands, but an alternative way of writing it is

$$\left[\left[\left\{(10{\cdot}03x + 7{\cdot}42)x + 6{\cdot}53\right\}x - 7{\cdot}67\right]x - 18{\cdot}32\right]x + 143{\cdot}21$$

which you can immediately verify is equivalent to the original expression. The process of doing this is called NESTING. An important advantage of this form, whatever aid you are using, is that the number of multiplications necessary is less than in the original. You can now go straight through the sequence of operations

$$\begin{array}{l} 10{\cdot}03 \\ \times 2{\cdot}36 \\ +7{\cdot}42 \\ \times 2{\cdot}36 \\ +6{\cdot}53 \\ \times 2{\cdot}36 \\ -7{\cdot}67 \quad \text{etc.} \end{array}$$

Provided you effectively have transfer facilities from the product register to the keyboard, no intermediate manual re-entering will be necessary. You must, however, be careful to look after your decimal point if this is not taken care of automatically. If you have a calculating aid available now carry out this calculation. (Assume that the numbers are sufficiently correct to be able to give the answer to three significant figures.)
****

1110

How would you rearrange $2x^6 + 3{\cdot}2x^4 + 2{\cdot}5x^2 + 13{\cdot}6x + 5{\cdot}2$ for ease of calculation? If you have access to a calculator, use your result to find the value of this expression when $x = 0{\cdot}46$, assuming the result can be given to three significant figures.
****

$$\left[\left\{(2x^2 + 3{\cdot}2)x^2 + 2{\cdot}5\right\}x + 13{\cdot}6\right]x + 5{\cdot}2$$
12·1

You will notice that the sequence of operations listed in FRAME 34 follows a simple repetitive form. A multiplication followed by an addition (which may be negative) forms the cycle. As repetitive

processes lend themselves to easy programming on to a computer, it
follows that nesting is a process which can be easily computerised.
When preparing a problem to put on to a computer, it is very important
that careful thought is given to the logic of the problem.   A great aid
to this end is the preparation of what is known as a flow diagram.   This
is simply a diagram that gives a graphic, easy to follow, description
of the process to be carried out.   Differently shaped containers are
used for the various instructions, but there is no single convention for
these shapes.   The ones used here are those adopted in the book
"Fortran and Algol" by A.C. Bajpai, H.W. Pakes, R.J. Clarke, J.M.
Doubleday and T.J. Stevens.   In your computing course, flow diagrams
are studied in more detail.   In the diagram that follows, a simple flow
chart is given for the calculation of the more general polynomial

$$a_0 x^n + a_1 x^{n-1} + \ldots + a_n$$

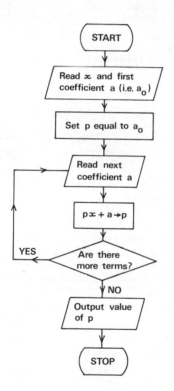

This will first be a_1, then a_2
the next time round, and so on.

Here $px + a$ is found and the result
taken as the new value of p.

The value of p is now that of
the required polynomial.

There is a variety of computing languages in use and we do not propose,
in general, to include computer programs in this book.   However, in case
you are unfamiliar with the form of such a program, a copy of a nesting
process print-out from a job run on the Loughborough University ICL
1904A computer is given on page 27.   It first lists the program,
written in the FORTRAN language, which can be used for a variety of
polynomials and different values of x.   In this particular run the data
supplied was that for the polynomial in FRAME 34, i.e.

$10 \cdot 03x^5 + 7 \cdot 42x^4 + 6 \cdot 53x^3 - 7 \cdot 67x^2 - 18 \cdot 32x + 143 \cdot 21$ with $x = 2 \cdot 36$, and the output of the calculations is shown below the program.   If the coefficients in the polynomial are not exact, the relevant number of significant figures can then be extracted from the computer solution. Other programs for the evaluation of a polynomial can be found in references (5), (8) and (9).

Referring to the calculation in FRAME 34, the link between the symbols in the flow diagram and the figures would be

$$x = 2 \cdot 36$$
$$a_0 = 10 \cdot 03$$
$$p = 10 \cdot 03$$

next $a = 7 \cdot 42$
 $p = 10 \cdot 03 \times 2 \cdot 36 + 7 \cdot 42 = 31 \cdot 0908$

next $a = 6 \cdot 53$
 $p = 31 \cdot 0908 \times 2 \cdot 36 + 6 \cdot 53 = 79 \cdot 904\ 285$

next $a = -7 \cdot 67$

What will be the expression for the next p?
****

36A

$79 \cdot 904\ 285 \times 2 \cdot 36 - 7 \cdot 67$

The process continues until there are no more a's to be read and the final value of p is the required answer.

The other type of formula that we will have a look at involves the subtraction of two nearly equal numbers.   It was seen in FRAME 24 that such subtraction can lead to a large relative error.   This is equivalent to a loss in the number of significant figures.   For example, suppose we have $2437 \cdot 2 - 2434 \cdot 0$.   The face value of this is $3 \cdot 2$.   Assuming that the original figures are the results of rounding, the maximum error in each is $0 \cdot 05$ and so the maximum possible error in $3 \cdot 2$ is $0 \cdot 1$.   Thus, whereas each of the original numbers is correct to five significant figures, you can only rely on one significant figure, i.e. 3, in the difference.

If this is going to occur when evaluating a formula, the formula can sometimes be modified to avoid it.   For example, the quadratic formula gives $x = \dfrac{24 \cdot 13 \pm 24 \cdot 01}{2}$ for the roots of $x^2 - 24 \cdot 13x + 1 \cdot 40 = 0$ and so the smaller root appears as $0 \cdot 06$.   But $\dfrac{-b - \sqrt{b^2 - 4ac}}{2a}$ can alternatively be written as $\dfrac{2c}{-b + \sqrt{b^2 - 4ac}}$ and the smaller root can now be calculated as $\dfrac{2 \cdot 80}{24 \cdot 13 + 24 \cdot 01} = 0 \cdot 0582$.   Thus a comparable number of significant figures can be given here as for the larger root ($24 \cdot 1$).

Now see if you can obtain, with four figure tables, a better value of

$\sqrt{n} - \sqrt{n-1}$ when $n = 1157$ than the $0 \cdot 02$ that such tables give directly.

****

38A

$$\sqrt{n} - \sqrt{n-1} = \frac{1}{\sqrt{n} + \sqrt{n-1}} = \frac{1}{68 \cdot 04} = 0 \cdot 0147$$

FRAME 39

Synthetic Division

In certain examples it is necessary to perform an algebraic long division. Although the amount of arithmetic cannot be altered, there are ways of setting out such a division so that the amount of writing involved is somewhat reduced. Start by finding the quotient and remainder when $x^5 - 2x^4 + 4x^2 + x - 7$ is divided by $x - 3$.

****

39A

The conventional layout is of the form

$$
\begin{array}{l}
x - 3)\ x^5 - 2x^4 \qquad\qquad + 4x^2 + x - 7(x^4 + x^3 + 3x^2 + 13x + 40 \\
\quad\ \ \underline{x^5 - 3x^4} \\
\qquad\ \ x^4 \\
\qquad\ \ \underline{x^4 - 3x^3} \\
\qquad\qquad 3x^3 \\
\qquad\qquad \underline{3x^3 - 9x^2} \\
\qquad\qquad\quad 13x^2 \\
\qquad\qquad\quad \underline{13x^2 - 39x} \\
\qquad\quad * \qquad\qquad 40x \\
\qquad\qquad\qquad \underline{40x - 120} \\
\qquad\qquad\qquad\quad 113
\end{array}
$$

giving quotient $x^4 + x^3 + 3x^2 + 13x + 40$ *and remainder 113.* *(The reason for the star is given in the next frame.)*

FRAME 40

If you examine this, you will see that quite a lot of the working need not be written down. For example, the $13x$ in the quotient automatically ensures that the starred line contains no x^2 term and so it is not really necessary to write down $13x^2$ in the line above it and do the actual subtraction. (Similar remarks apply at the other stages in the working.) Also it is unnecessary to keep on writing in the powers of x as these are all the same in any one column and are only really needed in the answer. If you examine the following scheme you will see that it contains all the essential working.

	1	-2	0	4	1	-7
3		3	3	9	39	120
	1	1	3	13	40	113

The quotient and the remainder are as before.
The following points should be noticed about it:

(i) The numbers in the first row are simply the coefficients of the powers of x in the dividend, zero being inserted where necessary.

(ii) To divide by $x - 3$, $+3$ is placed to the extreme left of the

second row.   This has the effect of replacing all the
subtractions in the working by additions.

(iii)   The numbers to the left of the vertical line in the third row
give the coefficients of the powers of x in the quotient and
the number to the right of this vertical line gives the
remainder.

(iv)   The powers of x in the quotient are not actually given in the
scheme but the first term is obviously x^4 (from x^5/x) and they
are then in descending order.

The working is done as follows, being indicated by the arrows.

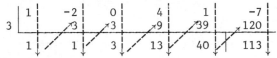

Each vertical arrow indicates an addition and each slanting arrow a
multiplication by 3, the working being carried out from left to right.
The arrows, of course, are not usually inserted.

Now use this method to divide $3x^6 + 2x^4 - 5x + 13$ by $x + 2$
    ****

40A

	3	0	2	0	0	−5	13
−2		−6	12	−28	56	−112	234
	3	−6	14	−28	56	−117	247

Quotient $3x^5 - 6x^4 + 14x^3 - 28x^2 + 56x - 117$, remainder 247.

With some calculators, the whole of the division can be carried out as
one continued operation, the relevant figures being recorded as they are
formed.   Negative − positive conversions can be carried out mentally
(for small figures) or on the machine.   In the first example (FRAME 40)
the sequence of operations would be

1	record
×3	
−2	record
×3	
+0	record
×3	
+4	record
×3	
+1	record
×3	
−7	record

Once again, a repetitive process is involved which makes it readily
programmable on to a computer.   The following is a simple flow chart
for the division of a polynomial of degree n by $x - \alpha$:

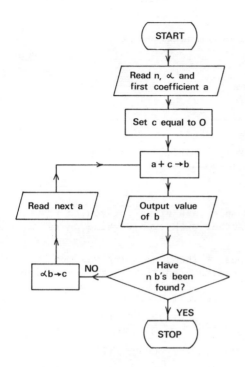

a and c are added, and
their sum is called b.

The first (n − 1) b's will be the
coefficients of the quotient. The
last b will be the remainder.

Again referring to the first example in FRAME 40, the numbers 1, −2, 0,
4 etc. in the first row are the successive values of a; 0 (left
blank), 3, 3, 9 etc. in the second row are the c's and 1, 1, 3, 13 etc.
in the third row are the b's.

FRAME 42

The process can easily be extended to division by quadratic factors, as
for example $(2x^5 - 3x^4 + x^2 + 10x - 17)/(x^2 - 2x + 5)$.

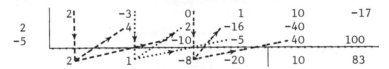

Quotient $2x^3 + x^2 - 8x - 20$, remainder $10x + 83$.
Note that the numbers at the front are here +2 and −5.
The three calculations indicated by ----→--- are carried out first,
i.e., the 2 is brought down and multiplied by 2 and −5.   The next
calculations are indicated by ·········→····· and are −3 + 4, 1 × 2,
1 × (−5).   The next set of calculations is indicated by — — —→ — —
i.e., 0 + 2 − 10, −8 × 2, −8 × (−5) and so on.   Unfortunately the
additional complications mean that you can no longer carry the working
through as one continued operation on a machine.

23

FRAME 42   (continued)

What are the quotient and remainder when $4x^6 - 2x^5 + 5x^3 - 4$ is divided by $x^2 + 3x + 1$?

****

42A

$$
\begin{array}{r|rrrrrrr}
 & 4 & -2 & 0 & 5 & 0 & 0 & -4 \\
-3 & & -12 & 42 & -114 & 285 & -841 & \\
-1 & & & -4 & 14 & -38 & 95 & -247 \\
\hline
 & 4 & -14 & 38 & -95 & 247 & -746 & -251
\end{array}
$$

Quotient $4x^4 - 14x^3 + 38x^2 - 95x + 247$, *remainder* $-746x - 251$.

FRAME 43

Miscellaneous Examples

In this frame, some miscellaneous examples are given for you to try. Answers are supplied in FRAME 44, together with such working as is considered helpful.

1. Evaluate the following expressions. Then, assuming maximum round-off error in each figure, estimate the limits between which your calculated value probably lies. Finally, quote each result to as many significant figures as are reliable.

    a) $\dfrac{2 \cdot 362 \times 1 \cdot 76}{1 \cdot 46 \times 0 \cdot 785}$ b) $7 \cdot 62 + \dfrac{54}{15 \cdot 3}$ c) $\dfrac{1 \cdot 32 - 0 \cdot 463}{16 \cdot 1 \times 2 \cdot 17}$

    In b) assume that there is no error in the number 54.

2. Evaluate the following expressions for the given values of x, first putting each in an appropriate form for the calculating aid you are using.

    a) $x^2 - 2 + \dfrac{1}{x^2}$ when $x = 1 \cdot 473$.

    b) $x^4 - 3 \cdot 17x^3 + 5 \cdot 29x - 17 \cdot 3$ when $x = 5 \cdot 12$.

    Estimate the approximate maximum error in each result, assuming that all integer coefficients are exact but that all decimals are subject to the maximum round-off error.

3. Find the quotient and remainder when

    i) $x^4 + 2x^3 + 3x^2 - 5x + 17$ is divided by $x + 1$,

    ii) $x^7 - 3x^3$ is divided by $x^2 + 2x - 1$.

4. Find, as accurately as your tables allow,

    $\ln(x - \sqrt{x^2 - 1})$ when $x = 27 \cdot 3$.

FRAME 44

Answers to Miscellaneous Examples

1. a) $3 \cdot 627$

    Numerator $\simeq 4 \cdot 157$

    Estimate of error in numerator $= 0 \cdot 0005 \times 1 \cdot 76 + 0 \cdot 005 \times 2 \cdot 362$

    $\simeq 0 \cdot 0009 + 0 \cdot 012 \simeq 0 \cdot 0129$

Denominator $\simeq 1 \cdot 146$

Estimate of error in denominator $= 0 \cdot 005 \times 0 \cdot 785 + 0 \cdot 0005 \times 1 \cdot 46$
$$\simeq 0 \cdot 004 + 0 \cdot 0007 = 0 \cdot 0047$$

Estimate of final error $= \left(\dfrac{0 \cdot 0129}{4 \cdot 157} + \dfrac{0 \cdot 0047}{1 \cdot 146} \right) \dfrac{4 \cdot 157}{1 \cdot 146}$
$$\simeq (0 \cdot 003 + 0 \cdot 004) 3 \cdot 6 \simeq 0 \cdot 025$$

Limits: $3 \cdot 602$ to $3 \cdot 652$
4

b) $11 \cdot 15$

Estimate of error in quotient $= \left(\dfrac{0 \cdot 05}{15 \cdot 3} \right) \dfrac{54}{15 \cdot 3} \simeq 0 \cdot 012$

Estimate of final error $\simeq 0 \cdot 005 + 0 \cdot 012 \simeq 0 \cdot 02$

Limits: $11 \cdot 13$ to $11 \cdot 17$
11

c) $0 \cdot 0245$

Estimate of error in difference $= 0 \cdot 005 + 0 \cdot 0005 = 0 \cdot 0055$

Estimate of error in product $\simeq 0 \cdot 05 \times 2 \cdot 17 + 0 \cdot 005 \times 16 \cdot 1 \simeq 0 \cdot 20$

Numerator $= 0 \cdot 857$ Denominator $= 34 \cdot 94$

Estimate of final error $\simeq \left(\dfrac{0 \cdot 0055}{0 \cdot 857} + \dfrac{0 \cdot 20}{34 \cdot 94} \right) \dfrac{0 \cdot 857}{34 \cdot 94} \simeq 0 \cdot 0003$

Limits: $0 \cdot 0242$ to $0 \cdot 0248$
$0 \cdot 02$

2.   a) $\dfrac{x^4 - 2x^2 + 1}{x^2} = \dfrac{(x^2 - 1)^2}{x^2}$ is a suggested form.

$\left[\left(x - \dfrac{1}{x} \right)^2 \text{ is another.} \right]$

$0 \cdot 631$

Error in $x^2 \simeq 2 \times 1 \cdot 473 \times 0 \cdot 0005 \simeq 0 \cdot 0015$

Error in $x^2 - 1 \simeq 0 \cdot 0015$.   This is the same as that in x^2 as the
-1 is exact.

Denominator $= 2 \cdot 170$ $x^2 - 1 = 1 \cdot 170$ Numerator $= 1 \cdot 369$

Estimate of error in numerator $= 2 \times 1 \cdot 369 \times 0 \cdot 0015 \simeq 0 \cdot 0041$

Estimate of error in final result $= \left(\dfrac{0 \cdot 0041}{1 \cdot 369} + \dfrac{0 \cdot 0015}{2 \cdot 170} \right) \dfrac{1 \cdot 369}{2 \cdot 170} \simeq 0 \cdot 002$

b) $\{(x - 3 \cdot 17)x^2 + 5 \cdot 29\}x - 1 \cdot 73$

$287 \cdot 1$

Approximate maximum error in $x^2 = 2 \times 5 \cdot 12 \times 0 \cdot 005 \simeq 0 \cdot 051$

$x^2 \simeq 26 \cdot 21$, $x - 3 \cdot 17 = 1 \cdot 95$, $(x - 3 \cdot 17)x^2 + 5 \cdot 29 = 51 \cdot 12$

Approximate error in $x - 3 \cdot 17 = 0 \cdot 005 + 0 \cdot 005 = 0 \cdot 01$

Approximate error in $(x - 3 \cdot 17)x^2 = 0 \cdot 01 \times 26 \cdot 21 + 1 \cdot 95 \times 0 \cdot 051$
$$\simeq 0 \cdot 451$$

Approximate error in $(x - 3 \cdot 17)x^2 + 5 \cdot 29 = 0 \cdot 451 + 0 \cdot 005 \simeq 0 \cdot 456$

Approximate error in $\{(x - 3 \cdot 17)x^2 + 5 \cdot 29\}x$
$$= 0 \cdot 456 \times 5 \cdot 12 + 0 \cdot 005 \times 51 \cdot 12 \simeq 2 \cdot 6$$

The subtraction of $1 \cdot 73$ will not affect this error to the degree of accuracy quoted.

3.   i) Quotient $x^3 + x^2 + 2x - 7$, remainder 24.
   ii) Quotient $x^5 - 2x^4 + 5x^3 - 12x^2 + 26x - 64$, remainder $154x - 64$.

4. $\ln (x - \sqrt{x^2 - 1})$ $= \ln \dfrac{1}{x + \sqrt{x^2 - 1}}$

$= \ln \dfrac{1}{54 \cdot 59}$,   using 4-figure tables

$= -\ln 54 \cdot 59$ $= -3 \cdot 998 \simeq -4 \cdot 00$

LOUGHBOROUGH UNIVERSITY COMPUTER CENTRE GEORGE 2L MK3F STREAM P RUN ON 16/05/75

JOB IMCPOLY,AM,IMC2084
HPFORTRAN
****
    SOFOR COMPILATION SYSTEM MARK 54A

```
      MASTER POLYNOMIAL
C     THIS PROGRAM GIVES INSTRUCTIONS FOR CALCULATING THE VALUE OF A POLYNOMIAL.
      READ(1,1)N,X,A
C     N IS THE DEGREE OF THE POLYNOMIAL. ITS VALUE WILL BE USED TO DECIDE
C     WHEN THE FINAL STAGE IN THE CALCULATION HAS BEEN REACHED.
    1 FORMAT(I3,F6.2,F8.2)
      WRITE(2,5)N,A
    5 FORMAT(40H FOR THE PARTICULAR POLYNOMIAL OF DEGREE,I3,18H WITH COE
     1FFICIENTS/1H ,F8.2)
      P=A
      I=0
    2 READ(1,3)A
    3 FORMAT(F8.2)
      WRITE(2,6)A
    6 FORMAT(1H ,F8.2)
      P=P*X+A
      I=I+1
C     THE VALUE OF I WILL BE INCREASED BY 1 UNTIL IT EQUALS N
      IF(I.LT.N)GO TO 2
      WRITE(2,4)X,P
    4 FORMAT(19H THE VALUE WHEN X =,F6.2,5H  IS ,F12.6)
      STOP
      END

      FINISH
```

```
FOR THE PARTICULAR POLYNOMIAL OF DEGREE  5 WITH COEFFICIENTS
    10.03
     7.62
     6.53
    -7.67
   -18.32
   143.21
THE VALUE WHEN X =  2.36  IS  1107.538385
```

Computer print-out for the evaluation of a polynomial, the details of which are given in FRAME 36.

Solution of Non-Linear Equations

Introduction

It is by no means uncommon for non-linear and transcendental equations to occur in practice. By the term 'non-linear equations' is meant those equations that involve powers other than the first, for example, $x^4 - x^3 + 5x - 7 = 0$. This is a quartic equation as it involves x^4. Transcendental equations are those which involve functions other than just powers of x, for example $e^x = \cos x + x^2$.

There are a few non-linear and transcendental equations that are readily solvable but these are definitely in the minority. Some examples of such equations are $2x^2 - 3x - 5 = 0$, $\sin x = 1/2$ and $e^x = 1$. Where no simple method exists for solving equations like these (only rather more complicated) numerical methods are frequently employed and it is the purpose of this programme to investigate some of these numerical techniques. Before doing so, however, let us have a look at a few practical examples where some equations requiring these techniques arise.

If you have read the programme "Partial Differential Equations for Technologists" in our book, "Mathematics for Engineers and Scientists, Vol. 2" in this series, you will have come across the equation $m \frac{\partial^2 y}{\partial t^2} + EI \frac{\partial^4 y}{\partial x^4} = 0$ for the transverse oscillations of a thin rod.

Under certain conditions, i.e., when the rod is clamped horizontally at both ends, it is found that vibrations take place only if the equation $\cos n\ell \cosh n\ell = 1$ is satisfied. In a similar way, if the beam is pinned at the end where $x = \ell$ instead of being clamped there, then the corresponding equation is $\tan n\ell = \tanh n\ell$.

In a certain crank mechanism, Freudenstein's equation $R_1 \cos \theta - R_2 \cos \phi + R_3 - \cos(\theta - \phi) = 0$, giving the relation between θ and ϕ, the input and output crank angles, must be satisfied.

Although it is possible to find ϕ as a function of θ directly from this equation, a numerical method of solution is sometimes preferred.

Taking off now into space, so to speak, it is found that if a missile is subjected to a thrust, then the equation $x^2(1 - \cos x \cosh x) - \gamma \sin x \sinh x = 0$ is associated with its flexural vibrations. γ is a constant which would be fixed in particular circumstances, and the solution for x required.

As a final example, if two space vehicles are describing the same orbit about the earth and a link-up between them is required, it is necessary to solve for e an equation of the form

$$\frac{\phi}{2} = \left(\frac{1 + e \cos \theta}{e^2 - 1} \right)^{3/2} \left[\frac{e \sqrt{e^2 - 1}}{1 + e \cos \theta} \sin \theta - \ln \left(\frac{\sqrt{e + 1} + \sqrt{e - 1} \tan \frac{1}{2}\theta}{\sqrt{e + 1} - \sqrt{e - 1} \tan \frac{1}{2}\theta} \right) \right]$$

e being the eccentricity of what is known as the transfer orbit. θ and ϕ are fixed in any particular problem.

Numerical techniques often make use of a process known as ITERATION. Before actually discussing non-linear equations themselves, we will have a look at what is meant by this process.

Iteration

First let us have a look at the formula $y = \frac{1}{2}\left(x + \frac{N}{x}\right)$ (4.1)

with $N = 1973$ and find the value of y when $x = 40$. It gives

$$y = \frac{1}{2}\left[40 + \frac{1973}{40}\right] \simeq \frac{1}{2}(40 + 49 \cdot 3) \simeq 44 \cdot 6$$

Now suppose $44 \cdot 6$ is substituted for x instead of 40. The result this time is

$$y = \frac{1}{2}\left[44 \cdot 6 + \frac{1973}{44 \cdot 6}\right] \simeq \frac{1}{2}(44 \cdot 6 + 44 \cdot 24) \simeq 44 \cdot 42$$

If now x is put equal to $44 \cdot 42$, the value of y is found to be $\frac{1}{2}(44 \cdot 42 + 44 \cdot 416\ 92) \simeq 44 \cdot 418\ 46$. Putting $x = 44 \cdot 418\ 46$ then gives $y = \frac{1}{2}(44 \cdot 418\ 46 + 44 \cdot 418\ 47) \simeq 44 \cdot 418\ 46$.

You will notice that x and N/x have been getting closer and closer to each other as we have proceeded, and at the last stage they were equal to one another to 6 significant figures. If the process is continued, you will find that they become equal to an even higher degree of accuracy. It is because of this gradual approach to equality that the number of significant figures has been gradually increased at each stage.

What can you say is the relationship between x and N when the difference between x and N/x is negligible?

****

$$x = \frac{N}{x} \quad \text{gives} \quad x = \sqrt{N}.$$

Now see what happens if you take $N = 57 \cdot 46$ and use the formula (4.1) as in the last frame, starting with $x = 8$. What is value of y when you have used it 3 times? Give your answer to 5 significant figures if possible, otherwise stop at 4 or even 3.

****

Successive values of y are $7 \cdot 59$, $7 \cdot 5802$, $7 \cdot 5802$. The fact that the last two are the same indicates that you have arrived at the value you are looking for.

$7 \cdot 5802$ is the square root of $57 \cdot 46$ correct to 5 significant figures. Continuing the process in this case very quickly leads to the more accurate result $7 \cdot 580\ 237\ 46$.

As this is a repetitive process, it is very convenient for programming on to a computer. A computer print out for this particular problem would give the values of successive calculations as

SOLUTION OF NON-LINEAR EQUATIONS

VALUE OF X USED	VALUE OF CORRESPONDING Y
	8.00000000
1ST	7.59125000
2ND	7.58024545
3RD	7.58023746
4TH	7.58023746

You will notice that all these are given to the same number of decimal places.   It is easier to let the computer do this than to program it to increase the number of decimal places as you go along.   The computer would be told to stop when two successive calculations give the same result to a pre-specified degree of accuracy.

The square root of any positive number can be found in this way.   It is necessary to make an initial guess as to the first value of x to take, but this is not difficult.   It helps if you take it reasonably close to the actual value.

FRAME 7

The process of finding successive approximations to a quantity, as was done in FRAMES 4 and 5, is called an ITERATIVE PROCESS.   Each use of the particular formula $\big($(4.1) in our examples$\big)$ is an ITERATION and each successive approximation is an ITERATE.

In FRAME 4, the first value of y was taken as the second value of x, the second value of y as the third value of x and so on.   In practice, however, a different notation is used, in order to avoid possible confusion.   The initial guess is often labelled x_0, the first value of y, i.e., the first iterate, x_1, the second x_2 and so on. Then (4.1) would lead to the successive equations

$$x_1 = \frac{1}{2}\left(x_0 + \frac{N}{x_0}\right) \qquad x_2 = \frac{1}{2}\left(x_1 + \frac{N}{x_1}\right) \qquad x_3 = \frac{1}{2}\left(x_2 + \frac{N}{x_2}\right) \qquad \text{etc.}$$

These can all be combined into the single equation

$$x_{n+1} = \frac{1}{2}\left(x_n + \frac{N}{x_n}\right) \qquad (7.1)$$

where n takes the values 0, 1, 2 .... in turn.
This notation is used in the flow diagram (illustrated on page 31) for the process of finding a square root by this method.

FRAME 8

The equation corresponding to (7.1) for finding a cube root is

$$x_{n+1} = \frac{1}{3}\left(2x_n + \frac{N}{x_n^{\,2}}\right)$$

$\left[\text{Don't worry for the time being how the formulae } x_{n+1} = \frac{1}{2}\left(x_n + \frac{N}{x_n}\right)\right.$
$\text{and } x_{n+1} = \frac{1}{3}\left(2x_n + \frac{N}{x_n^{\,2}}\right) \text{ are found.   You will discover this later.}\Big]$

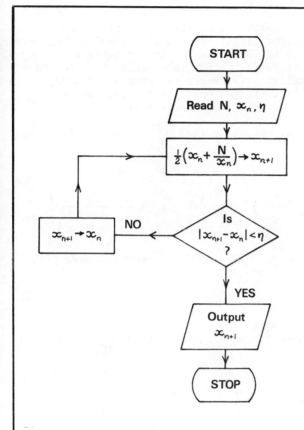

The value of x_n here will be x_0, the initial approximation to $\sqrt{N}$.

η is the pre-specified degree of accuracy mentioned in FRAME 6. As the computer may be working to more decimal places than shown in the print-out, η could be set at, say, $0\cdot000\ 000\ 001$. The computer would then cease the calculation when two successive iterates differ by less than this amount.

Flow Diagram for FRAME 7.

$\Big($ Programs for finding the square root of a number may be found in references (1) and (3). $\Big)$

FRAME 8   (continued)

By taking a suitable value for x_0, use this formula to find $\sqrt[3]{70\cdot13}$ to 5 significant figures.   (Again, stop at fewer significant figures if you can't manage 5.)

****

8A

As $4^3 = 64$ and $5^3 = 125$, $x_0 = 4$ is a reasonable initial guess. Successive iterates are

$$x_1 = 4\cdot13, \qquad x_2 = 4\cdot1238, \qquad x_3 = 4\cdot1238.$$

SOLUTION OF NON-LINEAR EQUATIONS

In the examples that have been taken so far, the approximations have got better and better as the number of iterations has increased. When this happens, the process is said to be CONVERGENT. In the cases taken, this convergence has been quite rapid. That this was to be expected can easily be shown theoretically. (If a process does not converge, it will be of no use, and if it converges only slowly, the numerical working involved is going to be considerably increased.)

Taking $x_{n+1} = \frac{1}{2}\left(x_n + \frac{N}{x_n}\right)$, let us assume that x_n is reasonably close to $\sqrt{N}$.

Then we can write $\sqrt{N} - x_n = \epsilon_n$, and so $x_n = \sqrt{N} - \epsilon_n$.

$$\therefore \quad x_{n+1} = \frac{1}{2}\left(\sqrt{N} - \epsilon_n + \frac{N}{\sqrt{N} - \epsilon_n}\right) = \frac{1}{2}\left\{\sqrt{N} - \epsilon_n + \frac{N}{\sqrt{N}\left(1 - \frac{\epsilon_n}{\sqrt{N}}\right)}\right\}$$

Now, as x_n is assumed to be near $\sqrt{N}$, ϵ_n will be small compared with $\sqrt{N}$ and so $\left|\frac{\epsilon_n}{\sqrt{N}}\right| < 1$.

By writing $\dfrac{1}{1 - \frac{\epsilon_n}{\sqrt{N}}}$ as $\left(1 - \frac{\epsilon_n}{\sqrt{N}}\right)^{-1}$ and expanding, find x_{n+1} as a

series in ϵ_n as far as the term in $\epsilon_n{}^3$.

****

9A

$$x_{n+1} = \frac{1}{2}\left\{\sqrt{N} - \epsilon_n + \sqrt{N}\left(1 + \frac{\epsilon_n}{\sqrt{N}} + \frac{\epsilon_n{}^2}{N} + \frac{\epsilon_n{}^3}{N\sqrt{N}} + \cdots\cdots\cdots\right)\right\}$$

$$= \sqrt{N} + \frac{1}{2}\frac{\epsilon_n{}^2}{\sqrt{N}} + \frac{1}{2}\frac{\epsilon_n{}^3}{N} + \cdots\cdots\cdots$$

Thus $\sqrt{N} - x_{n+1} \simeq -\frac{1}{2}\frac{\epsilon_n{}^2}{\sqrt{N}}$. This means that if the error in the nth iterate is of magnitude $|\epsilon_n|$, then $|\epsilon_{n+1}|$, the magnitude of the error in the (n+1)th iterate is of magnitude $\frac{1}{2}\frac{\epsilon_n{}^2}{\sqrt{N}}$. As $\left|\frac{\epsilon_n}{\sqrt{N}}\right| < 1$, this error is smaller in magnitude than the previous one. Similarly the error in the next iterate will be smaller than this one, and so on. A process in which the error at one stage is approximately a multiple of the square of the error at the previous stage is said to be a SECOND ORDER PROCESS.

Now take the formula $x_{n+1} = \dfrac{1}{3}\left(2x_n + \dfrac{N}{x_n^2}\right)$ and proceed similarly to

show that the method based on this formula for finding $N^{1/3}$ is also a second order process.

****

10A

Let $N^{1/3} - x_n = \varepsilon_n$ then $x_n = N^{1/3} - \varepsilon_n$ and

$$x_{n+1} = \frac{1}{3}\left\{2(N^{1/3} - \varepsilon_n) + \frac{N}{(N^{1/3} - \varepsilon_n)^2}\right\} = \frac{1}{3}\left\{2(N^{1/3} - \varepsilon_n) + N^{1/3}\left(1 - \frac{\varepsilon_n}{N^{1/3}}\right)^{-2}\right\}$$

$$= \frac{1}{3}\left\{2(N^{1/3} - \varepsilon_n) + N^{1/3}\left(1 + \frac{2\varepsilon_n}{N^{1/3}} + \frac{3\varepsilon_n^2}{N^{2/3}} + \dots\right)\right\} \simeq N^{1/3} + \frac{\varepsilon_n^2}{N^{1/3}}$$

i.e. $N^{1/3} - x_{n+1} \simeq -\dfrac{\varepsilon_n^2}{N^{1/3}}$

Solution of Non-Linear Equations by means of the Iteration Formula

$$x_{n+1} = F(x_n)$$

Having seen the basic ideas behind the process of iteration, the next thing is to consider how this can be applied to find the solution of an equation for which we do not know a simple analytical technique.

For a first example, let us take the equation

$$x^3 - 4x^2 + x - 10 = 0$$

Now the two formulae that were used earlier were

$$x_{n+1} = \frac{1}{2}\left(x_n + \frac{N}{x_n}\right) \qquad \text{and} \qquad x_{n+1} = \frac{1}{3}\left(2x_n + \frac{N}{x_n^2}\right)$$

i.e. they were of the form $x_{n+1} = F(x_n)$.   Furthermore, in the later stages, x_{n+1} was very nearly equal to x_n.   This suggests that it might be a help to rearrange $x^3 - 4x^2 + x - 10 = 0$ so that it is in the form $x = F(x)$.   There is obviously more than one way in which this can be done.   Some possibilities are

$$x = 10 - x^3 + 4x^2, \quad x = \tfrac{1}{2}\sqrt{x^3 + x - 10} \quad \text{(or } -\tfrac{1}{2}\sqrt{x^3 + x - 10}),$$

$$x = \frac{4x^2 - x + 10}{x^2}, \quad x = \frac{x^3 + x - 10}{4x}, \quad x = \sqrt[3]{4x^2 - x + 10}$$

Taking the first of these would suggest $x_{n+1} = 10 - x_n^3 + 4x_n^2$,   the second $x_{n+1} = \tfrac{1}{2}\sqrt{x_n^3 + x_n - 10}$ and so on.

Now with an iteration process, it is necessary to find, by some other means, an initial starting point.   Can you suggest any ways in which this can be done?

****

33

SOLUTION OF NON-LINEAR EQUATIONS

Various methods are possible, such as:

i) *A rough sketch of the curve* $y = x^3 - 4x^2 + x - 10$. x_0 *would be taken close to where this curve crosses the x-axis.*

ii) *Rough sketches, on the same axes, of* $y = x^3$ *and* $y = 4x^2 - x + 10$. x_0 *would be taken close to the value of* x *where they cross.*

iii) *A table of values of* x *and* y *where* $y = x^3 - 4x^2 + x - 10$. *At some point in the table the sign of* y *will change.* x_0 *would be taken close to where this happens.*

You should have met at least some of these methods at school.

FRAME 12

Using any one of the methods in 11A, see if you can find a suitable value for x_0. (You should find the required value lies between -2 and 6.)

****

12A

$x_0 = 4$

(*iii*) *A table of values of* y *against* x *is*

x	-2	-1	0	1	2	3	4	5	6
y	-36	-16	-10	-12	-16	-16	-6	20	68

(*i*) *The table in (iii) gives rise to the sketch in Figure (i).*
(*ii*) *New sets of values would be required and Figure (ii) would result.*

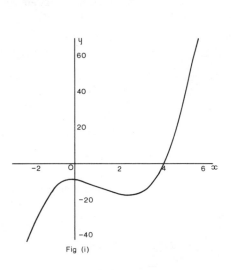

Fig (i)

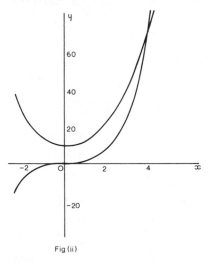

Fig (ii)

Note that this cubic equation only has one real root. For the time being, equations having more than one real root will not be introduced.

Now, taking $x_{n+1} = 10 - x_n^3 + 4x_n^2$ and $x_0 = 4$ gives

$$x_1 = 10, \qquad x_2 = -590, \qquad x_3 = -203\,986\,590,$$

and the whole process is obviously going wild.

But if we take $x_{n+1} = \dfrac{4x_n^2 - x_n + 10}{x_n^2}$ and $x_0 = 4$, then

$x_1 = 4 \cdot 4$, $x_2 = 4 \cdot 29$, $x_3 = 4 \cdot 310$, $x_4 = 4 \cdot 3063$, $x_5 = 4 \cdot 3070$, $x_6 = 4 \cdot 3069$, $x_7 = 4 \cdot 3069$

The fact that the last two results are the same indicates that the solution has now been obtained correct to 5 significant figures.

Two points are of interest here:

1) Why did the first formula for x_{n+1} lead to a ridiculous result, or rather, to no result?

2) Rather more steps had to be taken (up to x_7) than when square and cube roots were being obtained, for a similar degree of accuracy.

However, before going on to consider these points, you might like to find the root of $x^3 - 5x^2 - 29 = 0$, correct to 4 significant figures.
****

13A

A table of values gives, if $y = x^3 - 5x^2 - 29$,

x	0	1	2	3	4	5	6	7
y	-29	-33	-41	-47	-45	-29	7	69

and this suggests $x_0 = 6$.

From $x^3 - 5x^2 - 29 = 0$, $x = \dfrac{5x^2 + 29}{x^2}$

and so one formula for x_{n+1} *is* $x_{n+1} = \dfrac{5x_n^2 + 29}{x_n^2}$

If you tried $x = \dfrac{x^3 - 29}{5x}$, *you will have found that this will not give a solution.*

Successive approximations are:

$x_1 = 5 \cdot 8$, $x_2 = 5 \cdot 86$, $x_3 = 5 \cdot 845$, $x_4 = 5 \cdot 849$, $x_5 = 5 \cdot 848$, $x_6 = 5 \cdot 848$.

5·848 is the required root.

To find the answer to point number 1) in the last frame, a method is adopted which is somewhat similar to that used in FRAME 9.

Let the equation to be solved be $f(x) = 0$, and assume that this has been written in the form $x = F(x)$ so that the iteration formula $x_{n+1} = F(x_n)$ is used. Now let $x = a$ be the actual value of the root sought. Then $f(a) = 0$ and also $a = F(a)$. If x_n is an approximation for a, then we can write

$$a - x_n = \varepsilon_n$$

where, if x_n is a reasonably good approximation, ε_n is small.   Thus $x_n = a - \varepsilon_n$ and so

$$x_{n+1} = F(a - \varepsilon_n)$$

The R.H.S. of this equation can now be expanded by Taylor's series in powers of ε_n.   By using this expansion as far as the term in ε_n, find an expression for ε_{n+1}, where $\varepsilon_{n+1} = a - x_{n+1}$.

****

14A

$$a - \varepsilon_{n+1} \simeq F(a) - \varepsilon_n F'(a) \qquad (14A.1)$$

$$\varepsilon_{n+1} \simeq \varepsilon_n F'(a) \quad as \quad a = F(a)$$

FRAME 15

The iteration process is based on the assumption that the further we go, the less becomes the magnitude of the difference between a and each successive iterate.   This means that $|\varepsilon_{n+1}| < |\varepsilon_n|$.   Your answer in 14A shows that this can be expected if $|F'(a)| < 1$.   The word 'expected' has been used here because (14A.1) is not exact as the powers of ε_n have been neglected.   When testing $F(x)$ to see whether $|F'(a)| < 1$ or not, you will realise that $F'(a)$ cannot be found until a is known.   But in order to find whether the iterations will converge to a the value of $F'(a)$ is required!   To avoid this difficulty $F'(x)$ is tested in the neighbourhood of the root and if $|F'(x)| < 1$ for values of x in this neighbourhood then the iteration process can be expected to converge.

Returning to the example in FRAME 11, the first expression used in FRAME 13 for $F(x)$ was $10 - x^3 + 4x^2$.   For this $F'(x) = -3x^2 + 8x$ which, when $x = 4$, becomes -16.   ($x = 4$ was the first approximation used for the root).   Thus $|F'(4)| = 16$ which is certainly not < 1.   As you saw in FRAME 13, the iterations using this formula for $F(x)$ did not converge.   Even if you take a value of x closer to the actual root than 4, $|F'(x)|$ will still not become < 1.

Using $F(x) = \dfrac{4x^2 - x + 10}{x^2}$, $F'(x) = \dfrac{1}{x^2} - \dfrac{20}{x^3}$ and now $F'(4) = -\dfrac{1}{4}$, so $|F'(4)| < 1$.   As you will remember, the process worked for this $F(x)$.

Three other possibilities for $F(x)$ were given in FRAME 11.   Would any of them be suitable for finding the root of the equation $x^3 - 4x^2 + x - 10 = 0$?

****

15A

Yes: $\sqrt[3]{4x^2 - x + 10}$.   *If this is used however, the convergence is not so rapid as for* $\dfrac{4x^2 - x + 10}{x^2}$.   *This is because* $|F'(x)|$, *although* < 1, *is larger for* $\sqrt[3]{4x^2 - x + 10}$ *than it is for* $\dfrac{4x^2 - x + 10}{x^2}$.

What is an estimate for the value of $|F'(x)|$ near the root of
$x^3 - 5x^2 - 29 = 0$ when $F(x) = \dfrac{5x^2 + 29}{x^2}$?   (You found the root of
$x^3 - 5x^2 - 29 = 0$ in 13A.)
****

$$F(x) = -\frac{58}{x^3} \ , \qquad |F'(6)| = \frac{58}{216}$$

In FRAME 13 the remark was made that in the example worked there it was
necessary to take more steps than was the case with the square and cube
roots evaluated earlier, for a similar degree of accuracy.   From the
work that has been done, can you suggest any theoretical reason why this
might be so?   (This is not an easy question so – HINT – examine very
closely the first paragraph of FRAME 10 and the working in 14A for a
clue.)
****

*In the case in FRAME 10, the error at one stage was approximately a
multiple of the square of the error at the previous stage.   In 14A, you
found that for the process in FRAME 13, the error at one stage was
approximately a multiple of the actual error at the previous stage, not
the square of the error.   For an error numerically less than one, the
square of the error will be smaller than the actual error itself and so
the convergence will be more rapid.   You will remember that the process
in FRAME 10 is called a second order process.   Similarly one where ε_{n+1}
is a multiple of ε_n is called a FIRST ORDER PROCESS.*

A simple flow chart for finding the root of an equation that has been put
in the form $x = F(x)$ is shown on page 38.

The Newton-Raphson Iteration Formula

You have seen that, given an equation $f(x) = 0$, it is possible to find
a solution if it can be put in the form $x = F(x)$, where $F(x)$ is such
that $|F'(x)| < 1$ in the neighbourhood of the root.   Further, the
smaller is $F'(x)$, the quicker will the solution be obtained.

Now let us return to the equation $x^2 = N$.   (Effectively this was the
equation that was being solved at the beginning of the programme when $\sqrt{N}$
was being found.)   Can this be written in the form $x = F(x)$ in such a
way that $|F'(x)| < 1$?   The only obvious ways of rewriting $x^2 = N$ are
$x = \sqrt{N}$ and $x = \dfrac{N}{x}$.   The former isn't much help where iteration is
concerned as $F(x)$ must really contain x, not just be a constant.   If
you try the latter, the iteration formula would be

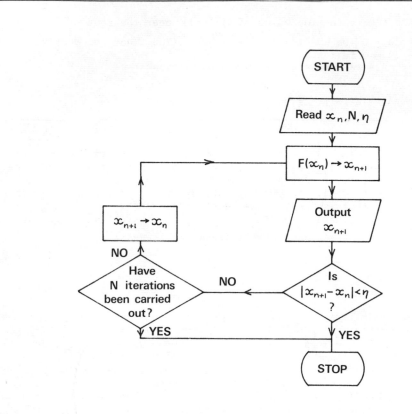

Flow Diagram for FRAME 18.

Note:- The purpose of N is to stop the computer carrying on for evermore should it meet a situation like that at the beginning of FRAME 13.   If the root, to the accuracy specified by η, has not been reached after N steps, the process is terminated.

(A program using the basic iteration method may be found in reference (6).)

FRAME 19   (continued)

$$x_{n+1} = \frac{N}{x_n}$$

and using the same figures as in FRAME 4, i.e. N = 1973 and $x_0 = 40$,

$$x_1 = \frac{1973}{40} \simeq 49 \cdot 3 \qquad\qquad x_2 = \frac{1973}{49 \cdot 3} \simeq 40$$

and all that will happen is an oscillation between 40 and 49·3, which isn't very helpful.   Now although it isn't obvious why this should be done, if, in this particular case the equation $x = \dfrac{N}{x}$ is written as

$$2x = x + \frac{N}{x} \quad \text{then we have} \quad x = \frac{1}{2}\left(x + \frac{N}{x}\right)$$

This leads to the iteration formula $x_{n+1} = \dfrac{1}{2}\left(x_n + \dfrac{N}{x_n}\right)$, which, as you have already seen, works.

The NEWTON-RAPHSON formula provides us with a means, in many cases, of writing down an iteration formula that enables us to get somewhere.

If you have read FRAMES 43-44, pages 1:190 - 1:191, of our book 'Mathematics for Engineers and Scientists, Vol. 1', you will have seen how the formula $x_2 = x_1 - \dfrac{f(x_1)}{f'(x_1)}$ can be obtained by using Taylor's Series. In a similar way, starting from an approximation x_n instead of x_1, the formula

$$x_{n+1} = x_n - \frac{f(x_n)}{f'(x_n)}$$

would result. Although the theoretical approach by Taylor's Series is important, an alternative, graphical, way of obtaining this equation is as follows:

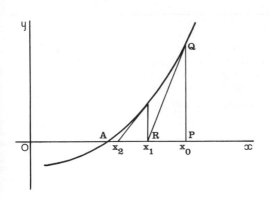

If the solution to the equation $f(x) = 0$ is required, a start is made by considering the graph of the function $y = f(x)$ in the neighbourhood of the solution. Let this be of the form shown. Also let x_0 be an approximate solution. Then if x_0 be substituted for x in $f(x)$, the result will be the ordinate PQ. At Q draw the tangent to the curve. Let this tangent meet Ox in R. Then, for the situation as shown, x_1 is a better approximation to the root than is x_0. If the process is repeated, x_2 will be a better approximation than x_1 and so on.

By working on $\triangle PQR$, can you obtain the result

$$x_1 = x_0 - \frac{f(x_0)}{f'(x_0)} \ ?$$

****

$$OR = OP - RP$$
$$= OP - PQ \cot P\hat{R}Q$$
$$= OP - \frac{PQ}{\tan P\hat{R}Q}$$
$$x_1 = x_0 - \frac{f(x_0)}{f'(x_0)}$$

In a similar way, $x_2 = x_1 - \dfrac{f(x_1)}{f'(x_1)}$ and generally $x_{n+1} = x_n - \dfrac{f(x_n)}{f'(x_n)}$.

What will this last formula give if

(i) $f(x) \equiv x^2 - N$ (ii) $f(x) \equiv x^3 - N$?

****

(i) $x_{n+1} = x_n - \dfrac{x_n^2 - N}{2x_n}$ *(ii)* $x_{n+1} = x_n - \dfrac{x_n^3 - N}{3x_n^2}$

       $= \dfrac{1}{2}\left(x_n + \dfrac{N}{x_n} \right)$ $= \dfrac{1}{3}\left(2x_n + \dfrac{N}{x_n^2} \right)$

These are the formulae used earlier when finding the square and cube roots of N.

In the case illustrated in FRAME 20, the sequence of numbers x_0, x_1, x_2 ..... was decreasing and tending to the actual solution. Now take x_0 on the other side of A, make a similar construction to that made at P previously and see what happens to x_1, x_2 etc.

****

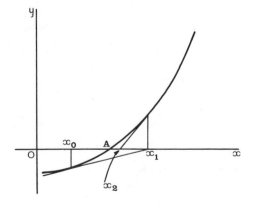

x_1 is on the opposite side of A from x_0. It may even be further away from A than x_0 is itself. After x_1 has been obtained, the ordinary process, described in FRAME 20, takes place.

In a similar way, the formula will give a method of finding the root of
$f(x) = 0$ when $y = f(x)$ takes on a form such as one of the following:

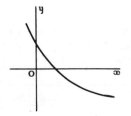
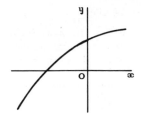
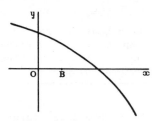

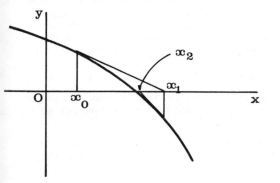

You can easily verify by
construction that the desired
result is going to be obtained
in any of these cases. Thus
taking x_0 at B in the last
figure, x_1, x_2 etc., are as
shown.

It may seem that these diagrams cover all conceivable cases. However,
as you will see later, difficulties with this method can occur, but for
the time being, no awkward examples will be taken.

The solution of $x^3 - 4x^2 + x - 10 = 0$ was found in FRAMES 11-13 by
rewriting this equation in the form $x = F(x)$, it being arranged so that
$|F'(x)| < 1$ in the neighbourhood of the root. This equation will now
be used to illustrate the Newton-Raphson process. Taking $f(x)$ as
$x^3 - 4x^2 + x - 10$, what will be the formula for x_{n+1} in terms of x_n?

****

24A

$$f'(x) = 3x^2 - 8x + 1$$

$$x_{n+1} = x_n - \frac{x_n^3 - 4x_n^2 + x_n - 10}{3x_n^2 - 8x_n + 1} \qquad (24A.1)$$

As in FRAME 11, it is necessary to find a value for x_0 by some other
means. Taking, as there, x_0 to be 4, we have

$$x_1 = 4 - \frac{4^3 - 4 \times 4^2 + 4 - 10}{3 \times 4^2 - 8 \times 4 + 1} \simeq 4 + 0 \cdot 35 = 4 \cdot 35$$

$$x_2 = 4 \cdot 35 - 0 \cdot 0424 = 4 \cdot 3076$$

$$x_3 = 4 \cdot 3076 - 0 \cdot 0007 = 4 \cdot 3069$$

$$x_4 = 4 \cdot 3069 - 0 \cdot 0000 = 4 \cdot 3069$$

Note that, as was suggested earlier in a similar situation, the number of figures in the working is increased as we proceed.   The equality of x_3 and x_4 indicates that the solution has been obtained accurate to 4 decimal places.

There are a few points to notice at this stage about the solution. Firstly, the R.H.S. of (24A.1) may either be left as it is, or put as a single fraction, i.e., $\dfrac{2x_n^{\ 3} - 4x_n^{\ 2} + 10}{3x_n^{\ 2} - 8x_n + 1}$.   Whether or not this is done, both numerator and denominator can be nested for the evaluation of the polynomials.   If the R.H.S. of (24A.1) is not combined into a single fraction, the numerator of the last term on the R.H.S. will tend to zero as the root is approached.   This is because it is of the same form as the left hand side of the original equation, which is zero at the root.

Secondly, you will notice that on the first application of the process we jumped across the root from 4 to $4 \cdot 35$.   The situation is similar to that illustrated in 22A.

Thirdly, the fact that x_4 and x_3 are the same indicates that the root has been obtained to the required degree of accuracy.   More about this later, however.

Fourthly, the root has been obtained after fewer iterations than were necessary in FRAME 13.

By now you will have realised that iterative processes are simply a series of repetitions.   As such they are very convenient for programming on to a computer.   A set of instructions to perform operations which lead to the solution of a problem is called an ALGORITHM.

Now use this process to solve, correct to 4 significant figures,

  (i) $x^3 - 5x^2 - 29 = 0$ (ii) $e^x = 10 - x$

The first of these you have already met and the second was mentioned in the previous programme.

****

i) $x_{n+1} = x_n - \dfrac{x_n^{\ 3} - 5x_n^{\ 2} - 29}{3x_n^{\ 2} - 10x_n}$.   *Taking, as before, $x_0 = 6$, leads*

to $x_1 = 5 \cdot 85$, $x_2 = 5 \cdot 848$, $x_3 = 5 \cdot 848$.

Notice that, once again, this process gives the result more quickly than that used in 13A.

ii) $f(x) = e^x + x - 10$, $\quad f'(x) = e^x + 1$, $\quad x_{n+1} = x_n - \dfrac{e^{x_n} + x_n - 10}{e^{x_n} + 1}$

x	0	1	2	3
$f(x)$	-9	$-6 \cdot 28$	$-0 \cdot 61$	$13 \cdot 09$

The table shows $x_0 = 2$ *to be a suitable starting point.*
$x_1 = 2 \cdot 073$, $\qquad x_2 = 2 \cdot 071$, $\qquad x_3 = 2 \cdot 071$

FRAME 27

The following is a flow diagram for finding a root of $f(x) = 0$ by the Newton-Raphson method.

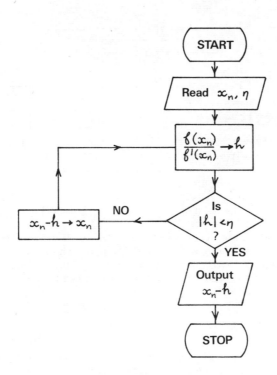

[Programs using the Newton-Raphson method may be found in references (1), (3), (4), (5), (6), (7) and (9).]

FRAME 28

In the examples taken so far, the iterations have converged quite rapidly to the root.   Is this coincidence, or is it to be expected?   To investigate this, a process similar to that in FRAME 14 is used.   With the same notation as adopted there,

$$x_{n+1} = a - \varepsilon_n - \frac{f(a - \varepsilon_n)}{f'(a - \varepsilon_n)}$$

Assuming that ε_n is small compared with a and remembering that $f(a) = 0$, show, by using Taylor's series to expand both $f(a - \varepsilon_n)$ and $f'(a - \varepsilon_n)$, and then the binomial series, that this gives

$$x_{n+1} \simeq a + \frac{1}{2} \varepsilon_n^2 \frac{f''(a)}{f'(a)} \qquad \text{Assume } f'(a) \neq 0.$$

$$************************************$$

28A

$$x_{n+1} = a - \varepsilon_n - \frac{f(a) - \varepsilon_n f'(a) + \frac{1}{2}\varepsilon_n^2 f''(a) \ldots\ldots\ldots}{f'(a) - \varepsilon_n f''(a) + \frac{1}{2}\varepsilon_n^2 f'''(a) \ldots\ldots}$$

$$= a - \varepsilon_n + \frac{\varepsilon_n \left\{ 1 - \frac{1}{2}\varepsilon_n \frac{f''(a)}{f'(a)} \ldots\ldots\ldots\ldots \right\}}{1 - \varepsilon_n \frac{f''(a)}{f'(a)} + \frac{1}{2}\varepsilon_n^2 \frac{f'''(a)}{f'(a)} \ldots\ldots}$$

$$= a - \varepsilon_n + \varepsilon_n \left\{ 1 - \frac{1}{2}\varepsilon_n \frac{f''(a)}{f'(a)} \ldots\ldots \right\} \left\{ 1 + \varepsilon_n \frac{f''(a)}{f'(a)} \ldots\ldots \right\}$$

$$= a - \varepsilon_n + \varepsilon_n \left\{ 1 - \frac{1}{2}\varepsilon_n \frac{f''(a)}{f'(a)} + \varepsilon_n \frac{f''(a)}{f'(a)} \ldots\ldots \right\}$$

$$\simeq a + \frac{1}{2}\varepsilon_n^2 \frac{f''(a)}{f'(a)}$$

Therefore $\quad a - x_{n+1} = -\frac{1}{2}\varepsilon_n^2 \frac{f''(a)}{f'(a)} \quad$ or $\quad \varepsilon_{n+1} = -\frac{1}{2}\varepsilon_n^2 \frac{f''(a)}{f'(a)} \qquad (29.1)$

Thus, the Newton-Raphson iteration formula is a second order process where the error at one stage is a multiple of the square of the error at the previous stage.   Unless the multiplying factor is high (which will occur if $f'(a)$ is small and/or $f''(a)$ is large) the convergence will therefore be quite rapid.

This means that very often one can tell that the root of an equation has been reached without getting two equal iterates.   To illustrate this, let us return to the equation $x^3 - 4x^2 + x - 10 = 0$, solved in FRAME 25.

The changes from x_0 to x_1, x_1 to x_2, etc., are $0 \cdot 35$, $-0 \cdot 0424$, $-0 \cdot 0007$, $-0 \cdot 0000$.   Taking into account magnitude only the second is about one third of the square of the first and the third, one third of the square of the second.   The fourth can therefore be expected to be in the region of $\frac{1}{3} \times 0 \cdot 0007^2$, which to four decimal places, is negligible.

Now examine the results obtained in 26A for the two equations you solved there and see whether you would have been justified in stopping at x_2 in each case.

$$************************************$$

i)   *The changes are* $-0 \cdot 15$, $-0 \cdot 002$, $0 \cdot 000$.

Second $\simeq \frac{1}{10} \times$ *square of first.   Third can be expected to be in the region of* $0 \cdot 002^2/10$, *i.e., negligible to 3 decimal places.*

ii) *The changes are* $0 \cdot 073$, $-0 \cdot 002$, $0 \cdot 000$.

Second $\simeq \frac{2}{5} \times$ *square of first.   Third can be expected to be in the region of* $\frac{2}{5} \times 0 \cdot 002^2$ *which is negligible to 3 decimal places.*

Hence in each case, a stop could have been made at x_2. *But this idea is only useful when you are performing the calculations manually, i.e. with tables or on a desk machine. When a computer is being used it is easier to let it carry on until it gets two equal iterates than to try and incorporate this more sophisticated test into a program.*

In practice, you will see that this means that if the changes are going down rapidly, you can stop when a change becomes very small, having regard to the accuracy to which you are working.

Now try the following two problems, which are examples of some applications of this work.

1. The equation $x - 0 \cdot 2 \sin x = 0 \cdot 5$ is a particular case of Kepler's equation which is used in computing the orbits of satellites.   Find the solution correct to 4 decimal places.

2. In the design of high voltage tubular electrical insulators, the equation $Q = \dfrac{\pi q^2 (x^2 - 1)}{(\ln x)^2}$ occurs.   Assuming q to be a constant, find the value of x (to 4 decimal places) for which Q is a minimum. (Note: $x = 1$ obviously makes the numerator of $\dfrac{dQ}{dx}$ zero but then the denominator is zero also.   The value of x required is the other one that makes the numerator of $\dfrac{dQ}{dx}$ zero.)

         ****

1. $0 \cdot 6155$ *(0, 1 or anything in between is a reasonable starting point.)*

2. $2 \cdot 2185$ *(2 is a reasonable starting point.)*

So far, only examples that have behaved themselves quite well have been taken to illustrate the use of the Newton-Raphson method.   Also, the first approximation has been taken reasonably close to the root and there has only been one real root.   The next examples to have a look at are some which have more than one real root, do not behave quite so nicely or where difficulties can arise if care is not taken.

SOLUTION OF NON-LINEAR EQUATIONS

The Newton-Raphson Formula applied to Examples with more than One Real Root

Some equations, of course, are satisfied by more than one value of x, for example, $x^4 - 6x^2 - 13x + 1 = 0$. To find first approximations, a table of values of $y = x^4 - 6x^2 - 13x + 1$ gives

x	-2	-1	0	1	2	3	4
y	19	9	1	-17	-33	-11	109

This table shows that there are roots near to 0 and 3. To find these roots more accurately, each of these values in turn is used as an x_0 in the Newton-Raphson formula, i.e., in

$$x_{n+1} = x_n - \frac{x_n^4 - 6x_n^2 - 13x_n + 1}{4x_n^3 - 12x_n - 13}$$

Using, first, $x_0 = 0$, successive iterations give $x_1 = 0 \cdot 077$, $x_2 = 0 \cdot 074\,38$, $x_3 = 0 \cdot 074\,37$, and so the root near to 0 is $0 \cdot 074\,37$.

In a similar way, taking $x_0 = 3$ leads to the root $3 \cdot 163\,73$.

Now find the roots of $x^4 - 1 \cdot 12x^3 - 2 \cdot 01x^2 - 3 \cdot 87x - 30 \cdot 72 = 0$, correct to 2 decimal places.

****

A table of values shows that there are two roots, at approximately -2 and 3. Taking x_0 as -2 leads to the root $-2 \cdot 13$ and as 3 to $3 \cdot 14$.

However, sometimes when solving polynomial equations, it is not necessary to use the Newton-Raphson method for all the roots. As an example, let us take the equation $x^3 - 4x^2 + 6 = 0$. Start by finding the largest root correct to 4 decimal places by the Newton-Raphson formula.

****

$3 \cdot 5141$.

The fact that there is a root $3 \cdot 5141$ means that the left hand side has a factor $x - 3 \cdot 5141$. When this factor is removed, we are left with the equation $x^2 - 0 \cdot 4859x - 1 \cdot 7075 = 0$. This is now an ordinary quadratic which can be solved in the usual way to give roots $1 \cdot 5720$ and $-1 \cdot 0862$.

From what has been said here, you will realise that an alternative way of solving the quartic in FRAME 32 is to find one root by Newton-Raphson and then take out the corresponding factor from the left hand side. This will leave a cubic equation to which Newton-Raphson can then be applied.

Now take the equation you solved in 32A, divide the left hand side by $x - 3 \cdot 14$ and apply Newton-Raphson to the resulting cubic. You will then be able to compare the amount of work involved by each method.

****

The cubic is $x^3 + 2 \cdot 02x^2 + 4 \cdot 33x + 9 \cdot 73 = 0$ *which with* $x_0 = -2$ *leads to the root* $-2 \cdot 13$.

The Choice of x_0 for the Newton-Raphson Process

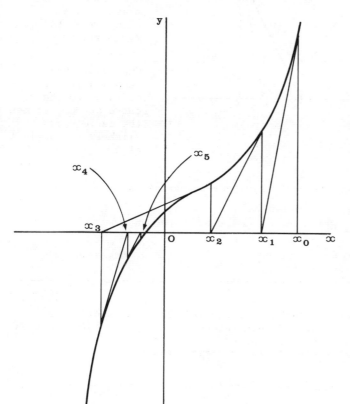

So far, not a great deal has been said about the value you take for x_0. It is more or less common sense to take it close to the root so that the number of iterations required to reach the root is relatively small. But what happens if you don't? Well, in some cases it doesn't make any difference apart from the number of iterations required. The diagram illustrates such a case.

You will notice that the second order property of the process does not hold until we are reasonably close to the root. You will realise that (29.1) is only an approximate formula, ignoring the higher powers of ε_n. Although this is permissible when ε_n is small, it certainly isn't when ε_n is not small.

The diagram shown on the next page illustrates a case where a poor choice of x_0 is absolutely disastrous.

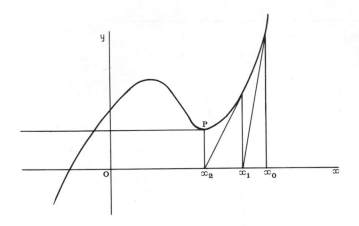

Here there is a minimum at P, $f'(x_2) = 0$ and so x_3 is now at infinity.

What would have happened if the ordinate to the curve at x_2 had met the curve very slightly to the left of P?

****

36A

x_3 *would have been larger than* x_2 *and could even have been larger than* x_0.

Thus, a small value (either +ve or −ve) of $f'(x_n)$ *at any stage can seriously upset the convergence of the process.*

FRAME 37

Returning for a moment to the example in FRAME 32   what would have happened if the larger root had been required and x_0 had been taken as zero, i.e. the simplest possible value?

****

37A

The iterations would have converged to the root $0 \cdot 074\ 37$, *not to the one required.*

This shows that, if an equation has more than one root, and a particular root is required, care must be taken that x_0 *is taken near to* that *root, not to another one.*

FRAME 38

Errors in the Calculation

Assuming now that x_0 has been well chosen and convergence is taking place, various things may happen if you make a mistake in your arithmetic at any point.   What consequences can you think of?   (Suppose the mistake occurs in calculating, say, x_3 from x_2.)

****

Any of the following may occur:

1) *The incorrect value of x_3 may be nearer the root than the correct value. You will have been very lucky as the mistake will have helped you on your way.*

2) *The incorrect value of x_3 may be further from the root but not very much. Further iterations will then probably converge to the root, but more slowly than if x_3 had been correct. This is the type of error referred to in the previous programme as being automatically taken care of. It is often referred to as a self-correcting error.*

3) *The incorrect x_3 may be near to another root of the equation, so that further iterations converge to this other root instead of to the one you are seeking.*

4) *The incorrect x_3 may be in such a position that in the end you get nowhere. For example, a situation similar to that in FRAME 36 might now occur.*

Equal or Nearly Equal Roots

If the graph of a function is of the form shown in the figure, there are two roots very close together in the neighbourhood of A.   If the local maximum shown there is actually on the x - axis, then these two roots become coincident.   Both cases lead to difficulty with the Newton-Raphson formula as dy/dx is small or zero in the neighbourhood of the roots.   Roots such as these occur less frequently than those we have been finding and the method of obtaining them is dealt with in APPENDIX A, rather than in the main part of the programme.

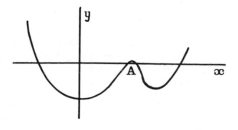

The Accuracy of the Result

As has been seen, the use of the Newton-Raphson formula enables a root to be obtained to a high degree of accuracy provided that the initial approximation is a reasonable one.   Even if it isn't, in some cases no trouble will arise apart from the necessity of using the formula a greater number of times.   However, we have so far assumed, in order not to introduce additional complications when dealing with the mechanics of the process, that the equations themselves are exact.   This means that all the working has been done on the assumption that all numerical coefficients in the equation are exact.   In practice, of course, such coefficients may be subject to round-off errors.

Another source of possible error occurs when solving equations involving

functions of x other than polynomials, for example, e^x, sin x, etc. If tables are used to find the values of such functions then each entry in the table is subject to round-off error and consequently this will affect the accuracy to which the solution can be obtained.

Although you are now aware of the problem, we think it is unlikely that you will normally have to worry about it.   If you do, you will find it treated in APPENDIX B at the end of this programme.

Complex Roots

So far, no complex roots have been found, although they have existed in some of the equations solved.   Thus, two roots of the quartic equation $x^4 - 6x^2 - 13x + 1 = 0$ were found in FRAME 32, but not the others.   In that case they would have presented no difficulty because once two roots have been found, a quartic can be resolved into two quadratic factors (or two linear and one quadratic).   However, some equations have no real roots, as, for example, $x^4 + 2x + 3 = 0$.   This equation can be solved by a method similar to that which we have been using but the arithmetic does become somewhat worse.   Alternatively, it is possible to resolve it directly into quadratic factors, from which the complex roots can be obtained.

Now you may be wondering why we should be interested in such roots.   A practical case where complex roots occur is stability theory.   There it is necessary to know where the complex roots of a polynomial forming the denominator of what is called the transfer function lie.   The ideas behind this are given in the programme "Further Laplace Transforms" in our book "Mathematics for Engineers and Scientists, Vol. 2".

If you need to know about the techniques involved for finding complex roots, you will find them described in APPENDIX C at the end of this programme.

The Secant Method and the Method of False Position

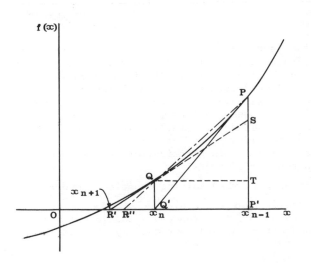

It has been seen that the advantage of the Newton-Raphson process over the straight-forward iteration considered in FRAMES 11-17 is its rapidity of convergence. However it requires the calculation of the derivative of the function.   The SECANT METHOD dispenses with the calculation of derivatives and, although not quite so rapidly convergent as Newton-Raphson, is better in this respect than the method of FRAMES 11-17.

The diagram shows the construction of x_n from x_{n-1} and of x_{n+1} from x_n when Newton-Raphson is used.   As you know, x_{n+1} is given by

$x_n - \dfrac{f(x_n)}{f'(x_n)}$ where $f'(x_n)$ is the slope of the line R'Q.   If R'Q is

produced to S, this slope is given by ST/QT.   The secant method replaces this slope by that of the chord QP, i.e., by PT/QT, giving R" for x_{n+1} instead of R'.   In the diagram as shown, there is a considerable difference between ST and PT, but the closer you get to the root, the less this difference becomes.   Also, as you know, the slope of the chord joining two points on a curve tends to the slope of the tangent, i.e., the derivative, as the two points become closer and closer together.   The net effect is that the convergence isn't quite so rapid as with Newton-Raphson, but the advantage of the method is that the calculation of the derivative, as remarked above, is no longer required.

What will the formula for x_{n+1} become when the secant method is used?

Give the result in terms of x_n, x_{n-1}, $f(x_n)$, and $f(x_{n-1})$.

****

43A

$$OR'' = OQ' - \frac{OQ'}{\tan Q\hat{R}''x} \qquad and \qquad Q\hat{R}''x = P\hat{Q}T$$

$$\therefore \quad x_{n+1} = x_n - \frac{f(x_n)}{\dfrac{f(x_{n-1}) - f(x_n)}{x_{n-1} - x_n}} \qquad (43A.1)$$

$$= x_n - \frac{(x_{n-1} - x_n)\, f(x_n)}{f(x_{n-1}) - f(x_n)} = \frac{x_{n-1}\, f(x_n) - x_n\, f(x_{n-1})}{f(x_n) - f(x_{n-1})}$$

FRAME 44

To reduce the amount of writing involved in formulae such as this, $f(x_n)$ is often denoted by f_n.   With this notation, the secant formula for x_{n+1} can be written as

$$x_{n+1} = \frac{x_{n-1}\, f_n - x_n\, f_{n-1}}{f_n - f_{n-1}} \qquad (44.1)$$

You will notice that, unlike the Newton-Raphson formula, it contains two functional values.   Thus, the calculation of x_{n+1} requires a knowledge of f_n and f_{n-1}.   At the beginning, assuming $n = 1$,

$$x_2 = \frac{x_0\, f_1 - x_1\, f_0}{f_1 - f_0} \qquad (44.2)$$

x_0 and x_1 are taken as two values of x for which the function is numerically small.   As you normally start a problem of this sort by

SOLUTION OF NON-LINEAR EQUATIONS

working out a table of values, two adjacent values in the table can usually be taken for x_0 and x_1.

To illustrate the method, examples will be taken that have previously been worked by other methods.   You will then be able to compare the various processes.

The equation $x^3 - 4x^2 + x - 10 = 0$ has already been solved by two methods.   In the first method the iteration formula

$$x_{n+1} = \frac{4x_n^2 - x_n + 10}{x_n^2}$$

was used and the second method was by the use of the Newton-Raphson formula.   From the table in 12A, you will see that we can take $x_0 = 4$ and $x_1 = 5$,   for which $f_0 = -6$, $f_1 = 20$. $\Big($These two values of x enclose the root and so it is reasonable to start with them.   An alternative selection would be 3 and 4 as these give rise to the two smallest values of $f(x)$ near the root.$\Big)$

Then $x_2 = \dfrac{4 \times 20 - 5 \times (-6)}{20 + 6}$ from (44.2)

$\simeq 4 \cdot 23$ and so $f_2 \simeq -1 \cdot 65$

Continuing $x_3 = \dfrac{x_1 f_2 - x_2 f_1}{f_2 - f_1}$ from (44.1) with $n = 2$

$= \dfrac{5 \times (-1 \cdot 65) - 4 \cdot 23 \times 20}{-1 \cdot 65 - 20}$ (44.3)

$\simeq 4 \cdot 289$ and so $f_3 \simeq -0 \cdot 395$

What will be the numerical expression giving x_4?
****

44A

$$x_4 = \frac{4 \cdot 23(-0 \cdot 395) - 4 \cdot 289(-1 \cdot 65)}{-0 \cdot 395 + 1 \cdot 65}$$

From this $x_4 = 4 \cdot 3076$ and then $f_4 = 0 \cdot 015\,25$.   Continuing gives $x_5 = 4 \cdot 3069$ and $f_5 = -0 \cdot 0003$. x_5 is the root correct to 4 decimal places.   Although x_5 is the root, the iteration formula has only been used four times.   The original method in FRAME 13 got the result after 6 iterations and Newton-Raphson after 3.

The working by this method can be exhibited in a table as follows:

4	−6
5	20
4·23	−1·65
4·289	−0·395
4·3076	0·015 25
4·3069	−0·0003

If you look at the working for any stage you will see how the table can be used.   Thus, suppose you have progressed as far as

$$
\begin{array}{ll}
4 & -6 \\
5 & 20 \\
4 \cdot 23 & -1 \cdot 65
\end{array}
$$

The numerator of (44.3) is the determinant $\begin{vmatrix} 5 & 20 \\ 4 \cdot 23 & -1 \cdot 65 \end{vmatrix}$ and the

denominator is simply the difference $-1 \cdot 65 - 20$ of the elements in the second column of this determinant.

Now use this method for the equations   (i) $x^3 - 5x^2 - 29 = 0$ and

(ii) $e^x = 10 - x$,   finding x to 4 significant figures.   (See 13A and 26A for tables of values.)

****

45A

i) Taking x_0 as 5 and x_1 as 6 gives

$$
\begin{array}{ll}
5 & -29 \\
6 & 7 \\
5 \cdot 81 & -1 \cdot 66 \\
5 \cdot 846 & -0 \cdot 087 \\
5 \cdot 848 & 0 \cdot 0008 \\
5 \cdot 848 &
\end{array}
$$

ii) Taking x_0 as 2 and x_1 as 3 gives

$$
\begin{array}{ll}
2 & -0 \cdot 61 \\
3 & 13 \cdot 09 \\
2 \cdot 04 & -0 \cdot 27 \\
2 \cdot 06 & -0 \cdot 09 \\
2 \cdot 070 & -0 \cdot 0052 \\
2 \cdot 071 & 0 \cdot 0038 \\
2 \cdot 071 &
\end{array}
$$

In this example, as the root is obviously near to 2, 2 and 2·1 could be taken as starting values.   The working would then be

$$
\begin{array}{ll}
2 & -0 \cdot 61 \\
2 \cdot 1 & 0 \cdot 27 \\
2 \cdot 07 & -0 \cdot 005 \\
2 \cdot 071 & 0 \cdot 0038 \\
2 \cdot 071 &
\end{array}
$$

giving the root that much quicker.

A variation of this method retains the x_0 and f_0 in (44.2) in place of the x_{n-1} and f_{n-1} in (44.1), giving

$$
x_{n+1} = \frac{x_0 f_n - x_n f_0}{f_n - f_0}
$$

This means that the slope of the chord PQ is replaced by the slope of the chord joining Q to the top of the first ordinate used.   In general, with this method, the iterations will converge to the root more slowly than previously.

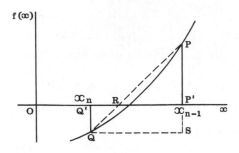

Now suppose the secant formula is used with x_{n-1} and x_n as shown here.   Where will be x_{n+1} on this diagram?   You can possibly state the answer immediately, but if not, obtain it geometrically by starting from the secant formula in the form (43A.1).   If you do it this way, express the various terms in (43A.1) as distances in the figure, i.e. x_{n-1} = OP', $f(x_n)$ = -QQ', etc.

****

46A

In FRAME 43, R'', the next approximation to the root, was the point where PQ produced met Ox.   In the present situation PQ actually intersects Ox and x_{n+1} is given by R, this point of intersection.

Alternatively,

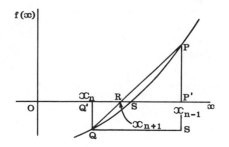

The right hand side of (43A.1) is equivalent to

$$OQ' - \frac{-QQ'}{PS/Q'P'}$$

$$= OQ' + \frac{QQ'}{PS/QS}$$

$$= OQ' + \frac{QQ'}{QQ'/Q'R}$$

$$= OQ' + Q'R = OR$$

The only difference between this process and the secant method is that, here, P' and Q' are taken specifically on opposite sides of S.   It follows that $f(x_{n-1})$ and $f(x_n)$ are of opposite sign.   One starts by choosing x_0 and x_1 so that these two points enclose the root, and using the secant formula to find x_2. x_2 is then taken in conjunction with either x_0 or x_1 and the formula used to find x_3.   The choice between x_0 or x_1 here is made by taking whichever gives a functional value opposite in sign to f_2.

Applying this process to the equation $x^3 - 4x^2 + x - 10 = 0$ and taking the same starting points as in FRAME 44, the first calculation is as before, giving the approximation 4·23.   For this value of x, f(x) is negative (-1·65).   As f(4) = -6 and f(5) = 20, 4·23 is now used in conjunction with 5, as f(4·23) and f(5) are opposite in sign.   The next approximation to the root is therefore

$$\frac{4 \cdot 23 \times 20 - 5 \times (-1 \cdot 65)}{20 - (-1 \cdot 65)} = 4 \cdot 289$$

Now $f(4 \cdot 289) = -0 \cdot 395$ and so using $4 \cdot 289$ in conjunction with a point previously used for which the functional value is positive, write down the numerical expression for the next approximation to the root.

****

47A

Using $4 \cdot 289$ with 5 gives $\frac{4 \cdot 289 \times 20 - 5 \times (-0 \cdot 395)}{20 - (-0 \cdot 395)}$.

This works out to be $4 \cdot 3028$.

FRAME 48

This gives a functional value of $-0 \cdot 09113$ and using this in conjunction with (5, 20) gives $4 \cdot 3060$ as the next approximation. Continuing the process leads to $4 \cdot 3067$ and then $4 \cdot 3069$, the value previously obtained.

Any of the variations described in the last few frames can be called a METHOD of FALSE POSITION. Unfortunately, various authors are not unanimous as to which method they call <u>the</u> rule of false position, but the majority define this as being the last of the processes we have considered. The rule of false position is also known as REGULA FALSI.

A flow diagram for this rule is shown on page 56.

FRAME 49

Simultaneous Non-Linear Equations

This programme has concentrated on methods of finding solutions of single non-linear equations. Some of these methods can be extended to simultaneous non-linear equations and are considered in APPENDIX D at the end of this programme. If your course includes this topic, you should read this appendix. Otherwise you can omit it as to do so will not affect your understanding of the rest of the book.

Various other methods for tackling non-linear equations have been devised. These have not been considered here as this is by no means intended to be an exhaustive treatment of the subject. The idea has been to give you an insight into the sort of approach used so that you can proceed further with a study of the subject, if, at a later stage, this becomes necessary.

FRAME 50

Miscellaneous Examples

In this frame a collection of miscellaneous examples is given for you to try. Answers are provided in FRAME 51, together with such working as is considered helpful.

1. By taking logarithms, find the solution of $x^x = 10$, correct to 4 significant figures, by (i) Newton-Raphson, (ii) Secant Method.

2. Find, correct to 4 significant figures, the positive root of $4x^4 = x + 8$.

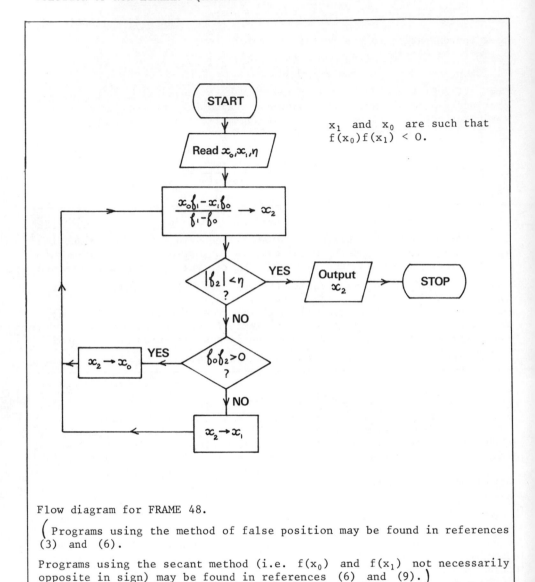

x_1 and x_0 are such that $f(x_0)f(x_1) < 0$.

Flow diagram for FRAME 48.

$\Big($ Programs using the method of false position may be found in references (3) and (6).

Programs using the secant method (i.e. $f(x_0)$ and $f(x_1)$ not necessarily opposite in sign) may be found in references (6) and (9). $\Big)$

FRAME 50   (continued)

3. The equation $\sin \omega t - e^{-at} = 0$ arises in the motion of a planetary gear system used in automatic transmission.   Determine the smallest root correct to 3 decimal places when $\omega = 0 \cdot 573$ and $a = 0 \cdot 01$.

4. Find the diameter D of the pipe which satisfies the flow equation $8820D^5 - 2 \cdot 31D - 0 \cdot 6465 = 0$, correct to 3 decimal places.

5. The equation $\tan x - x = 0 \cdot 01$ arises in the motion of helical gears.

Find the smallest positive root of this equation correct to 2 decimal places.

6. Show, by sketching the graphs of e^x and $(x + 1)/(x - 1)$, that the equation $x + 1 = (x - 1)e^x$ has only two real roots.   By using Newton's method find the position of the positive root correct to 2 significant figures.   (L.U.)
   (Note: Here the term 'Newton's method' has been used to mean the Newton-Raphson process.)

7. In FRAME 2, it was mentioned that in a certain vibration problem it is necessary to solve $\cos n\ell \cosh n\ell = 1$, which, if $n\ell$ is put equal to ϕ, becomes $\cos \phi \cosh \phi = 1$.   An obvious solution of this equation is $\phi = 0$.   Show that the smallest positive value of ϕ satisfying this equation is approximately $0 \cdot 018$ in excess of $3\pi/2$.

Answers to Miscellaneous Examples

1. $2 \cdot 506$ 2. $1 \cdot 233$ 3. $2 \cdot 362$

4. $0 \cdot 163$ 5. $0 \cdot 31$ 6. $1 \cdot 5$

7. A rough sketch of $\cos x$ and $\operatorname{sech} x$ shows the smallest positive root to be slightly in excess of $3\pi/2$.   One application of Newton-Raphson gives the excess as approximately $0 \cdot 018$.

SOLUTION OF NON-LINEAR EQUATIONS

APPENDIX A

Equal or Nearly Equal Roots

Given the equation $x^4 - 5 \cdot 6x^3 + 10 \cdot 84x^2 - 16 \cdot 8x + 23 \cdot 52 = 0$, the following table for $y = x^4 - 5 \cdot 6x^3 + 10 \cdot 84x^2 - 16 \cdot 8x + 23 \cdot 52$ can be formed.

x	-1	0	1	2	3	4
y	57·76	23·52	12·96	4·48	0·48	27·36

If the graph of this function is sketched, one cannot be quite sure whether, in the neighbourhood of $x = 3$, it is

like 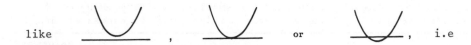 , or , i.e

whether the original equation has no root, two coincident roots or two nearly equal roots. But whatever happens, as there is a minimum in this neighbourhood, $\frac{dy}{dx}$ is going to pass through zero, and as you have seen, this can lead to difficulties with the Newton-Raphson formula.

However, there should be no difficulty in determining where $\frac{dy}{dx} = 0$. This will give us the x-coordinate for the minimum point on the curve. If, then, the y-coordinate is found, it will be immediately obvious which of the cases depicted is the actual one for this function.

Find where $\frac{dy}{dx} = 0$ and the value of y there.

****

$\frac{dy}{dx} = 4x^3 - 16 \cdot 8x^2 + 21 \cdot 68x - 16 \cdot 8$

This is zero in the neighbourhood of $x = 3$, $\therefore$ *let* $x_0 = 3$

$$x_{n+1} = x_n - \frac{4x_n^3 - 16 \cdot 8x_n^2 + 21 \cdot 68x_n - 16 \cdot 8}{12x_n^2 - 33 \cdot 6x_n + 21 \cdot 68}$$

$x_1 = 2 \cdot 83,$ $x_2 = 2 \cdot 801,$ $x_3 = 2 \cdot 800$

The value $x = 2 \cdot 8$ makes $y = 0$. There is a minimum here and hence two coincident roots.

Next, taking the equation $x^4 - 5 \cdot 6x^3 + 10 \cdot 84x^2 - 16 \cdot 8x + 23 \cdot 62 = 0$, the corresponding y function will be $y = x^4 - 5 \cdot 6x^3 + 10 \cdot 84x^2 - 16 \cdot 8x + 23 \cdot 62$. What can you say about the minimum value of y this time and the roots of the new equation?

****

y will have a minimum value of 0·10 at x = 2·8 and the new equation will have no real root in the vicinity of 2·8.

On the other hand, $y = x^4 - 5 \cdot 6x^3 + 10 \cdot 84x^2 - 16 \cdot 8x + 23 \cdot 42$ will have a minimum y of $-0 \cdot 10$ at $x = 2 \cdot 8$ and so $x^4 - 5 \cdot 6x^3 + 10 \cdot 84x^2 - 16 \cdot 8x + 23 \cdot 42 = 0$ will have two roots, both near to $x = 2 \cdot 8$.

Why can't the Newton-Raphson formula be used to find either of these roots, taking $x_0 = 2 \cdot 8$?

****

As f'(x) = 0, the formula breaks down.

A first estimate of the distances of the roots away from $2 \cdot 8$ is obtained as follows:

Let $x = a$ be one of the roots sought and b the value of x for which $f'(x) = 0$. Then $f(a) = 0$ and $f'(b) = 0$. In our example, $b = 2 \cdot 8$, but this will not be used at the moment, so that a general formula can be obtained, rather than one which will only be valid for this particular example. Also let $a = b + \varepsilon$.

As $f(a) = 0$, $f(b + \varepsilon) = 0$

$$\therefore \quad f(b) + \varepsilon f'(b) + \tfrac{1}{2}\varepsilon^2 f''(b) + \ldots\ldots\ldots\ldots = 0 \qquad (A4.1)$$

If the higher powers of ε are ignored (on the assumption that ε is small) then, remembering that $f'(b) = 0$, $\varepsilon^2 = -2f(b)/f''(b)$.

Thus, provided $f(b)$ and $f''(b)$ are of opposite sign, which occurs in the particular situation being considered, two equal and opposite values, η and $-\eta$, say, of ε will be obtained. $b + \eta$ and $b - \eta$ will then give first estimates of the two roots sought. They will not be exactly the roots as high powers of ε were ignored in (A4.1).

What will be these first estimates for the equation

$$x^4 - 5 \cdot 6x^3 + 10 \cdot 84x^2 - 16 \cdot 8x + 23 \cdot 42 = 0?$$

****

As $x^4 - 5 \cdot 6x^3 + 10 \cdot 84x^2 - 16 \cdot 8x + 23 \cdot 52 = 0$ when $x = 2 \cdot 8$,
$x^4 - 5 \cdot 6x^3 + 10 \cdot 84x^2 - 16 \cdot 8x + 23 \cdot 42 = -0 \cdot 10$ and so $f(b) = -0 \cdot 10$.
$f''(x) = 12x^2 - 33 \cdot 6x + 21 \cdot 68$, $\therefore f''(b) = 21 \cdot 68$
  $\varepsilon^2 = -2(-0 \cdot 10)/21 \cdot 68$ $\varepsilon \simeq 0 \cdot 1$

  $\therefore$ First estimates of roots are 2·7 and 2·9.

Now using the ordinary Newton-Raphson formula with $x_0 = 2 \cdot 7$ gives $x_1 = 2 \cdot 701$ which is correct to 3 decimal places. Then using the ordinary formula with $x_0 = 2 \cdot 9$ gives $x_1 = 2 \cdot 894$ which is again correct to 3 decimal places.

SOLUTION OF NON-LINEAR EQUATIONS

What are the roots, to 4 decimal places, of

$$f(x) \equiv 3x^4 - 18 \cdot 84x^3 + 30 \cdot 577x^2 - 6 \cdot 28x + 9 \cdot 859 = 0?$$
****

The table of values of f(x)

x	-2	-1	0	1	2	3	4	5
$f(x)$	343·4	68·6	9·9	18·3	16·9	0·5	36·2	262·9

suggests a minimum in the neighbourhood of $x = 3$.   If this minimum is negative there will be two roots in this vicinity.

$$f'(x) = 12x^3 - 56 \cdot 52x^2 + 61 \cdot 154x - 6 \cdot 28$$

Putting $f'(x) = 0$ and taking $x_0 = 3$ leads to a minimum at $3 \cdot 1402$ of value $-0 \cdot 018\ 35$.

$$f''(x) = 36x^2 - 113 \cdot 04x + 61 \cdot 154$$

Using $\varepsilon^2 = -2f(x)/f''(x)$,   with $x = 3 \cdot 1402$, gives $\varepsilon = \pm 0 \cdot 0245$.

First approximations to roots: $3 \cdot 1157$, $3 \cdot 1647$

Using ordinary Newton-Raphson formula, $3 \cdot 1157$ leads to $3 \cdot 1155$ and $3 \cdot 1647$ to $3 \cdot 1645$.

APPENDIX B

The Accuracy of the Result

To obtain an estimate of the error due to inexact coefficients in an equation, suppose there are, say, four of these, c_1, c_2, c_3 and c_4, in the equation $f = 0$. Instead of regarding f as a function of x only, it is now considered to be a function of x, c_1, c_2, c_3 and c_4. Then

$$df = \frac{\partial f}{\partial x} dx + \frac{\partial f}{\partial c_1} dc_1 + \frac{\partial f}{\partial c_2} dc_2 + \frac{\partial f}{\partial c_3} dc_3 + \frac{\partial f}{\partial c_4} dc_4 \qquad (B1.1)$$

Now, as $f = 0$ at the root $x = a$ say, $df = 0$ there also. Thus the right hand side of (B1.1) is zero when $x = a$.

$$\therefore \quad \left(\frac{\partial f}{\partial x} dx \right)_{x=a} = -\left[\frac{\partial f}{\partial c_1} dc_1 + \frac{\partial f}{\partial c_2} dc_2 + \frac{\partial f}{\partial c_3} dc_3 + \frac{\partial f}{\partial c_4} dc_4 \right]_{x=a}$$

When $x = a$, a small change dx in x can be regarded as a small change in the root a of the equation, i.e. as da.

$$\therefore \quad \left(\frac{\partial f}{\partial x} \right)_{x=a} da = -\left[\frac{\partial f}{\partial c_1} dc_1 + \frac{\partial f}{\partial c_2} dc_2 + \frac{\partial f}{\partial c_3} dc_3 + \frac{\partial f}{\partial c_4} dc_4 \right]_{x=a}$$

from which da follows immediately.

When dealing with round-off errors, the actual errors are, of course, unknown. The only information available will be the magnitude of the maximum error in each c. Thus all that can be obtained is an estimate of the magnitude of the maximum error in a, i.e., the magnitude of the maximum value of da.

In the first programme in this Unit, it was seen that when additions or subtractions are performed, the magnitude of the maximum possible error is the sum of the magnitudes of the separate errors.

Thus we can write

$$\left| \frac{\partial f}{\partial x} \right|_{x=a} |da| \leqslant \left\{ \left| \frac{\partial f}{\partial c_1} \right| |dc_1| + \left| \frac{\partial f}{\partial c_2} \right| |dc_2| + \left| \frac{\partial f}{\partial c_3} \right| |dc_3| + \left| \frac{\partial f}{\partial c_4} \right| |dc_4| \right\}_{x=a} (B1.2)$$

It will be obvious how to proceed if the number of c's is different from 4.

If round-off errors also occur due to the use of tables, each such error can be treated as an additional dc.

Applying this to the root $3 \cdot 1645$ in the example done in A5A and assuming that the '3' in $3x^4$ is correct, we can write

$$f = 3x^4 + c_1 x^3 + c_2 x^2 + c_3 x + c_4$$

Then $\dfrac{\partial f}{\partial x} = 12x^3 + 3\dot{c}_1 x^2 + 2c_2 x + c_3,$

$$\frac{\partial f}{\partial c_1} = x^3, \quad \frac{\partial f}{\partial c_2} = x^2, \quad \frac{\partial f}{\partial c_3} = x, \quad \frac{\partial f}{\partial c_4} = 1.$$

61

SOLUTION OF NON-LINEAR EQUATIONS

Thus, (B1.2) gives

$$1 \cdot 52 \; |da| \leqslant 31 \cdot 7 \times 0 \cdot 005 + 10 \cdot 0 \times 0 \cdot 0005 + 3 \cdot 2 \times 0 \cdot 005 + 1 \times 0 \cdot 0005$$
$$\simeq 0 \cdot 18$$
$$|da| \simeq 0 \cdot 12$$

As this is only an estimate, great accuracy is not required and so approximate values have been taken for $x(3 \cdot 2)$, $x^2(10 \cdot 0)$ and $x^3(31 \cdot 7)$.

Subject to these errors, then, the root can lie anywhere between, approximately, the limits $3 \cdot 04$ and $3 \cdot 28$.

Now estimate the accuracy of the root $-2 \cdot 13$ in the example you did in 32A assuming that only the coefficient of x^4 is correct.

****

At $x = -2 \cdot 13$, $f'(x) \simeq -49$

Maximum possible error in each $c = 0 \cdot 005$

Partial derivatives w.r.t. the c's are again x^3, x^2, x and 1.

$$49 \; |da| \simeq 9 \cdot 7 \times 0 \cdot 005 + 4 \cdot 5 \times 0 \cdot 005 + 2 \cdot 1 \times 0 \cdot 005 + 1 \times 0 \cdot 005$$
$$|da| \simeq 0 \cdot 02$$

Limits are approximately $-2 \cdot 11$ to $-2 \cdot 15$.

Notice that in the case of nearly equal roots the error in da can be expected to be larger than would otherwise be the case. This is due to the division by $\left| \dfrac{\partial f}{\partial x} \right|_{x=a}$ which is small for nearly equal roots.

As an example of a case where a table of functions is used, the equation $e^x + x - 10 = 0$ (solved in 26A) can be taken. Assuming that the value of e^x, taken from a set of tables correct to 4 decimal places, is the only source of error, let $f = e^x + x - 10$, then $\dfrac{\partial f}{\partial x} = e^x + 1$.

For the purpose of considering the error in e^x, let $e^x = c_1$, then $f = c_1 + x - 10$ and so $\dfrac{\partial f}{\partial c_1} = 1$, $|dc_1| = 0 \cdot 000\,05$

Thus $\left| e^{2 \cdot 071} + 1 \right| |da| \leqslant 0 \cdot 000\,05$, from which, approximately, $|da| < 0 \cdot 000\,007$. This will have no effect on $2 \cdot 071$, the value obtained for the root.

APPENDIX C

Solution of Polynomial Equations with no Real Root

Extension of Newton's Method to Complex Roots

The equation $x^4 + 2x + 3 = 0$ was quoted in FRAME 42 as an example of a polynomial equation that has no real root. If, however, this is not known at the start, it will be discovered when a table of values is formed or when the graphs of $y = x^4$ and $y = -2x - 3$ are plotted. There are various ways of solving such an equation, of which just two will be considered here. One method is a direct application of Newton-Raphson, use being made of the formula

$$x_{n+1} = x_n - \frac{f(x_n)}{f'(x_n)}$$

where, in this case, $f(x) = x^4 + 2x + 3$, $f'(x) = 4x^3 + 2$.

Thus
$$x_{n+1} = x_n - \frac{x_n^4 + 2x_n + 3}{4x_n^3 + 2} \qquad (C1.1)$$

A suitable starting value, x_0, is now required. This <u>must</u> be complex, because if a real value is taken, the iteration formula will never give

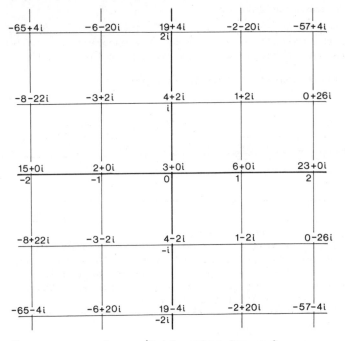

rise to a complex number. To find x_0, mark on an Argand diagram the values of $f(x)$ at a series of points.

In the diagram this has been done for several integer values of a and b where $x = a + ib$.

Now, for a real root, a suitable x_0 is where $f(x)$ is numerically small. For a complex root, a suitable x_0 is where $f(x)$ is again numerically small, i.e. where $|f(x)|$ is small.

Can you suggest a suitable value for x_0?

****

63

SOLUTION OF NON-LINEAR EQUATIONS

The smallest value of $|f(x)|$ for the points marked is 2 at -1.
However x here is real, and as was remarked in the last frame, such a
value is of no use in the present situation. The next smallest value
for the points marked is $\sqrt{5}$ at $1 + i$ and $1 - i$. Either of these is
therefore suitable.

Taking $x_0 = 1 + i$, (C1.1) gives

$$x_1 = 1 + i - \frac{1 + 2i}{4(1 + i)^3 + 2} = 1 + i - \frac{1 + 2i}{-6 + 8i} = 0 \cdot 9 + 1 \cdot 2i$$

What will be the value of x_2 to 2 decimal places?

****

$$x_2 = 0 \cdot 9 + 1 \cdot 2i - \frac{(0 \cdot 9 + 1 \cdot 2i)^4 + 2(0 \cdot 9 + 1 \cdot 2i) + 3}{4(0 \cdot 9 + 1 \cdot 2i)^3 + 2}$$

$$= 0 \cdot 9 + 1 \cdot 2i + 0 \cdot 05 - 0 \cdot 01i = 0 \cdot 95 + 1 \cdot 19i$$

Continuing in the same way gives:

$x_3 = 0 \cdot 951 + 1 \cdot 197i$, $\quad x_4 = 0 \cdot 9511 + 1 \cdot 1961i$, $\quad x_5 = 0 \cdot 9513 + 1 \cdot 1961i$,
$x_6 = 0 \cdot 9513 + 1 \cdot 1961i$

$x = 0 \cdot 9513 + 1 \cdot 1961i$ is therefore one root correct to 4 decimal places.
If necessary, it can be improved further.

Without doing any working, can you write down the value of a second root
of the equation? (If you haven't previously met the theory that enables
you to do this, go straight to the answer frame without attempting the
question.)

****

$0 \cdot 9513 - 1 \cdot 1961i$

You may have met before the fact that complex roots of a polynomial
equation with real coefficients occur in conjugate pairs — you are
certainly familiar with this result if the polynomial is a quadratic.
We shall not prove it here for higher order equations but it will be
obvious by the time you have finished this appendix.

Two factors of $x^4 + 2x + 3$ are therefore very nearly
$x - 0 \cdot 9513 - 1 \cdot 1961i$ and $x - 0 \cdot 9513 + 1 \cdot 1961i$. Multiplied together
these give the real quadratic factor $x^2 - 1 \cdot 9026x + 2 \cdot 3356$. Dividing
this quadratic factor into $x^4 + 2x + 3$ gives the second quadratic
factor $x^2 + 1 \cdot 9026x + 1 \cdot 2843$. Putting this equal to zero and solving
gives $-0 \cdot 9513 \pm 0 \cdot 6159i$ for the other two roots.

In order to keep the iteration part of the work neat and tidy, you will probably find it helpful to set this out in the form of a table.   The table below shows the beginning of a suggested layout.

Row	Formation		$n = \rightarrow$	0	1	2	3	4	5
a		x_n	R	1	0.9	0.95	0.951	0.9511	
b			I	1	1.2	1.19	1.197	1.1961	
c	$a^2 - b^2$	x_n^2	R	0	-0.63	-0.51	-0.528		
d	2ab		I	2	2.16	2.26	2.277		
e	$ac - bd$	x_n^3	R	-2	-3.16	-3.17	-3.228		
f	$bc + da$		I	2	1.19*	1.54	1.533		
g	$c^2 - d^2$	x_n^4	R	-4	-4.27	-4.85	-4.906		
h	2cd		I	0	-2.72	-2.31	-2.405		
k	$g + 2a + 3$	$f(x_n)$	R	1	0.53	0.05	-0.004		
ℓ	$h + 2b$		I	2	-0.32	0.07	-0.011		
m	$4e + 2$	$f'(x_n)$	R	-6	-10.64	-10.68	-10.912		
p	4f		I	8	4.76	6.16	6.132		
q	$km + \ell p$	Numerator of $f(x_n)/f'(x_n)$	R	10	-7.16	-0.10	-0.024		
r	$\ell m - pk$		I	-20	0.88	-1.06	0.145		
s	$m^2 + p^2$	Denominator of $f(x_n)/f'(x_n)$	R	100	135.87	152.01	156.673		
t	q/s	$f(x_n)/f'(x_n)$	R	0.1	-0.05	-0.001	-0.0001		
u	r/s		I	-0.2	0.01	-0.007	0.0009		
v	$a - t$	x_{n+1}	R	0.9	0.95	0.951	0.9511		
w	$b - u$		I	1.2	1.19	1.197	1.1961		

Labelling each row and writing down how it is formed will help you with the arithmetic.

Here, for example, 1·19 (starred) is 1·2 × (−0·63) + 2·16 × 0·9.   The entries in its column in rows a, b, c and d are combined according to the formula bc + da.

The completion of each column forms a loop in the complete calculation. The process is therefore very suitable for programming on to a computer.

Bairstow's Method

It was mentioned in FRAME 42 that an alternative method of solving an equation such as $x^4 + 2x + 3 = 0$ directly is to resolve its left hand side into quadratic factors.   When this has been done, the roots can immediately be found.   The process is applicable to a polynomial equation of any even degree.   However, if any real roots exist, it is advisable to remove the corresponding factors from the left hand side before the technique is applied.   In the case of a polynomial equation of odd degree there is always one real root and so the left hand side can be factorised into a linear factor and a factor of even degree. Although the roots of $x^4 + 2x + 3 = 0$ will eventually be obtained by this method, the more general polynomial equation   p = 0   (p a function of   x) will be taken in order to derive the theory.

When using the ordinary Newton-Raphson technique, an approximate root is first found.   Similarly it is necessary now to find, or guess at, an initial quadratic factor to start the process.   Suppose this factor is $x^2 + ax + b$.   Then we can write

$$p = (x^2 + ax + b)q + rx + s \qquad (C7.1)$$

where   q   is the quotient when   p   is divided by $x^2 + ax + b$ and rx + s   is the remainder.

If   a   and   b   are now varied, q, r and s   will all change.   The idea is to find what changes in   a   and   b   will make   r   and   s   zero.

Now suppose that changes   δa   and   δb   are made in   a   and   b   respectively and that these cause changes   δr   and   δs   in   r   and   s.   Assuming that the powers and products of   δa   and   δb   are sufficiently small to be ignored, what will be the approximate expressions for   δr   and   δs   as functions of δa   and   δb?

****

$$\delta r \simeq \frac{\partial r}{\partial a}\,\delta a + \frac{\partial r}{\partial b}\,\delta b \qquad\qquad \delta s \simeq \frac{\partial s}{\partial a}\,\delta a + \frac{\partial s}{\partial b}\,\delta b$$

Now it is hoped that the change   δr   in   r   will make this coefficient zero, i.e., it is hoped that   δr   is such that   r + δr = 0.   This will not, of course, happen exactly for the value of   δr   as given in C8A as

66

the powers and products of δa and δb have been ignored.   However, with
any luck, $r + \delta r$ will be nearer to zero than is r.   In order to
proceed it is assumed that $r + \delta r = 0$ and so $\delta r = -r$.   Similarly,
with s, $\delta s = -s$.

The equations in C8A then become

$$\frac{\partial r}{\partial a}\, \delta a + \frac{\partial r}{\partial b}\, \delta b = -r \qquad\qquad \frac{\partial s}{\partial a}\, \delta a + \frac{\partial s}{\partial b}\, \delta b = -s \qquad (C9.1)$$

To be precise, each = should be $\simeq$, but although not theoretically
true, = signs are normally used here.

The two equations (C9.1) are now taken as two simultaneous equations in
δa and δb.   In order to solve for δa and δb, all that is now

necessary are the values of $\dfrac{\partial r}{\partial a}$, $\dfrac{\partial r}{\partial b}$, $\dfrac{\partial s}{\partial a}$ and $\dfrac{\partial s}{\partial b}$.

As was remarked in FRAME C7, q, r and s all change as a and b vary
and so can be regarded as functions of a and b.   (q, of course, is a
function of x as well.)   The partial derivatives required will
therefore become involved if (C7.1) is differentiated w.r.t. a and b.
What equations will result when these differentiations are carried out?
(Remember from FRAME C6 that the equation being solved is $p = 0$.)
****

$$0 = xq + (x^2 + ax + b)\, \frac{\partial q}{\partial a} + x\, \frac{\partial r}{\partial a} + \frac{\partial s}{\partial a} \qquad (C9A.1)$$

$$0 = q + (x^2 + ax + b)\, \frac{\partial q}{\partial b} + x\, \frac{\partial r}{\partial b} + \frac{\partial s}{\partial b} \qquad (C9A.2)$$

Note that as $p = 0$, $\dfrac{\partial p}{\partial a} = \dfrac{\partial p}{\partial b} = 0$

(C9A.2) can be rewritten as

$$q = (x^2 + ax + b) \left(- \frac{\partial q}{\partial b}\right) - x\, \frac{\partial r}{\partial b} - \frac{\partial s}{\partial b} \qquad (C10.1)$$

which can be regarded as the statement that when q is divided by
$x^2 + ax + b$, the quotient is $-\dfrac{\partial q}{\partial b}$ and the remainder is $-\dfrac{\partial r}{\partial b}\, x - \dfrac{\partial s}{\partial b}$.
If this division is therefore carried out, the terms in the remainder
will provide two of the partial derivatives required.

Can you now see how $\dfrac{\partial r}{\partial a}$ and $\dfrac{\partial s}{\partial a}$ can be obtained?
****

They will appear in the remainder when xq *is divided by* $x^2 + ax + b$ *as*

$$xq = (x^2 + ax + b) \left(- \frac{\partial q}{\partial a}\right) - x\, \frac{\partial r}{\partial a} - \frac{\partial s}{\partial a} \qquad (C10A.1)$$

SOLUTION OF NON-LINEAR EQUATIONS

As all the partial derivatives appear in (C10.1) and (C10A.1) with minus signs, it is better to use equations (C9.1) in the form

$$- \frac{\partial r}{\partial a} \delta a - \frac{\partial r}{\partial b} \delta b = r \qquad\qquad - \frac{\partial s}{\partial a} \delta a - \frac{\partial s}{\partial b} \delta b = s$$

Having solved these equations for δa and δb, it is hoped that $x^2 + (a + \delta a)x + (b + \delta b)$ is a better quadratic factor than was $x^2 + ax + b$. The process is then repeated to obtain a better factor still and so on. Unfortunately, however, the new factor may not be an improvement on the old, especially in the early stages. You will remember that with the Newton-Raphson process, sometimes an iterate appears that is further away from the root than is the preceeding one. You will also remember that a small value of $f'(x)$ could lead to this. Here, a somewhat analogous situation arises if the determinant

$$\begin{vmatrix} \dfrac{\partial r}{\partial a} & \dfrac{\partial r}{\partial b} \\[2ex] \dfrac{\partial s}{\partial a} & \dfrac{\partial s}{\partial b} \end{vmatrix}$$

is small.

If the process does appear to be going haywire, you can do one of two things. One alternative is to continue with your iterations in the hope that they eventually settle down, i.e. converge. The other alternative is to start with a completely new trial factor.

To illustrate this method, the example already worked by Newton-Raphson will be taken, i.e. $x^4 + 2x + 3 = 0$. The starting value used there was $x_0 = 1 + i$. Now if $1 + i$ is an approximate root of this polynomial equation then so also is $1 - i$. Two approximate linear factors are therefore $x - (1 + i)$ and $x - (1 - i)$. An approximate quadratic factor is the product of these, i.e., $(x - \overline{1 + i})(x - \overline{1 - i}) = x^2 - 2x + 2$ and this will be taken as the initial trial factor.

Having decided on a trial factor, the next thing is to divide it into the expression $x^4 + 2x + 3$ to obtain q, r and s. This can be done by the synthetic division process illustrated in FRAME 42, page 23.

What do you get for q, r and s when you do this?

****

		1	0	0	2	3
2			2	4	4	
-2				-2	-4	-4
		1	2	2	2	-1

$q = x^2 + 2x + 2,\qquad r = 2,\qquad s = -1.$

68

xq is therefore $x^3 + 2x^2 + 2x$ and it is now necessary to divide both xq and q by the original factor $x^2 - 2x + 2$. The division of xq is

$$
\begin{array}{r|rrrr}
 & 1 & 2 & 2 & 0 \\
2 & & 2 & 8 & \\
-2 & & & -2 & -8 \\
\hline
 & 1 & 4 & 8 & -8 \\
\end{array}
$$

and from this $- \dfrac{\partial r}{\partial a} = 8,$ $\qquad - \dfrac{\partial s}{\partial a} = -8$

Similarly the division of q is

$$
\begin{array}{r|rrr}
 & 1 & 2 & 2 \\
2 & & 2 & \\
-2 & & & -2 \\
\hline
 & 1 & 4 & 0 \\
\end{array}
$$

and hence $- \dfrac{\partial r}{\partial b} = 4,$ $\qquad - \dfrac{\partial s}{\partial b} = 0.$

The equations for δa and δb are therefore

$$
\begin{aligned}
8\ \delta a + 4\ \delta b &= 2 \\
-8\ \delta a \qquad\quad &= -1
\end{aligned}
$$

from which $\delta a = 0 \cdot 12,$ $\quad \delta b = 0 \cdot 25.$

The second trial factor is then $x^2 + (-2 + 0 \cdot 12)x + (2 + 0 \cdot 25)$, i.e., $x^2 - 1 \cdot 88x + 2 \cdot 25.$

The complete working of the next stage can be set out as follows:

$$
\begin{array}{r|rrrrr}
 & 1 & 0 & 0 & 2 & 3 \\
1 \cdot 88 & & 1 \cdot 88 & 3 \cdot 53 & 2 \cdot 41 & \\
-2 \cdot 25 & & & -2 \cdot 25 & -4 \cdot 23 & -2 \cdot 88 \\
\hline
 & 1 & 1 \cdot 88 & 1 \cdot 28 & 0 \cdot 18 & 0 \cdot 12 \\
\end{array}
$$

$$
\begin{array}{r|rrrr}
 & 1 & 1 \cdot 88 & 1 \cdot 28 & 0 \\
1 \cdot 88 & & 1 \cdot 88 & 7 \cdot 07 & \\
-2 \cdot 25 & & & -2 \cdot 25 & -8 \cdot 46 \\
\hline
 & 1 & 3 \cdot 76 & 6 \cdot 10 & -8 \cdot 46 \\
\end{array}
\qquad
\begin{array}{r|rrr}
 & 1 & 1 \cdot 88 & 1 \cdot 28 \\
1 \cdot 88 & & 1 \cdot 88 & \\
-2 \cdot 25 & & & -2 \cdot 25 \\
\hline
 & 1 & 3 \cdot 76 & -0 \cdot 97 \\
\end{array}
$$

$$
\begin{aligned}
6 \cdot 10\ \delta a + 3 \cdot 76\ \delta b &= 0 \cdot 18 \\
-8 \cdot 46\ \delta a - 0 \cdot 97\ \delta b &= 0 \cdot 12
\end{aligned}
$$

$$
\delta a = -0 \cdot 024 \qquad \delta b = 0 \cdot 087
$$
$$
x^2 - 1 \cdot 904x + 2 \cdot 337
$$

Now do the next iteration completely.

****

SOLUTION OF NON-LINEAR EQUATIONS

$$
\begin{array}{c|ccccc}
 & 1 & 0 & 0 & 2 & 3 \\
1 \cdot 904 & & 1 \cdot 904 & 3 \cdot 625 & 2 \cdot 452 & \\
-2 \cdot 337 & & & -2 \cdot 337 & -4 \cdot 450 & -3 \cdot 010 \\
\hline
 & 1 & 1 \cdot 904 & 1 \cdot 288 & 0 \cdot 002 & -0 \cdot 010
\end{array}
$$

$$
\begin{array}{c|cccc}
 & 1 & 1 \cdot 904 & 1 \cdot 288 & 0 \\
1 \cdot 904 & & 1 \cdot 904 & 7 \cdot 250 & \\
-2 \cdot 337 & & & -2 \cdot 337 & -8 \cdot 899 \\
\hline
 & 1 & 3 \cdot 808 & 6 \cdot 201 & -8 \cdot 899
\end{array}
\qquad
\begin{array}{c|ccc}
 & 1 & 1 \cdot 904 & 1 \cdot 288 \\
1 \cdot 904 & & 1 \cdot 904 & \\
-2 \cdot 337 & & & -2 \cdot 337 \\
\hline
 & 1 & 3 \cdot 808 & -1 \cdot 049
\end{array}
$$

$$
6 \cdot 201 \, \delta a + 3 \cdot 808 \, \delta b = 0 \cdot 002
$$
$$
-8 \cdot 899 \, \delta a - 1 \cdot 049 \, \delta b = -0 \cdot 010
$$
$$
\delta a = 0 \cdot 0013 \qquad \delta b = -0 \cdot 0016
$$
$$
x^2 - 1 \cdot 9027 x + 2 \cdot 3354
$$

The next stage is now:

$$
\begin{array}{c|ccccc}
 & 1 & 0 & 0 & 2 & 3 \\
1 \cdot 9027 & & 1 \cdot 9027 & 3 \cdot 6203 & 2 \cdot 4448 & \\
-2 \cdot 3354 & & & -2 \cdot 3354 & -4 \cdot 4436 & -3 \cdot 0008 \\
\hline
 & 1 & 1 \cdot 9027 & 1 \cdot 2849 & 0 \cdot 0012 & -0 \cdot 0008
\end{array}
$$

$$
\begin{array}{c|cccc}
 & 1 & 1 \cdot 9027 & 1 \cdot 2849 & 0 \\
1 \cdot 9027 & & 1 \cdot 9027 & 7 \cdot 2405 & \\
-2 \cdot 3354 & & & -2 \cdot 3354 & -8 \cdot 8871 \\
\hline
 & 1 & 3 \cdot 8054 & 6 \cdot 1900 & -8 \cdot 8871
\end{array}
\qquad
\begin{array}{c|ccc}
 & 1 & 1 \cdot 9027 & 1 \cdot 2849 \\
1 \cdot 9027 & & 1 \cdot 9027 & \\
-2 \cdot 3354 & & & -2 \cdot 3354 \\
\hline
 & 1 & 3 \cdot 8054 & -1 \cdot 0505
\end{array}
$$

$$
6 \cdot 1900 \, \delta a + 3 \cdot 8054 \, \delta b = 0 \cdot 0012
$$
$$
-8 \cdot 8871 \, \delta a - 1 \cdot 0505 \, \delta b = -0 \cdot 0008
$$
$$
\delta a = 0 \cdot 0001, \qquad \delta b = 0 \cdot 0002
$$
$$
x^2 - 1 \cdot 9026 x + 2 \cdot 3356
$$

Further iterations will give the factor to more decimal places, if
required.   The two complex roots corresponding to this factor are
obtained by solving $x^2 - 1 \cdot 9026x + 2 \cdot 3356 = 0$, giving
$x = 0 \cdot 9513 \pm 1 \cdot 1961i$.   The other two roots are obtained as before.

A flow diagram for Bairstow's method is shown on page 71.

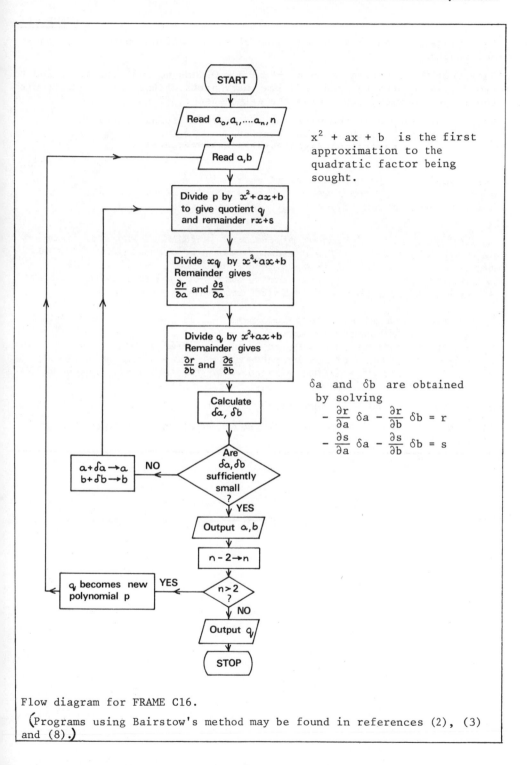

$x^2 + ax + b$ is the first approximation to the quadratic factor being sought.

δa and δb are obtained by solving

$$-\frac{\partial r}{\partial a}\, \delta a - \frac{\partial r}{\partial b}\, \delta b = r$$

$$-\frac{\partial s}{\partial a}\, \delta a - \frac{\partial s}{\partial b}\, \delta b = s$$

Flow diagram for FRAME C16.

(Programs using Bairstow's method may be found in references (2), (3) and (8).)

SOLUTION OF NON-LINEAR EQUATIONS

The following example occurred in an old examination paper set by London University.

Show that the equation $x^4 + x^2 - x + 1 = 0$ has no real roots. Find to two decimal places the real and imaginary parts of the complex roots with the larger modulus, explaining the method used.

As a final example on the work covered in this appendix, try this question by each of the methods discussed.

****

C17A

The equation can be written as $x^4 + (x - \frac{1}{2})^2 + \frac{3}{4} = 0$. The left hand side is the sum of terms which are positive (or zero) for real x and hence no real x can make the left hand side zero.

By Newton-Raphson: A diagram of the type in FRAME C1 suggests i or $-i$ as a starting point. If you started with i you should have obtained the root $-0 \cdot 55 + 1 \cdot 12i$, from which a second root, $-0 \cdot 55 - 1 \cdot 12i$, immediately follows. If you started with $-i$, you should have obtained $-0 \cdot 55 - 1 \cdot 12i$ and then $0 \cdot 55 + 1 \cdot 12i$ immediately follows.

By Bairstow: The starting points i and $-i$ in the Newton-Raphson method suggest the trial quadratic factor $(x - i)(x + i) = x^2 + 1$. Successive iterations then give $x^2 + x$, $x^2 + 0 \cdot 5x + 0 \cdot 5$, $x^2 + 0 \cdot 99x + 2 \cdot 00$ which can be taken as $x^2 + x + 2$, $x^2 + x + 1 \cdot 50$, $x^2 + 1 \cdot 1x + 1 \cdot 55$, $x^2 + 1 \cdot 095x + 1 \cdot 556$. No further change occurs to 3 decimal places.

$x^2 + 1 \cdot 095x + 1 \cdot 556 = 0$ then gives $x = -0 \cdot 55 \pm 1 \cdot 12i$.

It is easy to verify that these two roots have larger moduli than the others.

APPENDIX D

Solution of Simultaneous Non-Linear Equations

Extension of Direct Iteration Method

In FRAMES 11-17 of the main part of this programme, you saw how the solution of an equation of the form $f(x) = 0$ could be found by rewriting the equation as $x = F(x)$ to give the iteration formula $x_{n+1} = F(x_n)$. In order for the iterations to converge to the root, it was necessary for $|F'(x)| < 1$ in the neighbourhood of the root. Also, from the way in which the rate of convergence depends on $|F'(x)|$, it is better for $f(x) = 0$ to be rearranged in such a way that $|F'(x)|$ is as small as possible.

We shall now illustrate how this method can be extended to simultaneous equations.

As an example, let us consider the simultaneous equations

$$2x = \sin \tfrac{1}{2}(x - y) \ \Bigg\} \quad \text{(D2.1)}$$
$$2y = \cos \tfrac{1}{2}(x + y) \ \Bigg\}$$

In order to use the method, it is necessary to rearrange the equations so that x is the subject of one of them and y is the subject of the other, i.e., one equation is written in the form $x = F(x, y)$ and the other in the form $y = G(x, y)$. Obviously the way in which this can be done is not unique, just as the form $x = F(x)$ wasn't unique in FRAMES 11-17. But, just as in that earlier situation it was necessary to choose $F(x)$ so that the iterations converged, so here it is necessary to choose F and G so that the process converges. It can be shown that if

$$\left|\frac{\partial F}{\partial x}\right| + \left|\frac{\partial G}{\partial x}\right| < 1 \ \Bigg\}$$
$$\text{and} \quad \left|\frac{\partial F}{\partial y}\right| + \left|\frac{\partial G}{\partial y}\right| < 1 \ \Bigg\} \quad \text{(D2.2)}$$

in the neighbourhood of the root, convergence takes place.

Another preliminary that was necessary before was finding an approximate root as a starting point x_0. Similarly here it is necessary to obtain a pair of such values, one for x and the other for y. These are often denoted by x_0 and y_0. If it is possible to sketch the two curves represented by the simultaneous equations, the point of intersection can be used to obtain x_0 and y_0. Fortunately in this particular example this can be done. (How to proceed otherwise will be indicated later.) Just for the purpose of sketching the curves, the two equations can be rewritten as

$$y = x - 2 \sin^{-1} 2x$$
$$x = 2 \cos^{-1} 2y - y$$

and a table of values of x and y constructed for each of them.

Can you suggest limits within which x and y must lie, assuming that only real values are being used? Use the equations in the form (D2.1) to do this. ***************************************

SOLUTION OF NON-LINEAR EQUATIONS

As $2x = sin \frac{1}{2}(x - y)$ *and* $-1 \leqslant sin \frac{1}{2}(x - y) \leqslant 1$, $-\frac{1}{2} \leqslant x \leqslant \frac{1}{2}$.
Similarly $-\frac{1}{2} \leqslant y \leqslant \frac{1}{2}$.

A table of values for $y = x - 2 sin^{-1}2x$ is then

x	-0·5	-0·4	-0·3	-0·2	-0·1	0	0·1	0·2	0·3	0·4	0·5
y	2·64	1·46	0·98	0·62	0·30	0	-0·30	-0·62	-0·98	-1·46	-2·64

Only values of x have been taken between -0·5 and +0·5.   But, as
when the two equations are taken simultaneously, y also lies within
those limits, only the section of the table enclosed is relevant.   The
principal values of the inverse sine have been taken as only small values
of y are possible.

Now start to construct a similar table for $x = 2 cos^{-1}2y - y$ by using
y = -0·5, -0·4 and -0·3.
****

y	-0·5	-0·4	-0·3
$x \Big\{$	-5·78	-4·60	-4·12
	6·78	-5·40	4·72

Two values of x have been given for each y as for each value of 2y
there are two values of $cos^{-1}2y$, both equally small.

FRAME D5

Continuing this table, we have

y	-0·5	-0·4	-0·3	-0·2	-0·1	0	0·1	0·2	0·3	0·4	0·5
$x\Big\{$	-5·78	-4·60	-4·12	-3·76	-3·44	-3·14	-2·84	-2·52	-2·16	-1·68	-0·50
	6·78	5·40	4·72	4·16	3·64	3·14	2·64	2·12	1·56	0·88	-0·50

The only portion of this table that is of use is right at the end, using
the lower line of values of x.

Inserting a few more values into the table at this end gives, for values
of x in the lower line,

y	0·42	0·44	0·46	0·48
x	0·72	0·54	0·34	0·08

The graphs of the two curves, over the relevant sections, are shown on
page 75.

FRAME D6

For the purpose of sketching the graphs, the original equations were
written as
$$x = 2 cos^{-1}2y - y \qquad \text{and} \qquad y = x - 2 sin^{-1}2x$$
But also they can obviously be written as
$$x = \tfrac{1}{2} sin \tfrac{1}{2}(x - y) \qquad \text{and} \qquad y = \tfrac{1}{2} cos \tfrac{1}{2}(x + y)$$
Both of these sets are of the form $x = F(x, y)$ and $y = G(x, y)$.   Which
pair do you suggest should be used for improving the accuracy of the root?
****

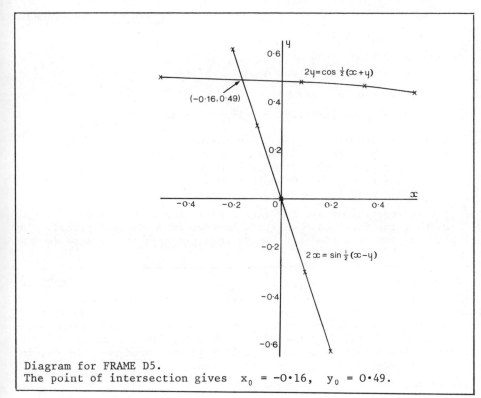

Diagram for FRAME D5.
The point of intersection gives $x_0 = -0.16$, $y_0 = 0.49$.

D6A

$x = \frac{1}{2} \sin \frac{1}{2}(x - y)$ and $y = \frac{1}{2} \cos \frac{1}{2}(x + y)$

In this case, $\dfrac{\partial F}{\partial x} = \frac{1}{4} \cos \frac{1}{2}(x - y)$, $\qquad \dfrac{\partial G}{\partial x} = -\frac{1}{4} \sin \frac{1}{2}(x + y)$

$\qquad\qquad\quad \dfrac{\partial F}{\partial y} = -\frac{1}{4} \cos \frac{1}{2}(x - y)$, $\qquad \dfrac{\partial G}{\partial y} = -\frac{1}{4} \sin \frac{1}{2}(x + y)$

Normally, one would now use $x = -0.16$, $y = 0.49$, and so

$$\left|\frac{\partial F}{\partial x}\right| + \left|\frac{\partial G}{\partial x}\right| = 0.27, \qquad\qquad \left|\frac{\partial F}{\partial y}\right| + \left|\frac{\partial G}{\partial y}\right| = 0.27$$

and both of these < 1. In this particular case however, it is unnecessary to use the specific values of x and y, as, for any pair of real values, each partial derivative has maximum numerical value of $\frac{1}{4}$.

The other pair do not satisfy the required conditions.

FRAME D7

We can thus take $x_{n+1} = \frac{1}{2} \sin \frac{1}{2}(x_n - y_n)$, $y_{n+1} = \frac{1}{2} \cos \frac{1}{2}(x_n + y_n)$.
However, as when y_{n+1} is being found, x_{n+1} is known and is expected to be more accurate than x_n, a slight alternative is to take

SOLUTION OF NON-LINEAR EQUATIONS

$y_{n+1} = \frac{1}{2} \cos \frac{1}{2}(x_{n+1} + y_n)$
$x_0 = -0 \cdot 16, \quad y_0 = 0 \cdot 49$ then gives
$$x_1 = \frac{1}{2} \sin \frac{1}{2}(-0 \cdot 16 - 0 \cdot 49) = -0 \cdot 1597$$
$x_1 = -0 \cdot 1597, \quad y_0 = 0 \cdot 49$ gives
$$y_1 = \frac{1}{2} \cos \frac{1}{2}(-0 \cdot 1597 + 0 \cdot 49) = 0 \cdot 4932$$

Then $x_2 = \frac{1}{2} \sin \frac{1}{2}(-0 \cdot 1597 - 0 \cdot 4932) = -0 \cdot 1603$
   $y_2 = \frac{1}{2} \cos \frac{1}{2}(-0 \cdot 1603 + 0 \cdot 4932) = 0 \cdot 4931$
   $x_3 = \frac{1}{2} \sin \frac{1}{2}(-0 \cdot 1603 - 0 \cdot 4931) = -0 \cdot 1605$
   $y_3 = \frac{1}{2} \cos \frac{1}{2}(-0 \cdot 1605 + 0 \cdot 4931) = 0 \cdot 4931$
   $x_4 = \frac{1}{2} \sin \frac{1}{2}(-0 \cdot 1605 - 0 \cdot 4931) = -0 \cdot 1605$

The fact that repetitions are now taking place for both x and y indicates that the root has been found correct to 4 decimal places.   By continuing the process, the root can be obtained accurate to more decimal places.

In FRAME D3 it was found that it was possible to construct tables of values of x and y for each equation.   This was because one equation could be solved for y in terms of x and the other for x in terms of y. If this cannot be done, an alternative method for finding x_0 and y_0 is necessary.   The method is somewhat similar to that used in FRAME C1, in that grids of function values are constructed.

To illustrate, let us take the equations

$$\left. \begin{array}{l} x + 2y = \dfrac{1}{5} \sin (x - y) \\[2mm] x - y \ \ = \dfrac{1}{5} \cos (x + y) \end{array} \right\} \qquad \text{(D8.1)}$$

Let $u = x + 2y - \dfrac{1}{5} \sin (x - y), \quad v = x - y - \dfrac{1}{5} \cos (x + y).$

On the first grid, values of u are given at the points indicated, and on the second, values of v.   (See page 77.)

A curve is then drawn on each grid passing through all points where the value of the tabulated function is zero.   You cannot expect to be able to do this accurately but we are only looking for a first estimate of the root anyway.   As a guide to drawing them, you would expect, for example, the u – curve to pass through a point approximately $\dfrac{1}{3}$ of the way down between A and B, as at A, $u = 0 \cdot 08$ and at B, $u = -0 \cdot 14$.   These two curves are now superimposed to give a first estimate of the root.

(See lower diagram on page 77.)

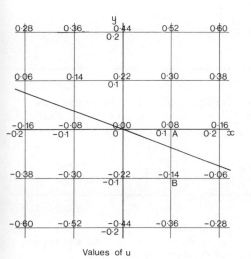

Values of u

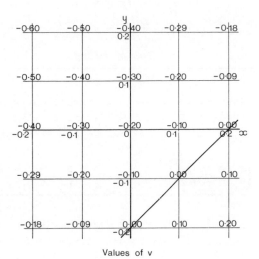

Values of v

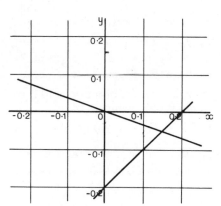

From this, we can take $x_0 = 0 \cdot 15$, $y_0 = -0 \cdot 05$.

Now try and rearrange the equations (D8.1) (or combinations of them) into the forms $x = F(x, y)$ and $y = G(x, y)$ so that conditions (D2.2) hold for x_0 and y_0.

****

D8A

The obvious won't work here, for neither

$$\left. \begin{array}{l} x = -2y + \dfrac{1}{5} \, sin \, (x - y) \\[2mm] y = x - \dfrac{1}{5} \, cos \, (x + y) \end{array} \right\} \quad nor \quad \left\{ \begin{array}{l} y = -\dfrac{1}{2}x + \dfrac{1}{10} \, sin \, (x - y) \\[2mm] x = y + \dfrac{1}{5} \, cos \, (x + y) \end{array} \right.$$

satisfy the conditions for convergence.

However, by eliminating y and x in turn from the L.H.S.'s of (D8.1),

SOLUTION OF NON-LINEAR EQUATIONS

D8A (continued)

$$\left. \begin{array}{l} x = \frac{1}{15} \ sin \ (x - y) + \frac{2}{15} \ cos \ (x + y) \\ y = \frac{1}{15} \ sin \ (x - y) - \frac{1}{15} \ cos \ (x + y) \end{array} \right\} \qquad (D8A.1)$$

and these are satisfactory.

FRAME D9

Working to 4 decimal places, carry out iterations using the equations (D8A.1) and $x_0 = 0 \cdot 15$, $y_0 = -0 \cdot 05$ until no further improvement can be obtained. ***

D9A

$\begin{cases} x_1 = 0 \cdot 1459 \\ y_1 = -0 \cdot 0534 \end{cases}$ $\begin{cases} x_2 = 0 \cdot 1460 \\ y_2 = -0 \cdot 0532 \end{cases}$ $\begin{cases} x_3 = 0 \cdot 1460 \\ y_3 = -0 \cdot 0532 \end{cases}$

FRAME D10

Extension of Newton–Raphson Method

Suppose two equations $f(x, y) = 0$ and $g(x, y) = 0$ have the solution $x = X$, $y = Y$ and that x_0 and y_0 are first approximations to X and Y. Then we can write $X - x_0 = \varepsilon_0$, $Y - y_0 = \eta_0$, and so $X = x_0 + \varepsilon_0$, $Y = y_0 + \eta_0$. Therefore, using Taylor's series

$$f(X, Y) = f(x_0 + \varepsilon_0, y_0 + \eta_0)$$
$$= f(x_0, y_0) + \varepsilon_0 f_x(x_0, y_0) + \eta_0 f_y(x_0, y_0) +$$

terms involving powers and products of ε_0 and η_0.

If the higher power terms are sufficiently small to be neglected, then, as $f(X, Y) = 0$, this equation reduces to

$$\varepsilon_0 f_x(x_0, y_0) + \eta_0 f_y(x_0, y_0) = -f(x_0, y_0) \qquad (D10.1)$$

This is one equation in ε_0 and η_0. What other equation will they satisfy? ***

D10A

$$\varepsilon_0 g_x(x_0, y_0) + \eta_0 g_y(x_0, y_0) = -g(x_0, y_0) \qquad (D10A.1)$$

FRAME D11

These two equations are used to obtain ε_0 and η_0. Second approximations to the solution are then $x_0 + \varepsilon_0$, $y_0 + \eta_0$. Why will these not be the exact values X and Y?
****

D11A

Because the higher powers of ε_0 and η_0 have been neglected.

FRAME D12

As you will remember, when using Newton–Raphson for an equation $f(x) = 0$ in a single variable, it is advisable for the first approximation to be reasonably good. Furthermore, difficulties can arise if $f'(x_0)$ is small. In a similar way, here it is advisable to start with reasonably good approximations and difficulties can arise if

$$\begin{vmatrix} f_x(x_0, y_0) & f_y(x_0, y_0) \\ g_x(x_0, y_0) & g_y(x_0, y_0) \end{vmatrix}$$

is small.

Having found new approximations x_1, y_1, say, the process can be repeated as many times as is necessary.

FRAME D13

Taking, as an example, the equations in FRAME D2, i.e.,

$$\begin{cases} 2x = \sin \tfrac{1}{2}(x - y) \\ 2y = \cos \tfrac{1}{2}(x + y) \end{cases}$$

we can write $f(x, y) = 2x - \sin \tfrac{1}{2}(x - y)$
$g(x, y) = 2y - \cos \tfrac{1}{2}(x + y)$

Then $f_x = 2 - \tfrac{1}{2} \cos \tfrac{1}{2}(x - y)$ $\qquad f_y = \tfrac{1}{2} \cos \tfrac{1}{2}(x - y)$
$g_x = \tfrac{1}{2} \sin \tfrac{1}{2}(x + y)$ $\qquad g_y = 2 + \tfrac{1}{2} \sin \tfrac{1}{2}(x + y)$

Using the same starting values, i.e. $x_0 = -0 \cdot 16$, $y_0 = 0 \cdot 49$, ε_0 and η_0 are given by $\left(\text{using (D10.1) and (D10A.1)}\right)$

$\{2 - \tfrac{1}{2} \cos \tfrac{1}{2}(-0 \cdot 65)\}\varepsilon_0 + \{\tfrac{1}{2} \cos \tfrac{1}{2}(-0 \cdot 65)\}\eta_0 = -\{2(-0 \cdot 16) - \sin \tfrac{1}{2}(-0 \cdot 65)\}$

$\{\tfrac{1}{2} \sin \tfrac{1}{2}(0 \cdot 33)\}\varepsilon_0 + \{2 + \tfrac{1}{2} \sin \tfrac{1}{2}(0 \cdot 33)\}\eta_0 = -\{2(0 \cdot 49) - \cos \tfrac{1}{2}(0 \cdot 33)\}$

i.e. $1 \cdot 5262\varepsilon_0 + 0 \cdot 4738\eta_0 = 0 \cdot 0007$, $\qquad 0 \cdot 0821\varepsilon_0 + 2 \cdot 0821\eta_0 = 0 \cdot 0064$

from which $\varepsilon_0 = -0 \cdot 0005$, $\qquad \eta_0 = 0 \cdot 0031$.

Next approximations to the root are therefore

$$x_1 = -0 \cdot 1605, \qquad y_1 = 0 \cdot 4931$$

Now, to 6 decimal places, find x_2 and y_2.
(Note: $\sin 0 \cdot 3268 = 0 \cdot 321\,014$, $\cos 0 \cdot 3268 = 0 \cdot 947\,074$,
$\sin 0 \cdot 1663 = 0 \cdot 165\,535$, $\cos 0 \cdot 1663 = 0 \cdot 986\,204$)
****

D13A

$x_2 = -0 \cdot 160\,510$ $\qquad\qquad y_2 = 0 \cdot 493\,102$

FRAME D14

The next application of the formula gives $x_3 = -0 \cdot 160\,509$, $y_3 = 0 \cdot 493\,102$.

Now use the same method for the equations (D8.1) with $x_0 = 0 \cdot 15$ and $y_0 = -0 \cdot 05$, working to 4 decimal places.
****

D14A

$$f = x + 2y - \frac{1}{5} \sin (x - y) \qquad g = x - y - \frac{1}{5} \cos (x + y)$$

$$f_x = 1 - \frac{1}{5} \cos (x - y) \qquad f_y = 2 + \frac{1}{5} \cos (x - y)$$

$$g_x = 1 + \frac{1}{5} \sin (x + y) \qquad g_y = -1 + \frac{1}{5} \sin (x + y)$$

SOLUTION OF NON-LINEAR EQUATIONS

The first iteration gives $x_1 = 0 \cdot 1459$, $y_1 = -0 \cdot 0532$ *and the next produces no change.*

Two points to finish up with:

i)   Only two simultaneous equations have been dealt with in this appendix.   The methods used for their solution can easily be extended to more than two equations.   You would, though, have much greater difficulty in finding a starting point.

ii)   A process used considerably these days is optimisation.   This involves finding the values of the variables for which a function – for example, a cost function – is a minimum.   The processes discussed in this appendix can be applied to this problem.   If it is required to minimise $\phi(x, y)$ say, it is necessary to find x and y for which ϕ_x and ϕ_y are both zero.   There are thus two simultaneous equations to be solved for x and y.

Finally, try these two questions, both of which have appeared in C.E.I. Examinations.

1.   Rearrange the equations

$$f(x, y) = \sin(xy) - \frac{y}{2\pi} - x = 0$$
$$g(x, y) = \left(1 - \frac{1}{4\pi}\right)\left(e^{2x} - e\right) + \frac{ey}{\pi} - 2ex = 0$$

in the form $x = F(x, y)$
                        $y = G(x, y)$

and state a sufficient condition for the convergence of a direct iteration method based on this arrangement.

Starting with initial values $x_0 = 0 \cdot 4$, $y_0 = 3 \cdot 0$, iterate on the above until the solution $x = \frac{1}{2}$, $y = \pi$ is obtained correct to two significant figures.

2.   Derive the Newton–Raphson method for solving the non-linear algebraic equations

$$f_i(x_1, x_2, \ldots\ldots\ldots, x_n) = 0, \quad i = 1, 2, \ldots\ldots\ldots\ldots, n.$$

Use the above to find a solution of the equations

$$f(x, y) = x^2 + y^2 - 1 = 0$$
$$g(x, y) = x^2 + 9y^2 - 4x + 18y + 9 = 0$$

correct to 2 decimal places in the neighbourhood of the point $(1, -\frac{1}{2})$.

With the aid of a sketch, or otherwise, deduce a second solution of the given equations.

****

1. $x = \sin xy - \dfrac{y}{2\pi}$ $y = \dfrac{\pi}{e}\{2ex - (1 - \dfrac{1}{4\pi})(e^{2x} - e)\}$

   Successive iterates are

   $\begin{cases} x_1 = 0 \cdot 455 \\ y_1 = 3 \cdot 107 \end{cases}$ $\begin{cases} x_2 = 0 \cdot 494 \\ y_2 = 3 \cdot 137 \end{cases}$ $\begin{cases} x_3 = 0 \cdot 501 \\ y_3 = 3 \cdot 141 \end{cases}$ $\begin{cases} x_4 = 0 \cdot 500 \\ y_4 = 3 \cdot 142 \end{cases}$

   and no further change occurs.   Required solution, $x = 0 \cdot 50$, $y = 3 \cdot 1$.

2. *The equations for ε and η are*

   $\qquad 2x\varepsilon + 2y\eta = -(x^2 + y^2 - 1)$
   $\qquad (2x - 4)\varepsilon + 18(y + 1)\eta = -(x^2 + 9y^2 - 4x + 18y + 9)$

   $\qquad x_0 = 1, \quad y_0 = -\tfrac{1}{2}, \quad \varepsilon_0 = -0 \cdot 10, \quad \eta_0 = 0 \cdot 06$
   $\qquad x_1 = 0 \cdot 90, \quad y_1 = -0 \cdot 44, \quad \varepsilon_1 = -0 \cdot 004, \quad \eta_1 = -0 \cdot 004$
   $\qquad x_2 = 0 \cdot 896, \quad y_2 = -0 \cdot 444$

   and no further changes occur.   Required solution, $x = 0 \cdot 90$, $y = -0 \cdot 44$.

   $x^2 + y^2 - 1 = 0$ *is the circle, centre $(0, 0)$, radius 1.*

   $x^2 + 9y^2 - 4x + 18y + 9 = 0$ *can be rearranged as*

   $\dfrac{(x - 2)^2}{4} + \dfrac{(y + 1)^2}{4/9} = 1$ *which is an ellipse, centre $(2, -1)$,*

   semiaxes 2, $\dfrac{2}{9}$.   A sketch shows the other point of intersection at $(0, -1)$.

Simultaneous Linear Equations

FRAME 1

Introduction

One or two remarks have already been made about simultaneous linear
equations in this book (see FRAME 3, page 1). Also, if you read
APPENDIX D to the last programme, you found that it was necessary there
to solve two such equations in ε and η. In that appendix, nothing
was said about the actual solution of such equations, it being assumed
that you used either determinants, matrices or a method that you learnt
at school. In the present programme, we shall consider methods of
dealing with a large number of such equations in an equally large number
of unknowns.

FRAME 2

Simultaneous linear equations arise in many practical situations, and
the following are a few illustrations.

In a certain automatic packaging machine it was found necessary for a
certain point on a mechanism link to move over a path of approximately
the shape illustrated below. To accomplish this, five points were taken

on the shape and a conic fitted to them. (Five
points were taken as the general conic has six
constants in its equation, one of which can be
chosen at random.) This process led to five
simultaneous equations, the solution of which
determined the conic.

In designing cam profiles, standard displacement curves are available for
yielding certain types of motion. These standard curves are not
altogether satisfactory where high operating speeds are necessary. In
such cases, an equation of the form

$$y = c_0 + c_1\theta + c_2\theta^2 + \ldots\ldots\ldots\ldots + c_n\theta^n$$

is used. Corresponding values of θ and y are measured and then
simultaneous equations can be formed for the c's. If the polynomial is
taken as being of degree n, it follows that $n + 1$ equations will be
needed for the $n + 1$ c's.

A technique that is discussed later on in this book is that of curve
fitting by the method of least squares. When doing this an equation
connecting the variables is assumed, this equation containing a number of
constants. These are determined by building up simultaneous equations
which they have to satisfy.

Other applications of this work occur in network analysis and surveying.

FRAME 3

Before having a look at the methods of solution frequently used there is
one other point that needs to be mentioned in connection with
simultaneous equations, namely the effect of errors. You will remember
from previous work that the effect of errors can vary greatly according
to where they occur and the operations that are carried out on
quantities that involve them.

If you have to solve a set of equations which have exact coefficients and an exact solution, e.g.

$$2x + 3y = -5$$
$$x - 4y = 14$$

with solution $(2, -3)$, then no error is introduced.   However this situation is quite likely not to occur in practice.   It is more likely that equations such as

$$2 \cdot 37x + 3 \cdot 06y = -5 \cdot 63$$
$$0 \cdot 93x - 3 \cdot 72y = -14 \cdot 78$$

arise where the numerical values quoted are only known correct to so many, here two, places of decimals.   This means that round-off errors have been introduced, and consequently the accuracy of the solution will be affected.   At this stage we shall just illustrate how small changes in the coefficients can affect the solutions.

First, let us have a look at the equations

$$\left. \begin{array}{l} 2x + 3y = -5 \\ -10 \cdot 0001x + 9y = -14 \end{array} \right\} \quad \text{and} \quad \left. \begin{array}{l} 2x + 3y = -5 \\ -9 \cdot 9999x + 9y = -14 \end{array} \right\} \quad (4.1)$$

The solution of each pair (to 5 decimal places) is $(-0 \cdot 062\,50, \; -1 \cdot 625\,00)$ and you see that the effect of a change of $0 \cdot 0002$ in the coefficient of x in the second equation has been nil as far as the first five decimal places in the answer are concerned.   However, a change of the same magnitude can have a very different effect on the solution of other equations, as is shown in the following sets.

$$\left. \begin{array}{l} 2x + 3y = -5 \\ 10 \cdot 0001x + 9y = -14 \end{array} \right\} \qquad \left. \begin{array}{l} 2x + 3y = -5 \\ 9 \cdot 9999x + 9y = -14 \end{array} \right\} \quad (4.2)$$

Solutions $(0 \cdot 249\,99, \; -1 \cdot 833\,33)$ and $(0 \cdot 250\,01, \; -1 \cdot 833\,34)$

$$\left. \begin{array}{l} 2x + 3y = -5 \\ 6 \cdot 1001x + 9y = -14 \end{array} \right\} \qquad \left. \begin{array}{l} 2x + 3y = -5 \\ 6 \cdot 0999x + 9y = -14 \end{array} \right\} \quad (4.3)$$

Solutions $(9 \cdot 99, \; -8 \cdot 33)$ and $(10 \cdot 01, \; -8 \cdot 34)$

$$\left. \begin{array}{l} 2x + 3y = -5 \\ 6 \cdot 0011x + 9y = -14 \end{array} \right\} \qquad \left. \begin{array}{l} 2x + 3y = -5 \\ 6 \cdot 0009x + 9y = -14 \end{array} \right\} \quad (4.4)$$

Solutions $(909 \cdot 1, \; -607 \cdot 7)$ and $(1111 \cdot 1, \; -742 \cdot 4)$

$$\left. \begin{array}{l} 2x + 3y = -5 \\ 6 \cdot 0001x + 9y = -14 \end{array} \right\} \qquad \left. \begin{array}{l} 2x + 3y = -5 \\ 5 \cdot 9999x + 9y = -14 \end{array} \right\} \quad (4.5)$$

Solutions $(10\,000, \; -6668 \cdot 3)$ and $(-10\,000, \; 6665)$

$$\left. \begin{array}{l} 2x + 3y = -5 \\ 6 \cdot 0002x + 9y = -14 \end{array} \right\} \qquad \left. \begin{array}{l} 2x + 3y = -5 \\ 6x + 9y = -14 \end{array} \right\} \quad (4.6)$$

Solution $(5000, \; -3335)$ No solution

On first sight there doesn't appear to be all that difference between the various sets of equations.   But whereas some sets give reasonably consistent results – for example (4.1) and (4.2) – others look as though

they have gone completely wild, for example (4.4) and (4.5).   To
investigate further, the graphs of the various sets are useful.   Taking
(4.1), the graphs of the left hand pair of equations are

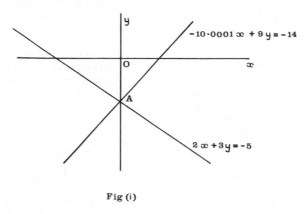

Fig (i)

As far as the right hand pair of equations (4.1) are concerned, $2x + 3y = -5$ is the same and $-9 \cdot 9999x + 9y = -14$ can be obtained by rotating $-10 \cdot 0001x + 9y = -14$ very slightly about its intercept A on the y-axis, as this intercept is the same for both equations. The effect on the point of intersection is negligible.   Note that A is not the point of intersection of the two lines.   An enlargement of the diagram around A is shown in Figure (ii).

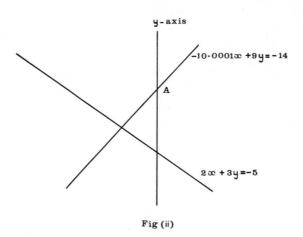

Fig (ii)

Now draw the graphs of the left hand pair of equations in (4.5) and think
what the effect on the point of intersection will be when a similar
rotation is carried out to give the right hand pair.
****

84

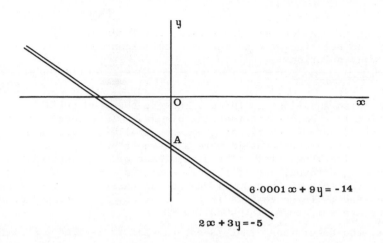

$$6 \cdot 0001x + 9y = -14$$

$$2x + 3y = -5$$

Here the two lines are almost (but not quite) parallel, and the point of intersection is way down in the bottom right hand corner. If the upper line is rotated very slightly anticlockwise about A to give 5·9999x + 9y = -14, it will pass through the position where it is parallel to 2x + 3y = -5 and as it turns further, the point of intersection will change to being a long way off in the upper left hand corner.

FRAME 6

You will thus appreciate that if the graphs of the two lines are nearly parallel, a very slight change in a coefficient will result in a considerably larger change in the position of the point of intersection. This does not happen if the lines are not nearly parallel.

Now go back to the sets of equations in FRAME 4 and work out the determinant of the coefficients of the two sets in each of (4.1) to (4.6).
****

6A

(4.1)	*48·0003*	*47·9997*
(4.2)	*−12·0003*	*−11·9997*
(4.3)	*−0·3003*	*−0·2997*
(4.4)	*−0·0033*	*−0·0027*
(4.5)	*−0·0003*	*0·0003*
(4.6)	*−0·0006*	*0*

FRAME 7

You will notice that although the change in the magnitude of the determinant is the same for each pair, the <u>relative</u> change is <u>very</u> different. Remembering that when you solve simultaneous equations by means of determinants, you have to divide by this determinant, you will see that a small change in this determinant, e.g. from 48·0003 to 47·9997, makes very little difference when the determinant is large. On the other hand it makes a very big difference when the value of the determinant is small, as is seen by the change from −0·0003 to +0·0003.

FRAME 7   (continued)

When the value of the determinant is zero, the lines are either parallel and there is no solution or else they are coincident and there are infinitely many solutions.

FRAME 8

If you have three equations in three unknowns, the geometrical interpretation is a set of intersecting planes. In general, these will meet in a point, but if any two of them are parallel, there is no solution. For more equations in more unknowns a geometrical interpretation is impossible. The determinant of the coefficients can still, however, be found, and if this is small in comparison with the coefficients themselves, then small changes in these coefficients can be expected to lead to large changes in the solution. Equations for which this happens are said to be ILL-CONDITIONED.

In practice, the determinant test is not used, as it is time consuming computerwise. To a certain extent, you can detect ill-conditioning if you find yourself dividing by very small quantities relative to the coefficients. If you find yourself attempting to divide by zero, there is no solution to the equations. When using a computer, you can instruct it to inform you if a very small divisor is encountered.   If it meets zero as a divisor, it will automatically boo at you anyway.   (It is, of course, assumed in all this work that the number of unknowns is the same as the number of equations.)

FRAME 9

Solution of Linear Algebraic Equations - Gaussian Elimination

You no doubt learnt, when you were at school, how to solve equations of this sort by elimination methods. You might be surprised to find that such methods are often still the most useful for solving sets of linear equations. However, as many equations are often involved, it is as well to adopt some systematic method of elimination instead of being haphazard about it, and this is one of the points we shall be discussing here.

The other main points involve (i) arranging our working so that the effect of errors in the equations is minimised and (ii) incorporating a check on our work.

But first let us have a look at a systematic arrangement of elimination without introducing these other points.

FRAME 10

As a first example, let us take the three equations

$$
\begin{aligned}
2 \cdot 37 x_1 + 3 \cdot 06 x_2 - 4 \cdot 28 x_3 &= 1 \cdot 76 & (10.1) \\
1 \cdot 46 x_1 - 0 \cdot 78 x_2 + 3 \cdot 75 x_3 &= 4 \cdot 69 & (10.2) \\
-3 \cdot 69 x_1 + 5 \cdot 13 x_2 + 1 \cdot 06 x_3 &= 5 \cdot 74 & (10.3)
\end{aligned}
$$

In practice, the number of equations and unknowns may be much larger, but small sets of equations serve just as well to illustrate the methods. However, bearing this point in mind we have used x_1, x_2 and x_3 for the unknowns as this notation can be easily extended.

The first step is to divide (10.1) by $2 \cdot 37$, so that the coefficient of x_1 is 1.   Thus

$$x_1 + 1 \cdot 2911x_2 - 1 \cdot 8059x_3 = 0 \cdot 7426 \qquad (10.4)$$

This equation is now multiplied by $-1 \cdot 46$, which is minus the coefficient of x_1 in (10.2), to give

$$-1 \cdot 46x_1 - 1 \cdot 8850x_2 + 2 \cdot 6366x_3 = -1 \cdot 0842 \qquad (10.5)$$

Adding this last equation to (10.2) gives

$$-2 \cdot 6650x_2 + 6 \cdot 3866x_3 = 3 \cdot 6058$$

The point of multiplying (10.4) by $-1 \cdot 46$ instead of $1 \cdot 46$ was to enable x_1 to be eliminated between (10.2) and (10.5) by an addition instead of a subtraction.

Now form from (10.4) another equation which, when added to (10.3), will give a second equation in x_2 and x_3 only.   What will this second equation be?   **

(10.4) multiplied by $3 \cdot 69$ gives

$$3 \cdot 69x_1 + 4 \cdot 7642x_2 - 6 \cdot 6638x_3 = 2 \cdot 7402 \qquad (10A.1)$$

(10.3) + (10A.1) $9 \cdot 8942x_2 - 5 \cdot 6038x_3 = 8 \cdot 4802$

We now have two equations in x_2 and x_3, i.e.,

$$-2 \cdot 6650x_2 + 6 \cdot 3866x_3 = 3 \cdot 6058 \qquad (11.1)$$
$$9 \cdot 8942x_2 - 5 \cdot 6038x_3 = 8 \cdot 4802 \qquad (11.2)$$

A similar procedure is now used to make the coefficient of x_2 in (11.1) 1 and then to eliminate x_2 between (11.1) and (11.2).   A single equation in x_3 will then be obtained.   Perform this stage of the elimination.
****

(11.1) ÷ (−2·6650)	$x_2 - 2 \cdot 3965x_3 = -1 \cdot 3530$	*(11A.1)*
(11A.1) × (−9·8942)	$-9 \cdot 8942x_2 + 23 \cdot 7115x_3 = 13 \cdot 3869$	*(11A.2)*
(11.2) + (11A.2)	$18 \cdot 1077x_3 = 21 \cdot 8671$	*(11A.3)*

The elimination process has now been completed and the equations (10.4), (11A.1) and (11A.3) $\div$ $18 \cdot 1077$ are used for the next part of the solution.   Repeating the first two of these equations and forming the third gives

$$x_1 + 1 \cdot 2911x_2 - 1 \cdot 8059x_3 = 0 \cdot 7426 \qquad (12.1)$$
$$x_2 - 2 \cdot 3965x_3 = -1 \cdot 3530 \qquad (12.2)$$
$$x_3 = 1 \cdot 2076 \qquad (12.3)$$

The value of x_3 is now known.   To find x_2, (12.3) is multiplied by $2 \cdot 3965$ and added to (12.2).   Effectively this substitutes the value of x_3 from (12.3) into (12.2).   Thus

SIMULTANEOUS LINEAR EQUATIONS

$$2 \cdot 3965 x_3 = 2 \cdot 8940$$
and $\quad\quad x_2 \quad\quad\quad = 1 \cdot 5410 \quad\quad\quad (12.4)$

What steps will be required to find x_1 in a similar way?

****

12A

(12.3) × 1·8059
(12.4) × (−1·2911)

The sum of these last two results and (12.1). $\left(\textit{Again, this}\atop\textit{effectively substitutes the values of }x_2\textit{ and }x_3\textit{ into (12.1).}\right)$

FRAME 13

Performing the operations listed in 12A,

$$
\begin{aligned}
1 \cdot 8059 x_3 &= 2 \cdot 1808 \\
-1 \cdot 2911 x_2 &= -1 \cdot 9896 \\
x_1 &= 0 \cdot 9338
\end{aligned}
$$

The solution is therefore $(0 \cdot 9338, 1 \cdot 5410, 1 \cdot 2076)$.

Substituting these results into the original equations gives

$$
\begin{aligned}
2 \cdot 37 x_1 + 3 \cdot 06 x_2 - 4 \cdot 28 x_3 &= 1 \cdot 7600 \\
1 \cdot 46 x_1 - 0 \cdot 78 x_2 + 3 \cdot 75 x_3 &= 4 \cdot 6899 \\
-3 \cdot 69 x_1 + 5 \cdot 13 x_2 + 1 \cdot 06 x_3 &= 5 \cdot 7397
\end{aligned}
$$

and these figures suggest that our solution is not unreasonable.

FRAME 14

The process of eliminating systematically the unknowns is called GAUSSIAN ELIMINATION.   Having obtained the three equations (12.1) − (12.3), the remainder of the procedure is known as BACK SUBSTITUTION.

There are several things to be said about this method and there are various modifications which will be looked at later.

The first thing you will notice is that all the working was done to more decimal places than were originally given.   This is often done and has the effect of reducing round-off errors that you yourself introduce. But to what extent can the solution be relied upon?   The answer to this requires a fair amount of theory and is given in the appendix rather than in the main part of the programme.

Next, you will appreciate that a considerable amount of arithmetic is involved and this leads to the possibility (at least, when you are doing the work) of arithmetical mistakes being made.   It would be helpful if some process could be incorporated into the solution so that such mistakes are detected as soon as possible.

Thirdly, while describing the method, all the equations were written out in full.   This together with the actual description itself, made the whole process somewhat lengthy.

A neat presentation of the whole calculation can be effected by the construction of a table.   This is done in the next frame and a current check on the arithmetic is also incorporated.   Furthermore a column headed δx has been added but you can ignore this now.   It will only be required if, later, you read the appendix.   You will appreciate, though, that such a table is only useful when you are doing the working manually.   A computer would just be told what to do and would then simply produce the final result.

Row	Formation	x_1	x_2	x_3	R.H.S.	s	δx
a		2·37	3·06	−4·28	1·76	2·91	1
b		1·46	−0·78	3·75	4·69	9·12	1
c		−3·69	5·13	1·06	5·74	8·24	1
d	a/2·37	1·0000	1·2911	−1·8059	0·7426	1·2278	0·42
e	−1·46d	−1·4600	−1·8850	2·6366	−1·0842	−1·7926	0·61
f	b + e		−2·6650	6·3866	3·6058	7·3274	1·61
g	3·69d	3·6900	4·7642	−6·6638	2·7402	4·5306	1·55
h	c + g		9·8942	−5·6038	8·4802	12·7706	2·55
i	f/(−2·6650)		1·0000	−2·3965	−1·3530	−2·7495	0·60
j	−9·8942i		−9·8942	23·7115	13·3869	27·2042	5·94
k	h + j			18·1077	21·8671	39·9748	8·49
ℓ	k/18·1077			1·0000	1·2076	2·2076	0·47
m	2·3965ℓ			2·3965	2·8940	5·2905	1·13
n	i + m		1·0000		1·5410	2·5410	1·73
p	1·8059ℓ			1·8059	2·1808	3·9867	0·85
q	−1·2911n		−1·2911		−1·9896	−3·2807	2·23
r	d + p + q	1·0000			0·9338	1·9338	3·50
Sol.		0·9338	1·5410	1·2076			

All the rows have been labelled in the first column, and the way in which each row after the first three has been formed is given in the second column.   In the columns headed x_1, x_2, x_3, the coefficients of these three quantities are given for each equation used.   The column headed R.H.S. speaks for itself and in the column labelled s the sum of the four numbers in columns x_1, x_2, x_3 and R.H.S. is given.   The solution row is obtained by extracting the relevant figures from the R.H.S. column.

The s column gives a current check on the arithmetic.   Having formed it by addition in the way indicated, it is then recalculated by using the formula given in the second column.   These two results should agree except for possible round-off errors.   Thus in row d,

$$1·0000 + 1·2911 - 1·8059 + 0·7426 = 1·2278$$
also
$$2·91 \div 2·37 = 1·2278$$
However, in row j,
$$-9·8942 + 23·7115 + 13·3869 = 27·2042$$
but
$$-9·8942 \times (-2·7495) = 27·2041$$

These two figures are sufficiently close to be acceptable.

A check like this is not necessary when a computer is being used, as this instrument is much too clever to make arithmetical mistakes.

Now check to see whether any four of the other figures in the s column
from rows d to r agree with those obtained by performing the
appropriate calculation as given in the 'Formation' column.

In the table in FRAME 15, each operation was performed singly and
recorded.   The amount of recording can be reduced by combining some of
the operations.   For example, referring to that table,

$$f = b + e = b - 1 \cdot 46d = b - 1 \cdot 46a/2 \cdot 37 = b - 0 \cdot 6160a$$

The following table is an abbreviation of that in FRAME 15.   It still,
however, contains all the essential working, although, as you will
realise, the process is modified slightly.   Rows showing the same
figures as previously carry different identifying letters for obvious
reasons.

Row	Formation	x_1	x_2	x_3	R.H.S.	s	δx
a		2·37	3·06	−4·28	1·76	2·91	1
b		1·46	−0·78	3·75	4·69	9·12	1
c		−3·69	5·13	1·06	5·74	8·24	1
d	b − (1·46/2·37)a = b − 0·6160a		−2·6650	6·3865	3·6058	7·3273	1·62
e	c + (3·69/2·37)a = c + 1·5570a		9·8944	−5·6040	8·4803	12·7707	2·58
f	e + $\dfrac{9 \cdot 8944}{2 \cdot 6650}$ d = e + 3·7127d			18·1072	21·8676	39·9748	8·59
g	f/18·1072			1·0000	1·2077	2·2077	0·47
h	d − 6·3865g		−2·6650		−4·1072	−6·7722	4·62
i	h/(−2·6650)		1·0000		1·5412	2·5412	1·73
j	a − 3·06i + 4·28g	2·3700			2·2129	4·5829	8·40
k	j/2·3700	1·0000			0·9337	1·9337	3·54
Sol.		0·9337	1·5412	1·2077			

If the results obtained this time are substituted into the original left
hand sides, the figures 1·7600, 4·6899, 5·7412 are obtained.

The last three equations in FRAME 13 can be written as

$$2 \cdot 37x_1 + 3 \cdot 06x_2 - 4 \cdot 28x_3 - 1 \cdot 76 = 0 \cdot 0000$$
$$1 \cdot 46x_1 - 0 \cdot 78x_2 + 3 \cdot 75x_3 - 4 \cdot 69 = -0 \cdot 0001$$
$$-3 \cdot 69x_1 + 5 \cdot 13x_2 + 1 \cdot 06x_3 - 5 \cdot 74 = -0 \cdot 0003$$

in which the original R.H.S.'s have been taken over to the left.

The three quantities remaining on the right are called the RESIDUALS.
If the equations have an exact solution and this is substituted into
them, the residuals will be zero.   If the solution is not quite exact

(as in our case) the residuals should be small.   After solving any set
of equations, you should check that this is so.   Unfortunately the
smallness of the residuals is not an absolute guarantee that you have an
accurate solution, as it is possible for an inaccurate solution to yield
small residuals if the equations are ill-conditioned.   That this is so
can be seen from the equations

$$5x_1 + 7x_2 + 6x_3 + 5x_4 = 23$$
$$7x_1 + 10x_2 + 8x_3 + 7x_4 = 32$$
$$6x_1 + 8x_2 + 10x_3 + 9x_4 = 33$$
$$5x_1 + 7x_2 + 9x_3 + 10x_4 = 31$$

which have exact solution (1, 1, 1, 1).   However the values (14·6,
-7·2, -2·5, 3·1) produce residuals 0·1, -0·1, -0·1, 0·1 and the
values (2·36, 0·18, 0·65, 1·21) produce residuals 0·01, -0·01, -0·01,
0·01.

What are the residuals for the solution to the three equations as
obtained in FRAME 16?   Why do you think this solution (and consequently
the residuals) are slightly different from previously?
****

0·0000,   -0·0001,   0·0012

*Due to round-off errors introduced during the calculation, some
arithmetical results become slightly different according to the order in
which the operations are performed.   For example,*
$3·06 \times \left(\dfrac{3·69}{2·37}\right) = 4·7644$ *but* $\left(\dfrac{3·06}{2·37}\right) \times 3·69 = 4·7642,$ *in each case
working to 4 decimal places.*

Having found the residuals, it is also possible to calculate the sum of
their squares.   Taking their squares eliminates the possibility of
positive and negative terms cancelling out on addition, thus giving a
false impression of accuracy.   If the solution is exact, this sum
(which can be denoted by ΣR^2) will be zero as each residual is itself
zero.   Otherwise it will be positive and then the smaller it is, the
better.   Thus, if the solution of a set of equations has been
calculated by two different processes, the better solution will be that
for which this sum is less.   Later in this book, we shall be considering
how to make the sum of a number of squares as small as possible - the so
called method of least squares.

Which of the two solutions obtained (0·9338, 1·5410, 1·2076) and
(0·9337, 1·5412, 1·2077) do you estimate is preferable for the equations
(10.1) - (10.3)?
****

For the first solution, $\Sigma R^2 = 10 \times 10^{-8}$ *and for the second,*
$\Sigma R^2 = 145 \times 10^{-8}.$ *The former is thus indicated as the better solution.*

Now take the set of equations

$$-2x_1 + 4x_2 + 3x_3 = 5$$
$$3x_1 + 2x_2 - 4x_3 = 7$$
$$4x_1 + 3x_2 + 5x_3 = 4$$

and construct the tables corresponding to those in FRAMES 15 and 16, working to 2 decimal places. Calculate the residuals and ΣR^2 in each case. (The coefficients have been chosen so that the arithmetic involved is relatively simple. They would not be so simple in a practical example, but working with numbers such as these will enable you to appreciate the method.)

****

The table corresponding to that in FRAME 15 is

Row	Formation	x_1	x_2	x_3	R.H.S.	s	δx
a		−2	4	3	5	10	1
b		3	2	−4	7	8	1
c		4	3	5	4	16	1
d	a/(−2)	1	−2	−1·5	−2·5	−5	0·5
e	−3d	−3	6	4·5	7·5	15	1·5
f	b + e		8	0·5	14·5	23	2·5
g	−4d	−4	8	6	10	20	2
h	c + g		11	11	14	36	3
i	f/8		1	0·06	1·81	2·87	0·31
j	−11i		−11	−0·66	−19·91	−31·57	3·41
k	h + j			10·34	−5·91	4·43	6·41
ℓ	k/10·34			1	−0·57	0·43	0·62
m	−0·06ℓ			−0·06	0·03	−0·03	0·04
n	i + m		1		1·84	2·84	0·35
p	1·5ℓ			1·5	−0·86	0·64	0·93
q	2n		2		3·68	5·68	0·70
r	d + p + q	1			0·32	1·32	2·13
Sol.		0·32	1·84	−0·57			

The residuals are 0·01, −0·08, −0·05. $\Sigma R^2 = 0·0090$

Again ignore the δx column.

The table corresponding to that in FRAME 16 is:

Row	Formation	x_1	x_2	x_3	R.H.S.	s	δx
a		-2	4	3	5	10	1
b		3	2	-4	7	8	1
c		4	3	5	4	16	1
d	$b + (3/2)a = b + 1·5a$		8	0·5	14·5	23	2·5
e	$c + (4/2)a = c + 2a$		11	11	14	36	3
f	$e - (11/8)d = e - 1·38d$			10·31	-6·01	4·30	6·45
g	$f/10·31$			1	-0·58	0·42	0·63
h	$d - 0·5g$		8		14·79	22·79	2·82
i	$h/8$		1		1·85	2·85	0·35
j	$a - 4i - 3g$	-2			-0·66	-2·66	4·29
k	$j/(-2)$	1			0·33	1·33	2·14
Sol.		0·33	1·85	-0·58			

The residuals are 0·00, 0·01, -0·03. $\Sigma R^2 = 0·0010$

You may be wondering why you were asked to do this example by the two methods.   The idea was simply to give you the feel of each so as to discover which you prefer.

<u>FRAME 20</u>

Gaussian Elimination with Partial Pivoting

If you look at the last table in 19A, you will see that during the elimination process, the multipliers 1·5, 2 and 1·38 were used.   In the first table in 19A, the same figures were effectively used, at least approximately as, for example, row a was divided by -2 and then the result multiplied by -3.   If the figures in row a are in error due, for example, to round-off, then the result of multiplying by 1·5 and 2 is to magnify this error.   If, however, it can be arranged that the multiplying factor is less than unity, then any such error will be decreased.   Returning to the example worked out in 19A, can you suggest any way of achieving this?
                    ****

<u>20A</u>

Instead of multiplying a by factors which will make the coefficient of x_1 equal to those in b and c, multiply c by factors that will make the coefficient of x_1 equal to those in a and b.

<u>FRAME 21</u>

When solving equations manually, they are sometimes rearranged so that the largest coefficient of x_1 appears in the first equation.   When a computer is being used, it would be told to search for the largest x_1 coefficient and then label it in such a way that the effect would be the same.

The three equations in FRAME 19 can be written

$$4x_1 + 3x_2 + 5x_3 = 4$$
$$-2x_1 + 4x_2 + 3x_3 = 5$$
$$3x_1 + 2x_2 - 4x_3 = 7$$

Again working to 2 decimal places, construct a table similar to either the first or the second (whichever you prefer) in 19A, using the equations in this new order.   What are the new residuals and ΣR^2?

****

21A

Either:

The table corresponding to the first of those in 19A is:

Row	Formation	x_1	x_2	x_3	R.H.S.	s
a		4	3	5	4	16
b		-2	4	3	5	10
c		3	2	-4	7	8
d	a/4	1	0·75	1·25	1	4
e	2d	2	1·50	2·50	2	8
f	b + e		5·50	5·50	7	18
g	-3d	-3	-2·25	-3·75	-3	-12
h	c + g		-0·25	-7·75	4	-4
i	f/5·50		1	1	1·27	3·27
j	0·25i		0·25	0·25	0·32	0·82
k	h + j			-7·50	4·32	-3·18
l	k/(-7·50)			1	-0·58	0·42
m	-l			-1	0·58	-0·42
n	i + m		1		1·85	2·85
p	-1·25l			-1·25	0·72	-0·53
q	-0·75n		-0·75		-1·39	-2·14
r	d + p + q	1			0·33	1·33
Sol.		0·33	1·85	-0·58		

The residuals are   0·00, 0·01, -0·03. $\Sigma R^2 = 0·0010$

_Or:_

The table corresponding to the second one in 19A is:

Row	Formation	x_1	x_2	x_3	R.H.S.	s
a		4	3	5	4	16
b		−2	4	3	5	10
c		3	2	−4	7	8
d	$b + (2/4)a = b + 0\cdot5a$		5·5	5·5	7	18
e	$c − (3/4)a = c − 0\cdot75a$		−0·25	−7·75	4	−4
f	$e + (0\cdot25/5\cdot5)d = e + 0\cdot05d$			−7·48	4·38	−3·1
g	$f/(−7\cdot48)$			1	−0·59	0·41
h	$d − 5\cdot5g$		5·5		10·24	15·74
i	$h/5\cdot5$		1		1·86	2·86
j	$a − 3i − 5g$	4			1·37	5·37
k	$j/4$	1			0·34	1·34
Sol.		0·34	1·86	−0·59		

The residuals are −0·01, −0·01, 0·10. $\Sigma R^2 = 0\cdot0102$

FRAME 22

If you look at the table you formed, you will find the two numbers 5·5 and −0·25 in the x_2 column. By chance, the first of these was bigger than the second and so at the next stage in the elimination process the multiplying factor was $\dfrac{0\cdot25}{5\cdot5}$. Had this not been the case these two rows would have been reversed to give a multiplying factor numerically less that 1.   The process whereby the multiplying factor is always arranged to be less than 1 during the elimination process is called PIVOTING and so we now have Gaussian Elimination with Pivoting.   (Actually the more correct term is PARTIAL PIVOTING, to distinguish it from FULL PIVOTING in which the columns are rearranged as well as the rows.)

Now use pivoting to solve the equations

$$0\cdot2x_1 + 0\cdot1x_2 + 0\cdot4x_3 = 0\cdot3$$
$$0\cdot4x_1 + 0\cdot3x_2 + 0\cdot6x_3 = 0\cdot2$$
$$0\cdot5x_1 + 0\cdot4x_2 + 0\cdot2x_3 = 0\cdot7$$

Work to 2 decimal places and calculate the residuals and ΣR^2.
****

95

SIMULTANEOUS LINEAR EQUATIONS

Either:

Row	Formation	x_1	x_2	x_3	R.H.S.	s
a		0·5	0·4	0·2	0·7	1·8
b		0·2	0·1	0·4	0·3	1·0
c		0·4	0·3	0·6	0·2	1·5
d	a/0·5	1	0·8	0·4	1·4	3·6
e	−0·2d	−0·2	−0·16	−0·08	−0·28	−0·72
f	−0·4d	−0·4	−0·32	−0·16	−0·56	−1·44
g	b + e		−0·06	0·32	0·02	0·28
h	c + f		−0·02	0·44	−0·36	0·06
j	g/(−0·06)		1	−5·33	−0·33	−4·66
k	0·02j		0·02	−0·11	−0·01	−0·10
ℓ	h + k			0·33	−0·37	−0·04
m	ℓ/0·33			1	−1·12	−0·12
n	5·33m			5·33	−5·97	−0·64
p	j + n		1		−6·30	−5·30
q	−0·4m			−0·4	0·45	0·05
r	−0·8p		−0·8		5·04	4·24
t	d + q + r	1			6·89	7·89
Sol.		6·89	−6·30	−1·12		

The residuals are 0·001, 0·000, −0·006. $\Sigma R^2 = 0·000\,037$

Note that at the stage *, it is necessary for the first of these
equations to have the larger coefficient of x_2.   This happens here,
but if not, rows g and h are interchanged.   This interchange can be
effected by repeating row g after row h.

Or:

Row	Formation	x_1	x_2	x_3	R.H.S.	s
a		0·5	0·4	0·2	0·7	1·8
b		0·2	0·1	0·4	0·3	1·0
c		0·4	0·3	0·6	0·2	1·5
d	b − (0·2/0·5)a = b − 0·4a		−0·06	0·32	0·02	0·28
e	c − (0·4/0·5)a = c − 0·8a		−0·02	0·44	−0·36	0·06
g	e − (0·02/0·06)d = e − 0·33d			0·33	−0·37	−0·04
h	g/0·33			1	−1·12	−0·12
i	e − 0·44h		−0·02		0·13	0·11
j	i/(−0·02)		1		−6·50	−5·5
k	a − 0·4j − 0·2h	0·5			3·52	4·02
ℓ	k/0·5	1			7·04	8·04
Sol.		7·04	−6·50	−1·12		

The residuals are   −0·004,   0·010,   −0·006. $\Sigma R^2 = 0·000\,152.$

96

Again, at †, row d would be repeated as row f had the coefficient of x_2 in row e been greater than that in row d.

The answers in the two tables here do not agree all that well.   This is due to the fact that you are only working to one more decimal place than in the original equations (in order not to get involved in too heavy arithmetic) and consequently the effect of round-off is considerable.

FRAME 23

Figures (1) and (2) are, respectively, flow charts for

a) Gaussian elimination without pivoting
b) Gaussian elimination with partial pivoting

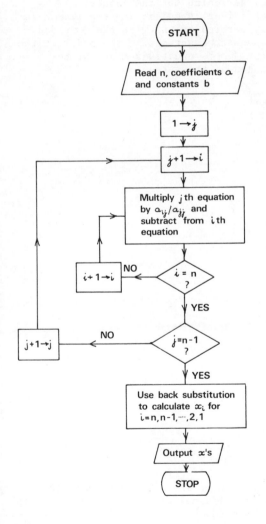

Equations are considered to be in the form

$$a_{11}x_1 + a_{12}x_2 + \ldots + a_{1n}x_n = b_1$$

$$a_{21}x_1 + a_{22}x_2 + \ldots + a_{2n}x_n = b_2 \quad \text{etc.}$$

The entries $\boxed{1 \to j}$ and $\boxed{j + 1 \to j}$ $\left(\text{and also } \boxed{j + 1 \to i} \text{ and } \boxed{i + 1 \to i}\right)$ are 'count' instructions. The first equation is originally called the jth equation and the 2nd equation called the ith equation.   The 3rd, 4th, 5th etc. equations up to the nth are then in turn labelled the ith equation.   After going through this sequence, the 2nd equation is called the jth equation and the 3rd to the nth equation in turn labelled the ith equation. The 3rd equation then becomes the jth equation and so on.

Fig. (1).

97

Equations are considered to be in the form

$$a_{11}x_1 + a_{12}x_2 + \ldots + a_{1n}x_n = b_1$$
$$a_{21}x_1 + a_{22}x_2 + \ldots + a_{2n}x_n = b_2 \quad \text{etc.}$$

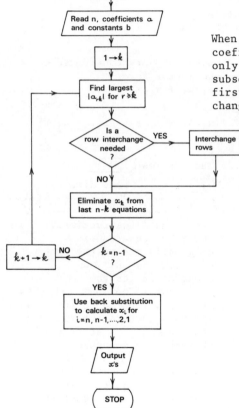

When searching for the largest coefficient in column k, the search is only carried out between the kth and subsequent equations. The order of the first k - 1 equations must not be changed at this stage.

Fig. (2)

(Programs using Gaussian elimination can be found in references (2), (5) and (9), while some using Gaussian elimination with partial pivoting can be found in references (4), (7) and (9).)

Gauss-Jordan Elimination

If you have read the programme 'Matrices I' of Unit 3 in our book
"Mathematics for Engineers and Scientists,Volume 1", you will realise
that a certain similarity exists between what we have just been doing
and the method of solution demonstrated in FRAMES 54-56 of the Matrices
programme.   Taking the last solution in 22A, the elimination process
changed the equations

$$0 \cdot 5x_1 + 0 \cdot 4x_2 + 0 \cdot 2x_3 = 0 \cdot 7$$
$$0 \cdot 2x_1 + 0 \cdot 1x_2 + 0 \cdot 4x_3 = 0 \cdot 3$$
$$0 \cdot 4x_1 + 0 \cdot 3x_2 + 0 \cdot 6x_3 = 0 \cdot 2$$

into

$$0 \cdot 5x_1 + 0 \cdot 4x_2 + 0 \cdot 2x_3 = 0 \cdot 7$$
$$-0 \cdot 02x_2 + 0 \cdot 44x_3 = -0 \cdot 36$$
$$0 \cdot 33x_3 = -0 \cdot 37$$

The augmented matrices for these are, respectively,

$$\begin{pmatrix} 0 \cdot 5 & 0 \cdot 4 & 0 \cdot 2 & \vdots & 0 \cdot 7 \\ 0 \cdot 2 & 0 \cdot 1 & 0 \cdot 4 & \vdots & 0 \cdot 3 \\ 0 \cdot 4 & 0 \cdot 3 & 0 \cdot 6 & \vdots & 0 \cdot 2 \end{pmatrix} \quad \text{and} \quad \begin{pmatrix} 0 \cdot 5 & 0 \cdot 4 & 0 \cdot 2 & \vdots & 0 \cdot 7 \\ 0 & -0 \cdot 02 & 0 \cdot 44 & \vdots & -0 \cdot 36 \\ 0 & 0 & 0 \cdot 33 & \vdots & -0 \cdot 37 \end{pmatrix}$$

This means that the coefficient matrix to the left of the dotted line has
been arranged to be upper triangular.

A variation of the ordinary Gaussian elimination process is the GAUSS-
JORDAN process which arranges the coefficient matrix to be diagonal.   If
you look at FRAMES 39-41 of the aforementioned programme on Matrices, you
will see that a similar process was adopted there for finding the inverse
of a matrix.   The Gauss-Jordan method can be used either with or without
pivoting.   In the example that follows, pivoting has been adopted.

To demonstrate the process, the equations you solved in FRAME 22 will be
used.   Again they are arranged so that the equation with the largest
coefficient of x_1 appears first.   Thus

$$0 \cdot 5x_1 + 0 \cdot 4x_2 + 0 \cdot 2x_3 = 0 \cdot 7 \quad (25.1)$$
$$0 \cdot 2x_1 + 0 \cdot 1x_2 + 0 \cdot 4x_3 = 0 \cdot 3 \quad (25.2)$$
$$0 \cdot 4x_1 + 0 \cdot 3x_2 + 0 \cdot 6x_3 = 0 \cdot 2 \quad (25.3)$$

The first steps are the same as those already carried out in the second
table in 22A for rows d and e.   Thus,

$(25.2) - \dfrac{0 \cdot 2}{0 \cdot 5} (25.1)$ i.e. $(25.2) - 0 \cdot 4(25.1)$ gives $-0 \cdot 06x_2 + 0 \cdot 32x_3$
$$= 0 \cdot 02$$

and

$(25.3) - \dfrac{0 \cdot 4}{0 \cdot 5} (25.1)$ i.e. $(25.3) - 0 \cdot 8(25.1)$ gives $-0 \cdot 02x_2 + 0 \cdot 44x_3$
$$= -0 \cdot 36$$

The three equations next used are

$$0 \cdot 5x_1 + 0 \cdot 4x_2 + 0 \cdot 2x_3 = 0 \cdot 7 \quad (25.1)$$
$$-0 \cdot 06x_2 + 0 \cdot 32x_3 = 0 \cdot 02 \quad (25.4)$$
$$-0 \cdot 02x_2 + 0 \cdot 44x_3 = -0 \cdot 36 \quad (25.5)$$

in this order.   If necessary the last two equations would have been
reversed so that the one with the larger coefficient of x_2 is written
down before the other.

Equation (25.4) is now used to eliminate the x_2 term from <u>both</u> (25.1) and (25.5). $\left(\text{In the previous method (25.1) would have been left unaltered at this stage.}\right)$

What operations will have to be carried out in order to achieve this? What will be the new equations replacing (25.1) and (25.5)?
****

25A

$(25.1) + \dfrac{0 \cdot 4}{0 \cdot 06} \ (25.4)$ *i.e.* $(25.1) + 6 \cdot 67 (25.4)$

$(25.5) - \dfrac{0 \cdot 02}{0 \cdot 06} \ (25.4)$ *i.e.* $(25.5) - 0 \cdot 33 (25.4)$

$$0 \cdot 5 x_1 \qquad\qquad -2 \cdot 33 x_3 \ = \ 0 \cdot 83$$
$$0 \cdot 33 x_3 \ = \ -0 \cdot 37$$

We next work on the three equations

$$0 \cdot 5 x_1 \qquad\qquad + 2 \cdot 33 x_3 \ = \ 0 \cdot 83 \qquad (26.1)$$
$$- 0 \cdot 06 x_2 + 0 \cdot 32 x_3 \ = \ 0 \cdot 02 \qquad (25.4)$$
$$0 \cdot 33 x_3 \ = \ -0 \cdot 37 \qquad (26.2)$$

The final step in the elimination process is to use (26.2) to eliminate the terms in x_3 from (26.1) and (25.4). To do this the operations

$$(26.1) - \dfrac{2 \cdot 33}{0 \cdot 33} \ (26.2) \quad i.e. \quad (26.1) - 7 \cdot 06 (26.2)$$

$$\text{and} \quad (25.4) - \dfrac{0 \cdot 32}{0 \cdot 33} \ (26.2) \quad i.e. \quad (25.4) - 0 \cdot 97 (26.2)$$

are carried out, giving

$$0 \cdot 5 x_1 \qquad\qquad = \ 3 \cdot 44$$
$$-0 \cdot 06 x_2 \qquad = \ 0 \cdot 38$$

and the solution is now obtained directly from

$$0 \cdot 5 x_1 \qquad\qquad = \ 3 \cdot 44 \left.\vphantom{\begin{matrix}1\\1\\1\end{matrix}}\right\}$$
$$-0 \cdot 06 x_2 \qquad = \ 0 \cdot 38$$
$$0 \cdot 33 x_3 \ = \ -0 \cdot 37 \qquad (26.3)$$

giving $x_1 = 6 \cdot 88,$ $x_2 = -6 \cdot 33,$ $x_3 = -1 \cdot 12.$

You will notice that if the augmented matrix for the equations (26.3) is written down

$$\begin{pmatrix} 0 \cdot 5 & 0 & 0 & : & 3 \cdot 44 \\ 0 & -0 \cdot 06 & 0 & : & 0 \cdot 38 \\ 0 & 0 & 0 \cdot 33 & : & -0 \cdot 37 \end{pmatrix}$$

then the coefficient matrix (to the left of the dotted line) is diagonal. In arriving at this matrix, some of the multiplying factors used were > 1, but this, unfortunately, cannot be helped. Even mathematicians can't have everything their own way!

What are the residuals for this solution and also what is ΣR^2?
****

-0·016   -0·005   -0·019   0·000 642

As with the previous method of solution, the calculation, when being done manually, is best set out in the form of a table.   Also, as before, it is advisable to incorporate a check column.

Row	Formation	x_1	x_2	x_3	R.H.S.	s
a		0·5	0·4	0·2	0·7	1·8
b		0·2	0·1	0·4	0·3	1·0
c		0·4	0·3	0·6	0·2	1·5
d	a	0·5	0·4	0·2	0·7	1·8
e	b − (0·2/0·5)a = b − 0·4a		−0·06	0·32	0·02	0·28
f	c − (0·4/0·5)a = c − 0·8a		−0·02	0·44	−0·36	0·06
g	d + (0·4/0·06)e = d + 6·67e	0·5		2·33	0·83	3·66
h	e		−0·06	0·32	0·02	0·28
i	f − (0·02/0·06)e = f − 0·33e			0·33	−0·37	−0·04
j	g − (2·33/0·33)i = g − 7·06i	0·5			3·44	3·94
k	h − (0·32/0·33)i = h − 0·97i		−0·06		0·38	0·32
ℓ	i			0·33	−0·37	−0·04
m	j/0·5	1			6·88	7·88
n	k/(−0·06)		1		−6·33	−5·33
p	ℓ/0·33			1	−1·12	−0·12
Sol.		6·88	−6·33	−1·12		

You will notice that several lines have been repeated in the table. This, of course, is not absolutely necessary but is serves to illustrate very nicely the matrix pattern that is being formed.   (Compare with FRAMES 40-41A of our programme Matrices I.)

Now use this method to solve the equations

$$2x_1 - 4x_2 - 2x_3 = 3$$
$$3x_1 + x_2 - 3x_3 = -4$$
$$4x_1 - 3x_2 + 2x_3 = 2$$

Work to 2 decimal places and arrange your working in tabular form.
****

Row	Formation	x_1	x_2	x_3	R.H.S.	s
a		4	−3	2	2	5
b		2	−4	−2	3	−1
c		3	1	−3	−4	−3
d	b − (2/4)a = b − 0·5a		−2·5	−3	2	−3·5
e	c − (3/4)a = c − 0·75a		3·25	−4·5	−5·5	−6·75
f	a	4	−3	2	2	5
g	e		3·25	−4·5	−5·5	−6·75
h	d		−2·5	−3	2	−3·5
i	f + (3/3·25)g = f + 0·92g	4	*	−2·14	−3·06	−1·20
j	g		3·25	−4·5	−5·5	−6·75
k	h + (2·5/3·25)g = h + 0·77g			−6·46	−2·24	−8·70
l	i − (2·14/6·46)k = i − 0·33k	4			−2·32	1·68
m	j − (4·5/6·46)k = j − 0·70k		3·25		−3·93	−0·68
n	k			−6·46	−2·24	−8·70
p	l/4	1			−0·58	0·42
q	m/3·25		1		−1·21	−0·21
r	n/(−6·46)			1	0·35	1·35
Sol.		−0·58	−1·21	0·35		

The residuals are −0·02, 0·00, 0·01 and $\Sigma R^2 = 0\cdot0005$.

Note that in certain places, as in previous examples, the formation of a row is designed to produce zeros. Due to round-off this will not actually always occur, but the particular number has always been taken as zero. For example, in the position indicated by *
f + (3/3·25)g = 0, but f + 0·92g = −0·01.

The Effect of Inaccurate Data

So far we have concentrated on the solution of the equations actually quoted. Mistakes made in the arithmetic will have been detected by the use of the s column and the reasonableness of the solution has been demonstrated by the smallness of the residuals. But as you know, unless exact data are given, the coefficients of the x's and the R.H.S.'s may themselves be in error due to round-off. How far can one rely on the solution if such errors are present? A rough estimate of this can be obtained and is given in the appendix at the end of this programme.

The Gauss-Seidel Iteration Method

Consider the circuit shown in diagram (1) on page 103.   The currents i_1 i_2 and i_3 are given by the equations

$$9i_1 \qquad\qquad - 5i_3 = 10$$
$$20i_2 - 12i_3 = -2$$
$$-5i_1 - 12i_2 + 20i_3 = 0$$

For the currents as shown, it is assumed that i_1 is flowing completely round the left hand branch of the network, i_2 round the upper right hand branch and i_3 round the lower right hand branch. The current in the wire AB, for example, is then $i_1 - i_2 \downarrow$.

Diagram (1)

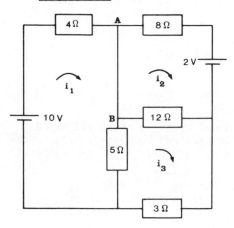

In matrix form, these equations are

$$\begin{pmatrix} 9 & 0 & -5 \\ 0 & 20 & -12 \\ -5 & -12 & 20 \end{pmatrix} \begin{pmatrix} i_1 \\ i_2 \\ i_3 \end{pmatrix} = \begin{pmatrix} 10 \\ -2 \\ 0 \end{pmatrix}$$

and you will notice that in each row of the coefficient matrix, the element on the leading diagonal is greater in magnitude than the sum of the magnitudes of the other elements in that row. Even if the equations had been written down in a different order, as is of course quite likely, they could still have been rearranged so that this property held. (A matrix which exhibits this property is said to be DIAGONALLY DOMINANT.)

A nearly similar situation arises in the case of the Wheatstone bridge illustrated. Here the equations can be arranged as

Diagram (2)

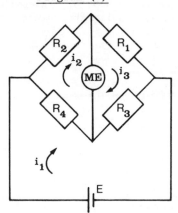

$$
\begin{array}{rcl}
(R_3 + R_4)i_1 & - R_4 i_2 \quad - R_3 i_3 & = E \\
R_4 i_1 - (R_2 + R_4)i_2 & & = 0 \\
R_3 i_1 & - (R_1 + R_3) i_3 & = 0
\end{array}
$$

or as

$$\begin{pmatrix} R_3 + R_4 & - R_4 & - R_3 \\ R_4 & - (R_2 + R_4) & 0 \\ R_3 & 0 & - (R_1 + R_3) \end{pmatrix} \begin{pmatrix} i_1 \\ i_2 \\ i_3 \end{pmatrix} = \begin{pmatrix} E \\ 0 \\ 0 \end{pmatrix}$$

As the R's are all essentially positive, the diagonal element in each row of the coefficient matrix is greater than or equal to the sum of the magnitudes of the other elements in the same row.

This is a situation that often arises in practice in scientific and engineering problems and the following method of solution, different from our previous elimination methods, can then often be adopted. In general it will be found that the more marked the diagonal dominance, the more likely is the method to succeed reasonably quickly.

The first set of equations in the last frame will be taken to illustrate the method, i.e.,

$$9i_1 \qquad\qquad - 5i_3 = 10$$
$$20i_2 - 12i_3 = -2$$
$$-5i_1 - 12i_2 + 20i_3 = 0$$

The equations, taken in this order, are solved for i_1, i_2 and i_3 respectively to give

$$i_1 = 1 \cdot 1111 + 0 \cdot 5556i_3 \qquad (31.1)$$
$$i_2 = -0 \cdot 1000 + 0 \cdot 6000i_3 \qquad (31.2)$$
$$i_3 = 0 \cdot 2500i_1 + 0 \cdot 6000i_2 \qquad (31.3)$$

if we work to 4 decimal places.

An initial "guess" is now made for a set of values of i_1, i_2 and i_3. For this "guess" the values (0, 0, 0) are often taken.

A second estimate of i_1 is now obtained by using these values of i_2 and i_3 (actually only that of i_3 is required here) in (31.1). Doing this gives $i_1 = 1 \cdot 11$. (At the beginning there is no need to retain a large number of decimal places.)

A second estimate of i_2 is found by putting $i_1 = 1 \cdot 11$, $i_3 = 0$ (in this particular example only i_3 is used) in (31.2). We thus get $i_2 = -0 \cdot 10$.

The values $i_1 = 1 \cdot 11$ and $i_2 = -0 \cdot 10$ are now used in (31.3) to give $0 \cdot 2500 \times 1 \cdot 11 - 0 \cdot 6000 \times 0 \cdot 10 = 0 \cdot 22$ as a second estimate of i_3.

The complete second estimate is therefore $(1 \cdot 11, -0 \cdot 10, 0 \cdot 22)$.

Equations (31.1), (31.2) and (31.3) are now used continuously in turn, the latest values obtained being inserted into the R.H.S.'s. As with iterative methods generally, it is hoped that eventually the values settle down to the actual solution of the equations.

The third estimates are:

$$i_1 = 1 \cdot 1111 + 0 \cdot 5556 \times 0 \cdot 22 \qquad = 1 \cdot 23$$
$$i_2 = -0 \cdot 1000 + 0 \cdot 6000 \times 0 \cdot 22 \qquad = 0 \cdot 03$$
$$i_3 = 0 \cdot 2500 \times 1 \cdot 23 + 0 \cdot 6000 \times 0 \cdot 03 = 0 \cdot 33$$

Now find the fourth and fifth estimates.

****

32A

The fourth estimates are

$$i_1 = 1 \cdot 1111 + 0 \cdot 5556 \times 0 \cdot 33 \qquad = 1 \cdot 29$$
$$i_2 = -0 \cdot 1000 + 0 \cdot 6000 \times 0 \cdot 33 \qquad = 0 \cdot 10$$
$$i_3 = 0 \cdot 2500 \times 1 \cdot 29 + 0 \cdot 6000 \times 0 \cdot 10 = 0 \cdot 38$$

The fifth estimates are

$$i_1 = 1 \cdot 1111 + 0 \cdot 5556 \times 0 \cdot 38 \qquad = 1 \cdot 32$$
$$i_2 = -0 \cdot 1000 + 0 \cdot 6000 \times 0 \cdot 38 \qquad = 0 \cdot 13$$
$$i_3 = 0 \cdot 2500 \times 1 \cdot 32 + 0 \cdot 6000 \times 0 \cdot 13 = 0 \cdot 41$$

Continuing in this way, further successive estimates are

$$
\begin{array}{lll}
(1\cdot34, & 0\cdot15, & 0\cdot42) \\
(1\cdot344, & 0\cdot152, & 0\cdot427) \\
(1\cdot348, & 0\cdot156, & 0\cdot431) \\
(1\cdot351, & 0\cdot159, & 0\cdot433) \\
(1\cdot3516, & 0\cdot1598, & 0\cdot4338) \\
(1\cdot3521, & 0\cdot1603, & 0\cdot4342) \\
(1\cdot3523, & 0\cdot1605, & 0\cdot4344) \\
(1\cdot3524, & 0\cdot1606, & 0\cdot4345) \\
(1\cdot3525, & 0\cdot1607, & 0\cdot4345)
\end{array}
$$

Further iterations produce no change in these values to 4 decimal places.

The residuals are $0\cdot0000$, $0\cdot0000$, $-0\cdot0009$ and $\Sigma R^2 = 81 \times 10^{-8}$.

This method of solution is known as the GAUSS-SEIDEL method.

Now use the method to find the solution of

$$
\begin{array}{l}
1\cdot3x_1 + 0\cdot5x_2 - 0\cdot1x_3 = 2\cdot4 \\
0\cdot4x_1 - 0\cdot1x_2 - 5\cdot5x_3 = 1\cdot7 \\
1\cdot7x_1 + 3\cdot5x_2 + 1\cdot5x_3 = 3\cdot2
\end{array}
$$

working to 3 decimal places.

****

33A

It is necessary to interchange the 2nd and 3rd equations.

Then
$$
\begin{array}{l}
x_1 = 1\cdot846 - 0\cdot385x_2 + 0\cdot077x_3 \\
x_2 = 0\cdot914 - 0\cdot486x_1 - 0\cdot429x_3 \\
x_3 = -0\cdot309 + 0\cdot073x_1 - 0\cdot018x_2
\end{array}
$$

Starting from (0, 0, 0), *successive estimates are*

$$
\begin{array}{lll}
(1\cdot8, & 0\cdot0, & -0\cdot2) \\
(1\cdot83, & 0\cdot11, & -0\cdot18) \\
(1\cdot790, & 0\cdot121, & -0\cdot181) \\
(1\cdot785, & 0\cdot124, & -0\cdot181) \\
(1\cdot784, & 0\cdot125, & -0\cdot181) \\
(1\cdot784, & 0\cdot125, & -0\cdot181)
\end{array}
$$

The residuals are $0\cdot000$, $-0\cdot003$, $-0\cdot001$ *and* $\Sigma R^2 = 10^{-5}$.

In the examples taken so far, the number of equations has been kept small. We will now look at just one, slightly more complicated, problem which will bring out some points not so obvious in smaller systems.

Suppose it is necessary to find the currents in the branches of the circuit shown on page 106:

FRAME 34   (continued)

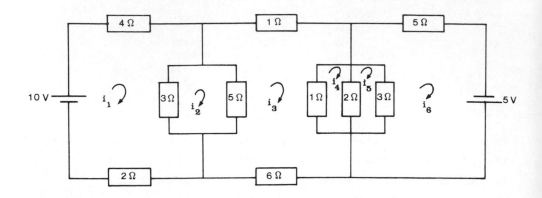

The equations for this circuit are:

$$
\begin{aligned}
9i_1 - 3i_2 &= 10 \\
-3i_1 + 8i_2 - 5i_3 &= 0 \\
-5i_2 + 13i_3 - i_4 &= 0 \\
-i_3 + 3i_4 - 2i_5 &= 0 \\
-2i_4 + 5i_5 - 3i_6 &= 0 \\
-3i_5 + 8i_6 &= 5
\end{aligned}
\right\} \quad (34.1)
$$

or, in matrix form,

$$
\begin{pmatrix}
9 & -3 & 0 & 0 & 0 & 0 \\
-3 & 8 & -5 & 0 & 0 & 0 \\
0 & -5 & 13 & -1 & 0 & 0 \\
0 & 0 & -1 & 3 & -2 & 0 \\
0 & 0 & 0 & -2 & 5 & -3 \\
0 & 0 & 0 & 0 & -3 & 8
\end{pmatrix}
\begin{pmatrix}
i_1 \\ i_2 \\ i_3 \\ i_4 \\ i_5 \\ i_6
\end{pmatrix}
=
\begin{pmatrix}
10 \\ 0 \\ 0 \\ 0 \\ 0 \\ 5
\end{pmatrix}
$$

You will notice that the diagonal element in each row is greater than
or equal to the sum of the magnitudes of the other elements in that row.
This suggests the Gauss–Seidel method of solution, although not very
strongly.   Furthermore you will notice that the coefficient matrix is a
band matrix.   (See page 8:32 in our second volume of Mathematics for
Engineers and Scientists if you have forgotten what a band matrix is.)
It often happens in practice that problems do give rise to coefficient
matrices that are band and/or sparse.

FRAME 35

From the equations (34.1),

$$
\begin{aligned}
i_1 &= 1 \cdot 111 + 0 \cdot 333 i_2 \\
i_2 &= 0 \cdot 375 i_1 + 0 \cdot 625 i_3 \\
i_3 &= 0 \cdot 385 i_2 + 0 \cdot 077 i_4 \\
i_4 &= 0 \cdot 333 i_3 + 0 \cdot 667 i_5 \\
i_5 &= 0 \cdot 400 i_4 + 0 \cdot 600 i_6 \\
i_6 &= 0 \cdot 625 + 0 \cdot 375 i_5
\end{aligned}
$$

As is now obvious, when the coefficient matrix is sparse, the number of terms on the R.H.S. of each equation is relatively small.

Starting from zero for all i's,

$$
\left.
\begin{aligned}
i_1 &= 1 \cdot 11 \\
i_2 &= 0 \cdot 375 \times 1 \cdot 11 = 0 \cdot 41 \\
i_3 &= 0 \cdot 385 \times 0 \cdot 41 = 0 \cdot 16 \\
i_4 &= 0 \cdot 333 \times 0 \cdot 16 = 0 \cdot 05 \\
i_5 &= 0 \cdot 400 \times 0 \cdot 05 = 0 \cdot 02 \\
i_6 &= 0 \cdot 625 + 0 \cdot 375 \times 0 \cdot 02 = 0 \cdot 63
\end{aligned}
\right\} \quad (35.1)
$$

The next estimates are

$$
\begin{aligned}
i_1 &= 1 \cdot 11 + 0 \cdot 333 \times 0 \cdot 41 = 1 \cdot 25 \\
i_2 &= 0 \cdot 375 \times 1 \cdot 25 + 0 \cdot 625 \times 0 \cdot 16 = 0 \cdot 57 \\
i_3 &= 0 \cdot 385 \times 0 \cdot 57 + 0 \cdot 077 \times 0 \cdot 05 = 0 \cdot 22 \\
i_4 &= 0 \cdot 333 \times 0 \cdot 22 + 0 \cdot 667 \times 0 \cdot 02 = 0 \cdot 09 \\
i_5 &= 0 \cdot 400 \times 0 \cdot 09 + 0 \cdot 600 \times 0 \cdot 63 = 0 \cdot 41 \\
i_6 &= 0 \cdot 625 + 0 \cdot 375 \times 0 \cdot 41 = 0 \cdot 78
\end{aligned}
$$

Now obtain the next two sets of estimates, giving the currents to 3 decimal places.

****

35A

$$
\begin{aligned}
i_1 &= 1 \cdot 111 + 0 \cdot 333 \times 0 \cdot 57 = 1 \cdot 301 \\
i_2 &= 0 \cdot 375 \times 1 \cdot 301 + 0 \cdot 625 \times 0 \cdot 22 = 0 \cdot 625 \\
i_3 &= 0 \cdot 385 \times 0 \cdot 625 + 0 \cdot 077 \times 0 \cdot 09 = 0 \cdot 248 \\
i_4 &= 0 \cdot 333 \times 0 \cdot 248 + 0 \cdot 667 \times 0 \cdot 41 = 0 \cdot 356 \\
i_5 &= 0 \cdot 400 \times 0 \cdot 356 + 0 \cdot 600 \times 0 \cdot 78 = 0 \cdot 610 \\
i_6 &= 0 \cdot 625 + 0 \cdot 375 \times 0 \cdot 610 = 0 \cdot 854
\end{aligned}
$$

$i_1 = 1 \cdot 319,$ $i_2 = 0 \cdot 650,$ $i_3 = 0 \cdot 278,$ $i_4 = 0 \cdot 499,$ $i_5 = 0 \cdot 712,$
$i_6 = 0 \cdot 892.$

FRAME 36

The process is now continued until no further changes occur in the estimates of the i's.   When not using a computer, the whole calculation is best set out in tabular form and this is done on page 108, starting with the values given in equations (35.1).

FRAME 37

The equations
$$
\begin{aligned}
3x_1 + x_2 - x_3 &= 1 \\
4x_1 - 2x_2 + 3x_3 &= 5 \\
5x_1 - x_2 - 2x_3 &= 4
\end{aligned}
$$

cannot be arranged so that the largest coefficients appear on the leading diagonal of the coefficient matrix.   (To 2 decimal places the equations have solution $0 \cdot 68$, $-0 \cdot 89$, $0 \cdot 16$.)   One way in which they can be 'solved' for x_1, x_2 and x_3 is

$$
\left.
\begin{aligned}
x_1 &= (1 - x_2 + x_3)/3 \\
x_2 &= (-5 + 4x_1 + 3x_3)/2 \\
x_3 &= (-4 + 5x_1 - x_2)/2
\end{aligned}
\right\} \quad (37.1)
$$

Table for FRAME 36.

i_1	i_2	i_3	i_4	i_5	i_6
1·11	0·41	0·16	0·05	0·02	0·63
1·25	0·57	0·22	0·09	0·41	0·78
1·301	0·625	0·248	0·356	0·610	0·854
1·319	0·650	0·278	0·499	0·712	0·892
1·327	0·671	0·297	0·574	0·765	0·912
1·334	0·686	0·308	0·613	0·792	0·922
1·339	0·695	0·315	0·633	0·806	0·927
1·342	0·700	0·318	0·643	0·813	0·930
1·344	0·703	0·320	0·649	0·818	0·932
1·345	0·704	0·321	0·652	0·820	0·932
1·345	0·705	0·322	0·654	0·821	0·933
1·346	0·706	0·322	0·655	0·822	0·933
1·346	0·706	0·322	0·656	0·822	0·933
1·346	0·706	0·322	0·656	0·822	0·933

The residuals are   −0·004, 0·000, 0·000, 0·002, −0·001, −0·002 and $\Sigma R^2 = 25 \times 10^{-6}$.

FRAME 37   (continued)

Taking this last set of equations, see what happens if you try to use the Gauss–Seidel method.   Find the first ten sets of estimates working to 2 decimal places. **

37A

x_1	x_2	x_3
0·33	−1·84	−0·26
0·86	−1·17	0·74
0·97	0·55	0·15
0·20	−1·88	−0·56
0·77	−1·80	0·82
1·21	1·15	0·45
0·10	−1·62	−0·94
0·56	−2·79	0·80
1·53	1·76	0·94
0·06	−0·97	−1·36

FRAME 38

What appears to be happening here? **

The form of the equations (37.1) is such that the iterations do not converge to the solution. If you think back to the previous programme, you will remember that iteration is not a suitable technique for every form into which an equation can be put. See FRAMES 11-13, pages 33 - 35 if you have forgotten about this.

We shall not fully consider conditions of convergence here. It can be shown that the Gauss-Seidel iterative method does converge if the coefficient matrix is diagonally dominant. This is a sufficient condition for convergence but some systems of equations which do not satisfy it also converge. In practice, if this condition is not satisfied one proceeds with caution, knowing that Gauss-Seidel may fail. In the example in FRAME 37 it was soon evident that this was the case. When using a computer, one would incorporate a test in the computer program to find whether convergence is taking place or not.

A flow diagram is shown, on page 110, for the solution of n simultaneous linear equations in n unknowns x_1, x_2, ... x_n using the Gauss-Seidel method.

The equations are taken as being in the form:

$a_{11}x_1 + a_{12}x_2 + \ldots + a_{1n}x_n = b_1$

$a_{21}x_1 + a_{22}x_2 + \ldots + a_{2n}x_n = b_2$ etc., the coefficient matrix being diagonally dominant.

For obtaining X_i, the new value for x_i, the ith equation gives

$X_i = (b_i - \Sigma_i)/a_{ii}$ where $\Sigma_i = \Sigma a_{ij}x_j$ with j taking all values from 1 to n except i.

Choleski's Method

Although matrix notation has been used in this programme, none of the methods used has involved the theory of matrices. Choleski's method however, does and so we shall leave consideration of it until the next programme which specifically deals with matrices.

Which Method is Best?

Various methods for solving simultaneous linear equations have been mentioned in this programme. Each method has its own advantages.

Elimination methods have the advantage that they can be used for any set of n equations in n unknowns provided the coefficient matrix is non-singular.

Iterative methods have the disadvantage that they can only be used for certain sets of equations, but practical problems often lead to equations that are amenable to this kind of treatment. When Gauss-Seidel can be used it has certain advantages over the elimination methods which makes it well worth while. Some of these advantages are:

Flow diagram
for FRAME 40.

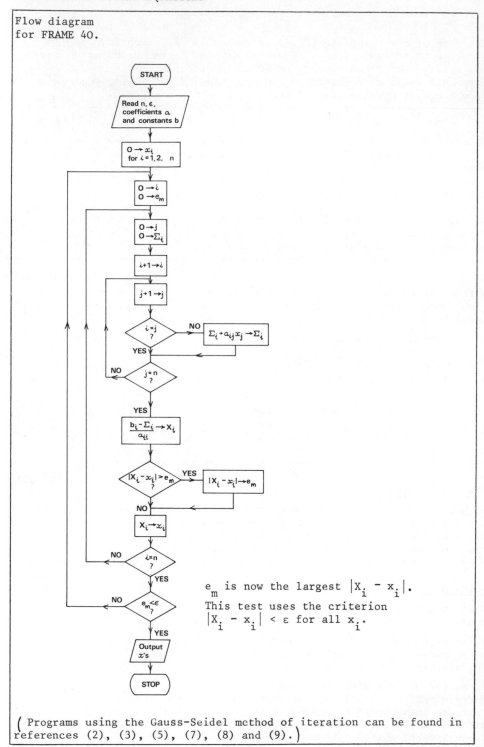

e_m is now the largest $|X_i - x_i|$.
This test uses the criterion
$|X_i - x_i| < \varepsilon$ for all x_i.

(Programs using the Gauss-Seidel method of iteration can be found in
references (2), (3), (5), (7), (8) and (9).)

i)   Fewer multiplications required for large systems.

ii)   Round-off error introduced has generally less effect.

iii)   The process is self-correcting.   This is only important though
        when calculations are being done manually.

iv)   When using a computer, may be more economical in core storage.

v)   Very often quicker and easier when coefficient matrix is sparse.

Miscellaneous Examples

In this frame a collection of miscellaneous examples is given for you to
try.   Answers are provided in FRAME 44, together with such working as
is considered helpful.   If you have access to a calculator, use the
questions marked A.   If not, modify the A questions as indicated in B.

1.A.   Working to 4 decimal places solve

$$0 \cdot 19x_1 + 0 \cdot 22x_2 + 0 \cdot 42x_3 = 0 \cdot 25$$
$$0 \cdot 27x_1 + 0 \cdot 34x_2 + 0 \cdot 56x_3 = 0 \cdot 18$$
$$0 \cdot 52x_1 + 0 \cdot 41x_2 + 0 \cdot 17x_3 = 0 \cdot 69$$

using Gaussian elimination in the form in FRAME 16   (i) without
pivoting (ii) with pivoting.

B.   Work to 2 decimal places only.

2.A.   Use the Gauss-Jordan method with pivoting to solve

$$0 \cdot 732x_1 - 5 \cdot 421x_2 + 1 \cdot 013x_3 = 4 \cdot 256$$
$$3 \cdot 491x_1 + 2 \cdot 203x_2 + 0 \cdot 782x_3 = -7 \cdot 113$$
$$0 \cdot 961x_1 - 1 \cdot 523x_2 + 4 \cdot 265x_3 = 3 \cdot 727$$

working to 4 decimal places.

B.   Work to 2 decimal places only, with each number in the equations
        rounded to 1 decimal place.

3.A.   Use the Gauss-Jordan method without pivoting to solve the set of
        equations in Question 2, working to 4 decimal places.

B.   Same modification as in 2.B.

4.A.   Solve the equations

$$10 \cdot 27x_1 - 1 \cdot 23x_2 + 0 \cdot 67x_3 = 4 \cdot 27$$
$$2 \cdot 39x_1 - 12 \cdot 65x_2 + 1 \cdot 13x_3 = 1 \cdot 26$$
$$1 \cdot 79x_1 + 3 \cdot 61x_2 + 15 \cdot 11x_3 = 12 \cdot 71$$

by each of the three methods

        a) Gaussian elimination with pivoting
        b) Gauss-Jordan elimination with pivoting
        c) Gauss-Seidel iteration,

    working to 4 decimal places.

    Compare the working involved in each case and the time you take.

111

FRAME 43   (continued)

B.   Work to 2 decimal places only, with each number in the equations
rounded to the nearest integer.

5.A.   The equilibrium of a certain framework requires the equations

$$10\,330\theta_1 + 1910\theta_2 + 434\theta_3 - 1736\theta_4 - 17\cdot78 = 0$$
$$1910\theta_1 + 10\,069\theta_2 - 1736\theta_3 + 694\theta_4 + 8\cdot89 = 0$$
$$434\theta_1 - 1736\theta_2 + 15\,718\theta_3 - 92\theta_4 - 23\cdot10 = 0$$
$$-1736\theta_1 + 694\theta_2 - 92\theta_3 + 15\,198\theta_4 + 10\cdot50 = 0$$

to be satisfied.   Working to 6 decimal places find θ_1, θ_2, θ_3
and θ_4.

B.   Work only to 5 decimal places with each θ coefficient rounded to
the nearest hundred and each constant term to the nearest integer.

6.A.

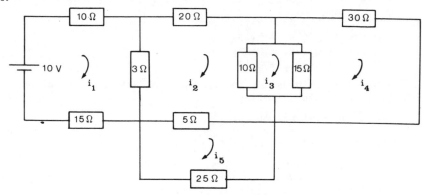

The currents in the above circuit are given by the equations:

$$28i_1 - 3i_2 = 10$$
$$-3i_1 + 38i_2 - 10i_3 - 5i_5 = 0$$
$$- 10i_2 + 25i_3 - 15i_4 = 0$$
$$- 15i_3 + 45i_4 = 0$$
$$- 5i_2 + 30i_5 = 0$$

Working to 4 decimal places find the values of the i's.

B.   Same as A.

FRAME 44

Answers to Miscellaneous Examples

You probably haven't found ΣR^2 in your answers but they have been
included here for comparison purposes.   If you have performed any of the
arithmetical operations in a different order to that which we used, you
may find a slight variation in the results.

1.A.   i)   4·9974, −5·0657, 0·9879.
            Residuals are 0·0000, 0·0002, −0·0003 and $\Sigma R^2 = 13 \times 10^{-8}$

       ii) 5·0025, −5·0716, 0·9887
            Residuals are 0·0000, 0·0000, 0·0000 and $\Sigma R^2 = 0$

B.   i)   4·63, −4·67, 0·94
          Residuals are 0·00, 0·01, −0·04 and ΣR^2 = 0·0017

     ii) 4·52, −4·38, 0·82
          Residuals are 0·00, −0·01, 0·01 and ΣR^2 = 0·0002

2.A.   −1·7244, −0·8380, 0·9632
          Residuals are 0·0003, 0·0002, 0·0002 and ΣR^2 = 17 × 10^{-8}

  B.   −1·72, −0·84, 0·97
          Residuals are 0·01, 0·00, 0·01 and ΣR^2 = 0·0002

3.A.   −1·7238, −0·8379, 0·9633
          Residuals are 0·0003, 0·0026, 0·0010 and ΣR^2 = 785 × 10^{-8}

  B.   −1·51, −0·84, 0·95
          Residuals are 0·13, 0·73, 0·14 and ΣR^2 = 0·5694

4.A.   a) and b) 0·3692, 0·0405, 0·7877
          Residuals −0·0004, 0·0002, 0·0008 and ΣR^2 = 84 × 10^{-8}

       c) 0·3693, 0·0405, 0·7878
          Residuals 0·0007, 0·0005, 0·0009 and ΣR^2 = 155 × 10^{-8}

  B.   a) 0·32, 0·04, 0·81
          Residuals are −0·03, −0·07, −0·05 and ΣR^2 = 0·0082

       b) 0·32, 0·03, 0·81
          Residuals are −0·02, 0·06, −0·09 and ΣR^2 = 0·0121

       c) 0·32, 0·04, 0·82
          Residuals are −0·02, −0·06, 0·10 and ΣR^2 = 0·0140

5.A.   0·001 770, −0·000 962, 0·001 312, −0·000 437

       Using the Gauss−Seidel method, the convergence was very rapid.
       You will notice that the diagonal dominance was very marked in
       this set of equations.

       The residuals are −0·005 280, 0·003 412, 0·000 432, −0·002 578
       and ΣR^2 = 46 × 10^{-6}.   The residuals may appear to be high, but
       when you consider that, for example, a change of 1 in the last
       figure of θ_1 will change the first residual by over 0·01 you
       will realise that they are not really as big as they seem.

  B.   0·001 81, −0·000 98, 0·001 31, −0·000 40
       Residuals are −0·015, 0·034, −0·003, 0·026 and ΣR^2 = 0·002 066 .

6.   0·3607, 0·0336, 0·0168, 0·0056, 0·0056
       Residuals are −0·0012, −0·0013, 0·0000, 0·0000, 0·0000.
       ΣR^2 = 3 × 10^{-6}.

SIMULTANEOUS LINEAR EQUATIONS

APPENDIX

The Effect of Inaccurate Data

Suppose there are three equations, i.e.,

$$a_{11}x_1 + a_{12}x_2 + a_{13}x_3 = b_1$$
$$a_{21}x_1 + a_{22}x_2 + a_{23}x_3 = b_2$$
$$a_{31}x_1 + a_{32}x_2 + a_{33}x_3 = b_3$$

in three unknowns. (If there are more, the extension of the process will be obvious.)

Then, approximately, if small changes are made in all quantities in the first equation,

$$a_{11}\delta x_1 + x_1\delta a_{11} + a_{12}\delta x_2 + x_2\delta a_{12} + a_{13}\delta x_3 + x_3\delta a_{13} = \delta b_1 \qquad (A1.1)$$

where as usual δa_{11} indicates the change in a_{11}, etc. Here the "change" will be the error introduced when the value of a_{11} is rounded off to so many decimal places. δx_1, etc., will then be the consequent errors in the terms in the solution. (Note: (A1.1) comes from an extension of the formula

$$\delta z \simeq \frac{\partial z}{\partial x}\,\delta x + \frac{\partial z}{\partial y}\,\delta y + \frac{\partial z}{\partial t}\,\delta t$$

if z is a function of the three independent variables x, y and t.)

Then from (A1.1)

$$a_{11}\delta x_1 + a_{12}\delta x_2 + a_{13}\delta x_3 = \delta b_1 - (x_1\delta a_{11} + x_2\delta a_{12} + x_3\delta a_{13}) \qquad (A1.2)$$

Now, remembering that δa_{11}, x_1, etc., may be positive or negative, what will be the maximum absolute value of the R.H.S. of (A1.2)?

****

$$|\delta b_1| + |x_1\delta a_{11}| + |x_2\delta a_{12}| + |x_3\delta a_{13}|$$

$$or \qquad |\delta b_1| + |x_1||\delta a_{11}| + |x_2||\delta a_{12}| + |x_3||\delta a_{13}|$$

Now suppose that the a's and b_1 are quoted correct to n decimal places. The actual values of the δa's and δb_1 will be unknown. But the maximum value of each of these will be $\frac{1}{2}$ in the last decimal place, i.e., $\frac{1}{2} \times 10^{-n} = \varepsilon$, say. Hence the maximum value of the expression in A1A is

$$\varepsilon + |x_1|\varepsilon + |x_2|\varepsilon + |x_3|\varepsilon = (1 + |x_1| + |x_2| + |x_3|)\varepsilon$$

Let this value be C. Then the maximum value of

$$|a_{11}\delta x_1 + a_{12}\delta x_2 + a_{13}\delta x_3|$$

is C.

Exactly similar operations may be carried out on the other two equations and you would find that the maximum value of the corresponding expression is the same in each case, i.e.

$$|a_{21}\delta x_1 + a_{22}\delta x_2 + a_{23}\delta x_3| < C$$

$$|a_{31}\delta x_1 + a_{32}\delta x_2 + a_{33}\delta x_3| < C$$

When a table such as that in FRAME 15 is used to solve a set of equations, linear combinations of the various equations are made, as indicated in the 'formation' column. For each new equation that is formed in this way there will be a maximum value of the expression in it corresponding to $|a_{11}\delta x_1 + a_{12}\delta x_2 + a_{13}\delta x_3|$. For example, in the table in FRAME 15, there will be a maximum value of

$$|1{\cdot}0000\ \delta x_1 + 1{\cdot}2911\ \delta x_2 - 1{\cdot}8059\ \delta x_3| \quad \text{for row d} \quad \text{and of}$$

$$|-9{\cdot}8942\ \delta x_2 + 23{\cdot}7115\ \delta x_3| \quad \text{for row j.}$$

In particular there will be a maximum value of

$$|1{\cdot}0000\ \delta x_3| \qquad \text{for row } \ell$$

$$|1{\cdot}0000\ \delta x_2| \qquad \text{for row n}$$

$$\text{and} \quad |1{\cdot}0000\ \delta x_1| \qquad \text{for row r}$$

and these, of course, will be the maximum values of $|\delta x_3|$, $|\delta x_2|$ and $|\delta x_1|$. These are the quantities that we are now looking for.

The maximum values of the respective expressions for rows a, b and c are each C, as found in FRAME A2. Then, as d is formed by dividing a by $2{\cdot}37$, the maximum value of

$$|1{\cdot}0000\ \delta x_1 + 1{\cdot}2911\ \delta x_2 - 1{\cdot}8059\ \delta x_3| \quad \text{is} \quad C/2{\cdot}37 \quad \text{i.e.} \quad 0{\cdot}42C.$$

e is formed by multiplying d by $-1{\cdot}46$. The maximum value of

$$|-1{\cdot}4600\ \delta x_1 - 1{\cdot}8850\ \delta x_2 + 2{\cdot}6366\ \delta x_3| \quad \text{is therefore}$$

$(0{\cdot}42C) \times 1{\cdot}46 = 0{\cdot}61C$. As we are here only dealing with absolute values $0{\cdot}42C$ is multiplied by <u>plus</u> $1{\cdot}46$, not minus $1{\cdot}46$.

f is formed by adding b and e. The maximum value of

$$|-2{\cdot}6650\ \delta x_2 + 6{\cdot}3866\ \delta x_3| \quad \text{is therefore} \quad C + 0{\cdot}61C = 1{\cdot}61C.$$

This process can be continued right down the table. As each quantity will be a multiple of C, it is only necessary to give the actual coefficient of C. This has been done in the column headed δx. Remember that in forming this column, all minus signs are ignored.

Extracting the relevant information at the end, the maximum value of

$$|\delta x_1| \qquad \text{is} \quad 3{\cdot}50C \qquad \text{from row r}$$

$$|\delta x_2| \qquad \text{is} \quad 1{\cdot}73C \qquad \text{from row n}$$

$$|\delta x_3| \qquad \text{is} \quad 0{\cdot}47C \qquad \text{from row } \ell$$

In this particular example $n = 2$ and so $\varepsilon = 0{\cdot}005$. (See FRAME A2.) Thus $C = (1 + 0{\cdot}93 + 1{\cdot}54 + 1{\cdot}21) \times 0{\cdot}005 = 0{\cdot}024$, the values of x_1, x_2 and x_3 in the formula for C being those obtained as the solution of the original equations. As only a rough estimate is being found here, they have been reduced to 2 decimal places. Finally, the maximum possible variations in x_1, x_2 and x_3, due to the worst possible combination of

115

maximum round-off errors in the a's are approximately

$$3 \cdot 50 \times 0 \cdot 024, \quad 1 \cdot 73 \times 0 \cdot 024 \quad \text{and} \quad 0 \cdot 47 \times 0 \cdot 024,$$

i.e. $0 \cdot 084$, $0 \cdot 042$ and $0 \cdot 011$.

Hence the limits within which x_1 can lie are $0 \cdot 9338 \pm 0 \cdot 084$, i.e., $0 \cdot 8498$ and $1 \cdot 0178$. Similarly for x_2 they are $1 \cdot 4990$ and $1 \cdot 5830$ and for x_3, $1 \cdot 1966$ and $1 \cdot 2186$.

If you now examine the table in FRAME 16, you will see that the corresponding extra column has been included there also.   Remember that in row d for example the formation <u>for this extra column only</u> is b + 0·62a.

Turning now to the equations in FRAME 19, add the δx column to your two solutions and use the values obtained in your second table to estimate the maximum variations in the x's due to maximum round-off errors in the a's.
           ****

The δx column has already been added to each table in 19A.

For the first table in 19A,

$n = 0, \quad \varepsilon = 0 \cdot 5, \quad C = (1 + 0 \cdot 34 + 1 \cdot 85 + 0 \cdot 57) \times 0 \cdot 5 = 1 \cdot 88$

$$|\delta x_1| = 2 \cdot 13 \times 1 \cdot 88 = 4 \cdot 00$$
$$|\delta x_2| = 0 \cdot 35 \times 1 \cdot 88 = 0 \cdot 66$$
$$|\delta x_3| = 0 \cdot 62 \times 1 \cdot 88 = 1 \cdot 17$$

For the second table in 19A,

$n = 0, \quad \varepsilon = 0 \cdot 5, \quad C = (1 + 0 \cdot 33 + 1 \cdot 85 + 0 \cdot 58) \times 0 \cdot 5 = 1 \cdot 88$

$$|\delta x_1| = 2 \cdot 14 \times 1 \cdot 88 = 4 \cdot 02$$
$$|\delta x_2| = 0 \cdot 35 \times 1 \cdot 88 = 0 \cdot 66$$
$$|\delta x_3| = 0 \cdot 63 \times 1 \cdot 88 = 1 \cdot 18$$

Matrix Algebra, Eigenvalues and Eigenvectors

Arithmetical Operations on Matrices

As you already know, there are four ordinary arithmetical operations that can be carried out on matrices:- addition, subtraction, multiplication and inversion. You will remember that inversion is the operation that is used in matrix work instead of the ordinary division of arithmetic.

Now as far as computation is concerned, addition, subtraction and multiplication present no difficulties and can easily be programmed on to a computer. Inversion, however, is a different story and the use of the definition $A^{-1} = \frac{1}{|A|}$ adj A requires the evaluation of a number of determinants. For small matrices you can do the arithmetic involved in this quite easily. However, larger matrices require a computer to handle them and the evaluation of determinants by computers is rather costly time-wise. Alternative methods are therefore preferable, if these can be found. The first part of this programme will be devoted to such methods.

Inversion by the Gauss-Jordan Process

Make a start by using the definition $A^{-1} = \frac{1}{|A|}$ adj A to find the inverse of

$$A = \begin{pmatrix} 2 & -1 & 3 \\ -3 & 4 & -5 \\ 1 & 3 & -6 \end{pmatrix}$$

****

2A

$|A| = -34$ $\qquad$ adj $A = \begin{pmatrix} -9 & 3 & -7 \\ -23 & -15 & 1 \\ -13 & -7 & 5 \end{pmatrix}$

$$A^{-1} = -\frac{1}{34} \begin{pmatrix} -9 & 3 & -7 \\ -23 & -15 & 1 \\ -13 & -7 & 5 \end{pmatrix} = \begin{pmatrix} 0 \cdot 2647 & -0 \cdot 0882 & 0 \cdot 2059 \\ 0 \cdot 6765 & 0 \cdot 4412 & -0 \cdot 0294 \\ 0 \cdot 3824 & 0 \cdot 2059 & -0 \cdot 1471 \end{pmatrix}$$

You will probably not have carried out the last step, but we shall want it later.

Now use the Gauss-Jordan process, without pivoting, that you learnt in the previous programme to solve the equations

$$\left. \begin{array}{rcl} 2x_1 - x_2 + 3x_3 &=& 1 \\ -3x_1 + 4x_2 - 5x_3 &=& 0 \\ x_1 + 3x_2 - 6x_3 &=& 0 \end{array} \right\} \qquad (3.1)$$

working to 4 decimal places.

****

Row	Formation	x_1	x_2	x_3	R.H.S.	s
a		2	−1	3	1	5
b		−3	4	−5	0	−4
c		1	3	−6	0	−2
d	a	2	−1	3	1	5
e	b + (3/2)a = b + 1·5a		2·5	−0·5	1·5	3·5
f	c − (1/2)a = c − 0·5a		3·5	−7·5	−0·5	−4·5
g	d + (1/2·5)e = d + 0·4e	2		2·8	1·6	6·4
h	e		2·5	−0·5	1·5	3·5
i	f−(3·5/2·5)e = f − 1·4e			−6·8	−2·6	−9·4
j	g + (2·8/6·8)i = g + 0·4118i	2			0·5293	2·5293
k	h − (0·5/6·8)i = h − 0·0735i		2·5		1·6911	4·1911
ℓ	i			−6·8	−2·6	−9·4
m	j/2	1			0·2646	1·2646
n	k/2·5		1		0·6764	1·6764
p	ℓ/(−6·8)			1	0·3824	1·3824
Sol.		0·2646	0·6764	0·3824		

The residuals are 0·0000, −0·0002, −0·0006. Remember that you should check that the residuals are small.

The equations (3.1) can be written in matrix form, i.e., as

$$\begin{pmatrix} 2 & -1 & 3 \\ -3 & 4 & -5 \\ 1 & 3 & -6 \end{pmatrix} \begin{pmatrix} x_1 \\ x_2 \\ x_3 \end{pmatrix} = \begin{pmatrix} 1 \\ 0 \\ 0 \end{pmatrix} \quad (4.1)$$

and you will notice that the coefficient matrix is the A of FRAME 2. Now if you examine the solution, you will see that, apart from very slight differences caused by round-off errors in the arithmetic, it is the same as the first column of the result for A^{-1} in 2A.

This solution can also be written in matrix form, using the augmented matrix

$$\begin{pmatrix} 2 & -1 & 3 & : & 1 \\ -3 & 4 & -5 & : & 0 \\ 1 & 3 & -6 & : & 0 \end{pmatrix}$$

The steps in the solution this way are as follows:

Operations $R_2 + (3/2)R_1$ and $R_3 − (1/2)R_1$ give

$$\begin{pmatrix} 2 & -1 & 3 & : & 1 \\ 0 & 2·5 & -0·5 & : & 1·5 \\ 0 & 3·5 & -7·5 & : & -0·5 \end{pmatrix}$$

Then $R_1 + (1/2 \cdot 5)R_2$ and $R_3 - (3 \cdot 5/2 \cdot 5)R_2$ give

$$\begin{pmatrix} 2 & 0 & 2 \cdot 8 & : & 1 \cdot 6 \\ 0 & 2 \cdot 5 & -0 \cdot 5 & : & 1 \cdot 5 \\ 0 & 0 & -6 \cdot 8 & : & -2 \cdot 6 \end{pmatrix}$$

Next $R_1 + (2 \cdot 8/6 \cdot 8)R_3$ and $R_2 - (0 \cdot 5/6 \cdot 8)R_3$ give

$$\begin{pmatrix} 2 & 0 & 0 & : & 0 \cdot 5293 \\ 0 & 2 \cdot 5 & 0 & : & 1 \cdot 6911 \\ 0 & 0 & -6 \cdot 8 & : & -2 \cdot 6 \end{pmatrix}$$

Finally $(1/2)R_1$, $(1/2 \cdot 5)R_2$,   and $-(1/6 \cdot 8)R_3$ give

$$\begin{pmatrix} 1 & 0 & 0 & : & 0 \cdot 2646 \\ 0 & 1 & 0 & : & 0 \cdot 6764 \\ 0 & 0 & 1 & : & 0 \cdot 3824 \end{pmatrix}$$

You will notice that the idea here is to perform various row operations with the aim of producing the unit matrix on the left of the dotted line.

Now use this approach to solve the equations

$$\left. \begin{array}{r} 2x_1 - x_2 + 3x_3 = 0 \\ -3x_1 + 4x_2 - 5x_3 = 1 \\ x_1 + 3x_2 - 6x_3 = 0 \end{array} \right\} \qquad (4.2)$$

****

4A

$$\begin{pmatrix} 2 & -1 & 3 & : & 0 \\ -3 & 4 & -5 & : & 1 \\ 1 & 3 & -6 & : & 0 \end{pmatrix} \qquad R_2 + \frac{3}{2} R_1, \quad R_3 - \frac{1}{2} R_1$$

$$\begin{pmatrix} 2 & -1 & 3 & : & 0 \\ 0 & 2 \cdot 5 & -0 \cdot 5 & : & 1 \\ 0 & 3 \cdot 5 & -7 \cdot 5 & : & 0 \end{pmatrix} \qquad R_1 + \frac{1}{2 \cdot 5} R_2, \quad R_3 - \frac{3 \cdot 5}{2 \cdot 5} R_2$$

$$\begin{pmatrix} 2 & 0 & 2 \cdot 8 & : & 0 \cdot 4 \\ 0 & 2 \cdot 5 & -0 \cdot 5 & : & 1 \\ 0 & 0 & -6 \cdot 8 & : & -1 \cdot 4 \end{pmatrix} \qquad R_1 + \frac{2 \cdot 8}{6 \cdot 8} R_3, \quad R_2 - \frac{0 \cdot 5}{6 \cdot 8} R_3$$

$$\begin{pmatrix} 2 & 0 & 0 & : & -0 \cdot 1765 \\ 0 & 2 \cdot 5 & -0 \cdot 5 & : & 1 \cdot 1029 \\ 0 & 0 & -6 \cdot 8 & : & -1 \cdot 4 \end{pmatrix} \qquad \frac{1}{2} R_1, \quad \frac{1}{2 \cdot 5} R_2, \quad -\frac{1}{6 \cdot 8} R_3$$

$$\begin{pmatrix} 1 & 0 & 0 & : & -0 \cdot 0882 \\ 0 & 1 & 0 & : & 0 \cdot 4411 \\ 0 & 0 & 1 & : & 0 \cdot 2059 \end{pmatrix}$$

and the solution is the second column of A^{-1}.   *(The residuals are* $0 \cdot 0002$, $-0 \cdot 0005$, $-0 \cdot 0003$.)

If you compare the row operations you carried out here with those performed in FRAME 4, you will see that they are exactly the same.

What equations do you think you would have to solve in order to obtain the third column of the inverse of

$$\begin{pmatrix} 2 & -1 & 3 \\ -3 & 4 & -5 \\ 1 & 3 & -6 \end{pmatrix} \quad ?$$

****

$$\left. \begin{array}{rcl} 2x_1 - x_2 + 3x_3 &=& 0 \\ -3x_1 + 4x_2 - 5x_3 &=& 0 \\ x_1 + 3x_2 - 6x_3 &=& 1 \end{array} \right\} \qquad (5A.1)$$

The row operations performed in solving these would again be exactly the same.

As the row operations performed in the solutions of the three sets (3.1), (4.2) and (5A.1), of simultaneous equations are the same, these solutions can be combined into a single process by the use of a larger augmented matrix. The coefficient part of this (to the left of the dotted line) is the same and the other part is extended to three columns, these three columns being the R.H.S.'s of (3.1), (4.2) and (5A.1) in that order. Thus we start with

$$\begin{pmatrix} 2 & -1 & 3 & : & 1 & 0 & 0 \\ -3 & 4 & -5 & : & 0 & 1 & 0 \\ 1 & 3 & -6 & : & 0 & 0 & 1 \end{pmatrix}$$

and perform the operations $R_2 + (3/2)R_1$, $R_3 - (1/2)R_1$.

The process then continues

$$\begin{pmatrix} 2 & -1 & 3 & : & 1 & 0 & 0 \\ 0 & 2 \cdot 5 & -0 \cdot 5 & : & 1 \cdot 5 & 1 & 0 \\ 0 & 3 \cdot 5 & -7 \cdot 5 & : & -0 \cdot 5 & 0 & 1 \end{pmatrix} \qquad \begin{array}{l} R_1 + (1/2 \cdot 5)R_2, \\ R_3 - (3 \cdot 5/2 \cdot 5)R_2 \end{array}$$

$$\begin{pmatrix} 2 & 0 & 2 \cdot 8 & : & 1 \cdot 6 & 0 \cdot 4 & 0 \\ 0 & 2 \cdot 5 & -0 \cdot 5 & : & 1 \cdot 5 & 1 & 0 \\ 0 & 0 & -6 \cdot 8 & : & -2 \cdot 6 & -1 \cdot 4 & 1 \end{pmatrix} \qquad \begin{array}{l} R_1 + (2 \cdot 8/6 \cdot 8)R_3, \\ R_2 - (0 \cdot 5/6 \cdot 8)R_3 \end{array}$$

$$\begin{pmatrix} 2 & 0 & 0 & : & 0 \cdot 5293 & -0 \cdot 1765 & 0 \cdot 4118 \\ 0 & 2 \cdot 5 & 0 & : & 1 \cdot 6911 & 1 \cdot 1029 & -0 \cdot 0735 \\ 0 & 0 & -6 \cdot 8 & : & -2 \cdot 6 & -1 \cdot 4 & 1 \end{pmatrix} \qquad \begin{array}{l} (1/2)R_1, \\ (1/2 \cdot 5)R_2, \\ -(1/6 \cdot 8)R_3 \end{array}$$

$$\begin{pmatrix} 1 & 0 & 0 & : & 0 \cdot 2646 & -0 \cdot 0882 & 0 \cdot 2059 \\ 0 & 1 & 0 & : & 0 \cdot 6764 & 0 \cdot 4411 & -0 \cdot 0294 \\ 0 & 0 & 1 & : & 0 \cdot 3824 & 0 \cdot 2059 & -0 \cdot 1471 \end{pmatrix}$$

The complete inverse appears to the right of the dotted line when the requisite row operations have been performed to give the unit matrix to the left of this line.   We have thus converted the augmented matrix

$$\begin{pmatrix} A & : & I \end{pmatrix}$$

into $\begin{pmatrix} I & : & A^{-1} \end{pmatrix}$

Now try this process on the matrix

$$\begin{pmatrix} -2 & -1 & 2 \\ 4 & 0 & -1 \\ -3 & 2 & 1 \end{pmatrix}$$

Then find its inverse by the formula $A^{-1} = \dfrac{1}{|A|} \text{adj } A$ and see if your results agree.

****

6A

$$\left(\begin{array}{ccc:ccc} -2 & -1 & 2 & 1 & 0 & 0 \\ 4 & 0 & -1 & 0 & 1 & 0 \\ -3 & 2 & 1 & 0 & 0 & 1 \end{array}\right)$$

$R_2 + 2R_1,$
$R_3 - (3/2)R_1$

$$\left(\begin{array}{ccc:ccc} -2 & -1 & 2 & 1 & 0 & 0 \\ 0 & -2 & 3 & 2 & 1 & 0 \\ 0 & 3 \cdot 5 & -2 & -1 \cdot 5 & 0 & 1 \end{array}\right)$$

$R_1 - (1/2)R_2,$
$R_3 + (3 \cdot 5/2)R_2$

$$\left(\begin{array}{ccc:ccc} -2 & 0 & 0 \cdot 5 & 0 & -0 \cdot 5 & 0 \\ 0 & -2 & 3 & 2 & 1 & 0 \\ 0 & 0 & 3 \cdot 25 & 2 & 1 \cdot 75 & 1 \end{array}\right)$$

$R_1 - (0 \cdot 5/3 \cdot 25)R_3,$
$R_2 - (3/3 \cdot 25)R_3$

$$\left(\begin{array}{ccc:ccc} -2 & 0 & 0 & -0 \cdot 3076 & -0 \cdot 7692 & -0 \cdot 1538 \\ 0 & -2 & 0 & 0 \cdot 1538 & -0 \cdot 6154 & -0 \cdot 9231 \\ 0 & 0 & 3 \cdot 25 & 2 & 1 \cdot 75 & 1 \end{array}\right)$$

$-(1/2)R_1, \quad -(1/2)R_2,$
$(1/3 \cdot 25)R_3$

$$\left(\begin{array}{ccc:ccc} 1 & 0 & 0 & 0 \cdot 1538 & 0 \cdot 3846 & 0 \cdot 0769 \\ 0 & 1 & 0 & -0 \cdot 0769 & 0 \cdot 3077 & 0 \cdot 4616 \\ 0 & 0 & 1 & 0 \cdot 6154 & 0 \cdot 5385 & 0 \cdot 3077 \end{array}\right)$$

Using the formula,   inverse $= \dfrac{1}{13}\begin{pmatrix} 2 & 5 & 1 \\ -1 & 4 & 6 \\ 8 & 7 & 4 \end{pmatrix}$

$$= \begin{pmatrix} 0 \cdot 1538 & 0 \cdot 3846 & 0 \cdot 0769 \\ -0 \cdot 0769 & 0 \cdot 3077 & 0 \cdot 4615 \\ 0 \cdot 6154 & 0 \cdot 5385 & 0 \cdot 3077 \end{pmatrix}$$

The second method is easier here but remember we are looking for a method which is a good one from a computer standpoint, the use of a computer being a must when large matrices are involved.

As the process is simply an extension of the ordinary Gauss-Jordan method for solving simultaneous equations, the working, when you are doing this manually, can be laid out in the form of a table and an s column introduced to give a check on the arithmetic.   Thus, the last example could be tabulated as shown on page 122.

Row	Formation							s
a		-2	-1	2	1	0	0	0
b		4	0	-1	0	1	0	4
c		-3	2	1	0	0	1	1
d	a	-2	-1	2	1	0	0	0
e	b + 2a	0	-2	3	2	1	0	4
f	$c - \frac{3}{2}a = c - 1\cdot5a$	0	3·5	-2	-1·5	0	1	1
g	$d - \frac{1}{2}e = d - 0\cdot5e$	-2	0	0·5	0	-0·5	0	-2
h	e	0	-2	3	2	1	0	4
i	$f + \frac{3\cdot5}{2}e = f + 1\cdot75e$	0	0	3·25	2	1·75	1	8
j	$g - \frac{0\cdot5}{3\cdot25}i = g - 0\cdot1538i$	-2	0	0	-0·3076	-0·7692	-0·1538	-3·2306
k	$h - \frac{3}{3\cdot25}i = h - 0\cdot9231i$	0	-2	0	0·1538	-0·6154	-0·9231	-3·3847
ℓ	i	0	0	3·25	2	1·75	1	8
m	j/(-2)	1	0	0	0·1538	0·3846	0·0769	1·6153
n	k/(-2)	0	1	0	-0·0769	0·3077	0·4616	1·6924
p	ℓ/3·25	0	0	1	0·6154	0·5385	0·3077	2·4616

You will realise that rows d, h and ℓ are simply repetitions of previous rows and can therefore be omitted.

Now use a tabular form to find the inverse of

$$\begin{pmatrix} 1\cdot5 & -0\cdot7 & 3\cdot5 \\ -1\cdot7 & 2\cdot5 & -1\cdot9 \\ 0\cdot7 & -1\cdot1 & 7\cdot1 \end{pmatrix} \qquad (7.1)$$

working to 2 decimal places.
                    ****

Row	Formation									s
a		1·5	−0·7	3·5	1	0	0			5·3
b		−1·7	2·5	−1·9	0	1	0			−0·1
c		0·7	−1·1	7·1	0	0	1			7·7
d	$b + (1·7/1·5)a = b + 1·13a$	0	1·71	2·06	1·13	1	0			5·90
e	$c − (0·7/1·5)a = c − 0·47a$	0	−0·77	5·46	−0·47	0	1			5·22
f	$a + (0·7/1·71)d = a + 0·41d$	1·5	0	4·34	1·46	0·41	0			7·71
g	$e + (0·77/1·71)d = e + 0·45d$	0	0	6·39	0·04	0·45	1			7·88
h	$f − (4·34/6·39)g = f − 0·68g$	1·5	0	0	1·43	0·10	−0·68			2·35
i	$e − (5·46/6·39)g = e − 0·85g$	0	−0·77	0	−0·50	−0·38	0·15			−1·50
k	$h/ 1·5$	1	0	0	0·95	0·07	−0·45			1·57
l	$i/(−0·77)$	0	1	0	0·65	0·49	−0·19			1·95
m	$g/6·39$	0	0	1	0·01	0·07	0·16			1·24

In this solution the redundant rows have been omitted. If you have left them in you should still, of course, have obtained the same answer.

FRAME 8

The solution can be checked by multiplying the original matrix and the inverse together (in either order) which should give **I**. This is equivalent to checking the solution of simultaneous equations. Here the result of the multiplication is

$$\begin{pmatrix} 1·00 & 0·01 & 0·02 \\ −0·01 & 0·97 & −0·01 \\ 0·02 & 0·01 & 1·03 \end{pmatrix}$$

FRAME 9

So far, the process described has worked for every example. This suggests that we should be able to justify it theoretically.

Returning to FRAME 2, if a general 3×3 matrix **A** is taken instead of

$\begin{pmatrix} 2 & −1 & 3 \\ −3 & 4 & −5 \\ 1 & 3 & −6 \end{pmatrix}$ the matrix equation corresponding to (4.1) is $AX = \begin{pmatrix} 1 \\ 0 \\ 0 \end{pmatrix}$.

The solution of this equation is $X = A^{-1} \begin{pmatrix} 1 \\ 0 \\ 0 \end{pmatrix}$.

If $A^{-1} = \begin{pmatrix} b_{11} & b_{12} & b_{13} \\ b_{21} & b_{22} & b_{23} \\ b_{31} & b_{32} & b_{33} \end{pmatrix}$, what is $A^{-1} \begin{pmatrix} 1 \\ 0 \\ 0 \end{pmatrix}$?

****

$$\begin{pmatrix} b_{11} \\ b_{21} \\ b_{31} \end{pmatrix}, \quad \textit{which is the first column of } A^{-1}.$$

This was the result noticed in the first paragraph in FRAME 4.

In a similar way $AX = \begin{pmatrix} 0 \\ 1 \\ 0 \end{pmatrix}$ will work out to give $X = \begin{pmatrix} b_{12} \\ b_{22} \\ b_{32} \end{pmatrix}$, the

second column of A^{-1} and $AX = \begin{pmatrix} 0 \\ 0 \\ 1 \end{pmatrix}$ will have the third column of A^{-1}

for its solution. Our process is therefore justified for a 3×3 matrix and it can similarly be shown to work for an $n \times n$ matrix.

The augmented matrix $\begin{pmatrix} a_{11} & a_{12} & a_{13} & \vdots & 1 & 0 & 0 \\ a_{21} & a_{22} & a_{23} & \vdots & 0 & 1 & 0 \\ a_{31} & a_{32} & a_{33} & \vdots & 0 & 0 & 1 \end{pmatrix}$ (11.1)

can be regarded as coming from the matrix equation $AX = I$ (11.2)

just as $\begin{pmatrix} a_{11} & a_{12} & a_{13} & \vdots & 1 \\ a_{21} & a_{22} & a_{23} & \vdots & 0 \\ a_{31} & a_{32} & a_{33} & \vdots & 0 \end{pmatrix}$

can be written down from the equation $AX = \begin{pmatrix} 1 \\ 0 \\ 0 \end{pmatrix}$

However in (11.2), X will have to be a 3×3 matrix. Now we have seen that when the augmented matrix (11.1) is manipulated so that the unit matrix appears to the left of the dotted line, then the inverse of A appears to the right, i.e., we get

$$\begin{pmatrix} 1 & 0 & 0 & \vdots & b_{11} & b_{12} & b_{13} \\ 0 & 1 & 0 & \vdots & b_{21} & b_{22} & b_{23} \\ 0 & 0 & 1 & \vdots & b_{31} & b_{32} & b_{33} \end{pmatrix}$$

What matrix equation will this be the augmented matrix for?
****

$IX = A^{-1}$, *which is, of course* $X = A^{-1}$.

But (11.2), when solved, gives $X = A^{-1} I$ and so this is another way of justifying the process theoretically. You will notice that it introduces the concept of a single matrix equation ($AX = I$ in this case) being used to represent <u>several</u> sets $\big((3.1), (4.2)$ and $(5A.1)$ in the numerical example taken earlier $\big)$ of simultaneous equations. Previously a matrix equation has only been used to represent <u>one</u> set of such equations.

A simple flow diagram for this method of matrix inversion is

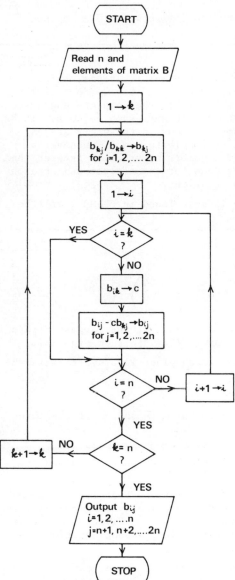

B is the augmented matrix $(A : I)$ where A is an $n \times n$ matrix and I is the $n \times n$ unit matrix.

In FRAME 6, the divisions by 2, 2·5 and −6·8 took place at the end. Alternatively it is possible to perform that by 2 right at the beginning and those by 2·5, −6·8 just before the other elements in the columns containing 2·5, −6·8 respectively, are converted to zero. The flow diagram has adopted this latter procedure.

The matrix B is now $(I : A^{-1})$ so the elements being output here are those of A^{-1}.

$\left(\text{Programs for matrix inversion by the Gauss-Jordan method can be found in references (3) and (8).}\right)$

125

What will happen to the Gauss-Jordan process, as it has been used so far, for finding the inverse of a matrix A if $a_{11} = 0$?

****

It will break down as it will be impossible to produce zeros in the positions occupied by a_{21} and a_{31}.

When solving a set of ordinary linear simultaneous equations, this difficulty can be avoided by pivoting. Question - Can we do a similar thing here? Answer - Yes, provided we do it to the complete augmented matrix. Furthermore, pivoting can be used in any example (so far it has not been used, to avoid additional complications in the early stages) where it is deemed advisable. It is advisable to use it whenever a multiplying factor can be changed from a number numerically greater than unity to one numerically less. It is done for exactly the same reasons as when solving ordinary simultaneous equations.

As an example, in the next frame, the inverse found in FRAME 7 will be reworked with pivoting.

The augmented matrix is

$$\begin{pmatrix} -2 & -1 & 2 & : & 1 & 0 & 0 \\ 4 & 0 & -1 & : & 0 & 1 & 0 \\ -3 & 2 & 1 & : & 0 & 0 & 1 \end{pmatrix}$$

In the table the first two rows are reversed so that the largest figure in the first column is also in the first row. (See table on page 127.)

In this solution repetition of rows has taken place in order to make the process quite clear.

You will notice the following points:

The unit matrix does <u>not</u> appear in columns 4, 5 and 6 of the three rows a, b and c.

As row g already contains a zero in column 2, row j is simply row g repeated. Normally some operation would be necessary to produce a zero in row j, column 2.

When rows d, e, f have been obtained, d <u>must</u> be left first. e and f are interchanged in position so that the numerically larger coefficient (2 here) in column 2 appears before the other. Pivoting has taken place at this stage just as when ordinary simultaneous equations were being solved.

No pivoting can take place after rows j, k and ℓ have been obtained otherwise the zeros in columns 1 and 2 will no longer be in the correct positions.

Now use pivoting to find the inverse of matrix (7.1).

Row	Formation	1	2	3	4	5	6	s
a		4	0	-1	0	1	0	4
b		-2	-1	2	1	0	0	0
c		-3	2	1	0	0	1	1
d	a	4	0	-1	0	1	0	4
e	$b + \frac{2}{4}a = b + 0 \cdot 5a$	0	-1	1·5	1	0·5	0	2
f	$c + \frac{3}{4}a = c + 0 \cdot 75a$	0	2	0·25	0	0·75	1	4
g	d	4	0	-1	0	1	0	4
h	f	0	2	0·25	0	0·75	1	4
i	e	0	-1	1·5	1	0·5	0	2
j	g	4	0	-1	0	1	0	4
k	h	0	2	0·25	0	0·75	1	4
ℓ	$i + \frac{1}{2}h = i + 0 \cdot 5h$	0	0	1·625	1	0·875	0·5	4
m	$j + \frac{1}{1 \cdot 625}\ell = j + 0 \cdot 6154\ell$	4	0	0	0·6154	1·5385	0·3077	6·4616
n	$k - \frac{0 \cdot 25}{1 \cdot 625}\ell = k - 0 \cdot 1538\ell$	0	2	0	-0·1538	0·6154	0·9231	3·3847
p	ℓ	0	0	1·625	1	0·875	0·5	4
q	m/4	1	0	0	0·1538	0·3846	0·0769	1·6154
r	n/2	0	1	0	-0·0769	0·3077	0·4616	1·6923
s	p/1·625	0	0	1	0·6154	0·5385	0·3077	2·4616

****

16A

The table for this solution is on page 128.

*Note that pivoting takes place in this example only at the first stage.
At the second stage only rows e and f can be reversed even although 2·5
is the largest figure in this column.   Having obtained two zeros in the
positions shown by the dashed rectangle they must not be destroyed.*

Row	Formation	1	2	3	4	5	6	s
a		-1·7	2·5	-1·9	0	1	0	-0·1
b		1·5	-0·7	3·5	1	0	0	5·3
c		0·7	-1·1	7·1	0	0	1	7·7
d	a	-1·7	2·5	-1·9	0	1	0	-0·1
e	$b + (1·5/1·7)a = b + 0·88a$	0	1·5	1·83	1	0·88	0	5·21
f	$c + (0·7/1·7)a = c + 0·41a$	0	-0·08	6·32	0	0·41	1	7·65
g	$d - (2·5/1·5)e = d - 1·67e$	-1·7	0	-4·96	-1·67	-0·47	0	-8·80
h	e	0	1·5	1·83	1	0·88	0	5·21
i	$f + (0·08/1·5)e = f + 0·05e$	0	0	6·41	0·05	0·45	1	7·91
j	$g + (4·96/6·41)i = g + 0·77i$	-1·7	0	0	-1·63	-0·12	0·77	-2·68
k	$h - (1·83/6·41)i = h - 0·29i$	0	1·5	0	0·99	0·75	-0·29	2·95
ℓ	i	0	0	6·41	0·05	0·45	1	7·91
m	$j/(-1·7)$	1	0	0	0·96	0·07	-0·45	1·58
n	$k/1·5$	0	1	0	0·66	0·50	-0·19	1·97
p	$ℓ/6·41$	0	0	1	0·01	0·07	0·16	1·24

FRAME 17

Multiply the result you have just obtained by the matrix (7.1) to check
how accurate it is.   Compare the new values with those in FRAME 8.
****

17A

$$\begin{pmatrix} 1·01 & 0·00 & 0·02 \\ 0·00 & 1·00 & -0·01 \\ 0·02 & 0·00 & 1·03 \end{pmatrix}$$

FRAME 18

The effect of Inaccurate Data

The process we have adopted to find the inverse of a matrix is effectively
the solution of sets of simultaneous equations.   The method of obtaining
the errors in the result due to inaccurate data will therefore be exactly
the same as that for simultaneous equations, dealt with in the appendix
to the previous programme, with the exception that all δb's will be zero
as the figures on the R.H.S.'s of the equations solved are either 0 or 1
and are exact.

FRAME 19

Choleski's Factorisation Process

What are the following matrix products?

i) $\begin{pmatrix} 1 & 0 & 0 \\ 2 & 1 & 0 \\ -3 & 4 & 1 \end{pmatrix} \begin{pmatrix} -2 & 3 & 1 \\ 0 & 5 & -4 \\ 0 & 0 & -1 \end{pmatrix}$ ii) $\begin{pmatrix} -2 & 0 & 0 \\ -4 & 5 & 0 \\ 6 & 20 & -1 \end{pmatrix} \begin{pmatrix} 1 & -\frac{3}{2} & -\frac{1}{2} \\ 0 & 1 & -\frac{4}{5} \\ 0 & 0 & 1 \end{pmatrix}$

iii) $\begin{pmatrix} 1 & 0 & 0 \\ 2 & 2 & 0 \\ -3 & 8 & 3 \end{pmatrix}\begin{pmatrix} -2 & 3 & 1 \\ 0 & \frac{5}{2} & -2 \\ 0 & 0 & -\frac{1}{3} \end{pmatrix}$ iv) $\begin{pmatrix} 1 & 0 & 0 \\ -2 & 1 & 0 \\ 4 & 3 & 1 \end{pmatrix}\begin{pmatrix} 1 & 2 & 3 \\ 0 & 4 & 5 \\ 0 & 0 & 6 \end{pmatrix}$

v) $\begin{pmatrix} 1 & 0 & 0 \\ -2 & 2 & 0 \\ 4 & 6 & 3 \end{pmatrix}\begin{pmatrix} 1 & 2 & 3 \\ 0 & 2 & \frac{5}{2} \\ 0 & 0 & 2 \end{pmatrix}$ vi) $\begin{pmatrix} 3 & 0 & 0 \\ -6 & 2 & 0 \\ 12 & 6 & 1 \end{pmatrix}\begin{pmatrix} \frac{1}{3} & \frac{2}{3} & 1 \\ 0 & 2 & \frac{5}{2} \\ 0 & 0 & 6 \end{pmatrix}$

****

19A

(i), (ii), (iii) all give $\begin{pmatrix} -2 & 3 & 1 \\ -4 & 11 & -2 \\ 6 & 11 & -20 \end{pmatrix}$

(iv), (v), (vi) all give $\begin{pmatrix} 1 & 2 & 3 \\ -2 & 0 & -1 \\ 4 & 20 & 33 \end{pmatrix}$

You will see from the above results that it is possible to write each matrix in 19A as a number of different products **AB** .   This is somewhat similar to the fact that 5 can be written as 1×5, $2 \times \frac{5}{2}$, $\frac{4}{7} \times \frac{35}{4}$ and so on.

Now look at the matrices in (i) - (vi) in the last frame and see if you can spot anything special about

(a)   the six left-hand matrices   (b)   the six right-hand matrices.
    ****

20A

(a) They are all lower triangular.   (b) They are all upper triangular.

These results suggest firstly that it might be possible to express a square matrix **A** as the product **LU** of two square matrices **L** and **U, L** being lower triangular and **U** upper triangular.   Secondly they suggest that the matrices **L** and **U** are not unique.

It is, in fact, possible to choose at will all the elements on the leading diagonal of either **L** or **U**.   If work is going to be carried out using these matrices, it is sensible to choose the elements on one of these leading diagonals to be as simple as possible.   Let us choose those in **L** to be all 1.   Then, given any square matrix **A** , we can write **LU = A** where **L** is lower triangular with each element on the leading diagonal 1 and **U** is upper triangular.

As an example let $A = \begin{pmatrix} 4 & 3 & -2 \\ 1 & 0 & 5 \\ 2 & -3 & -4 \end{pmatrix}$. Taking the forms for L and U

suggested above, let $L = \begin{pmatrix} 1 & 0 & 0 \\ \ell_{21} & 1 & 0 \\ \ell_{31} & \ell_{32} & 1 \end{pmatrix}$ and $U = \begin{pmatrix} u_{11} & u_{12} & u_{13} \\ 0 & u_{22} & u_{23} \\ 0 & 0 & u_{33} \end{pmatrix}$

It is now necessary to find the ℓ's and the u's so that $LU = A$, i.e.,

$$\begin{pmatrix} 1 & 0 & 0 \\ \ell_{21} & 1 & 0 \\ \ell_{31} & \ell_{32} & 1 \end{pmatrix} \begin{pmatrix} u_{11} & u_{12} & u_{13} \\ 0 & u_{22} & u_{23} \\ 0 & 0 & u_{33} \end{pmatrix} = \begin{pmatrix} 4 & 3 & -2 \\ 1 & 0 & 5 \\ 2 & -3 & -4 \end{pmatrix}$$

Can you suggest what u_{11} must be?

****

23A

4. *Taking the sum of the products*
(elements of row 1 in L) × (corresponding elements of column 1 in U)
gives only u_{11} and this must therefore be a_{11}, i.e., 4. Here the
standard notation for the elements of matrix A is being used.

By continuing the multiplication of L and U, we get $u_{12} = a_{12} = 3$
and $u_{13} = a_{13} = -2$ when the correct combination of elements in L and U
is taken to give a_{12} and a_{13}.

The formula giving a_{21} is $\ell_{21}u_{11}$ and hence, as $u_{11} = 4$, $4\ell_{21} = 1$ i.e.
$\ell_{21} = \frac{1}{4}$. Similarly, a_{31} is given by $\ell_{31}u_{11}$ and so $4\ell_{31} = 2$, or
$\ell_{31} = \frac{1}{2}$.

Continue in this way to find u_{22}, u_{23}, ℓ_{32} and u_{33} in this order. Then
write down the resulting matrix equation $A = LU$.
****

24A

$\ell_{21}u_{12} + u_{22} = a_{22}$ *i.e.* $u_{22} = -\dfrac{3}{4}$

$\ell_{21}u_{13} + u_{23} = a_{23}$ *i.e.* $u_{23} = \dfrac{11}{2}$

$\ell_{32} = 6$ $u_{33} = -36$

$$\begin{pmatrix} 4 & 3 & -2 \\ 1 & 0 & 5 \\ 2 & -3 & -4 \end{pmatrix} = \begin{pmatrix} 1 & 0 & 0 \\ \frac{1}{4} & 1 & 0 \\ \frac{1}{2} & 6 & 1 \end{pmatrix} \begin{pmatrix} 4 & 3 & -2 \\ 0 & -\frac{3}{4} & \frac{11}{2} \\ 0 & 0 & -36 \end{pmatrix}$$

Now express $\begin{pmatrix} 2 & 0 & 1 \\ -3 & 4 & -2 \\ 1 & 7 & -5 \end{pmatrix}$ in this form.

HINT: It is suggested that you start with

$$\begin{pmatrix} 1 & 0 & 0 \\ & 1 & 0 \\ & & 1 \end{pmatrix}\begin{pmatrix} & & \\ 0 & & \\ 0 & 0 & \end{pmatrix} = \begin{pmatrix} 2 & 0 & 1 \\ -3 & 4 & -2 \\ 1 & 7 & -5 \end{pmatrix}$$

and fill in the blanks as you go along.   Find the unknown elements in the order u_{11}, u_{12}, u_{13}, ℓ_{21}, ℓ_{31}, u_{22}, u_{23}, ℓ_{32}, u_{33}.

****

$$\begin{pmatrix} 1 & 0 & 0 \\ -\frac{3}{2} & 1 & 0 \\ \frac{1}{2} & \frac{7}{4} & 1 \end{pmatrix}\begin{pmatrix} 2 & 0 & 1 \\ 0 & 4 & -\frac{1}{2} \\ 0 & 0 & -\frac{37}{8} \end{pmatrix} = \begin{pmatrix} 2 & 0 & 1 \\ -3 & 4 & -2 \\ 1 & 7 & -5 \end{pmatrix}$$

This process is very easy to program for a computer.   From

$$\begin{pmatrix} 1 & 0 & 0 \\ \ell_{21} & 1 & 0 \\ \ell_{31} & \ell_{32} & 1 \end{pmatrix}\begin{pmatrix} u_{11} & u_{12} & u_{13} \\ 0 & u_{22} & u_{23} \\ 0 & 0 & u_{33} \end{pmatrix} = \begin{pmatrix} a_{11} & a_{12} & a_{13} \\ a_{21} & a_{22} & a_{23} \\ a_{31} & a_{32} & a_{33} \end{pmatrix}$$

we have the equations

$$\begin{cases} u_{11} = a_{11} \\ u_{12} = a_{12} \\ u_{13} = a_{13} \end{cases} \qquad \begin{cases} \ell_{21}u_{11} = a_{21} \\ \ell_{31}u_{11} = a_{31} \end{cases} \qquad \begin{cases} \ell_{21}u_{12} + u_{22} = a_{22} \\ \ell_{21}u_{13} + u_{23} = a_{23} \end{cases}$$

$$\begin{cases} \ell_{31}u_{12} + \ell_{32}u_{22} = a_{32} \\ \ell_{31}u_{13} + \ell_{32}u_{23} + u_{33} = a_{33} \end{cases}$$

from which

$$\begin{cases} u_{11} = a_{11} \\ u_{12} = a_{12} \\ u_{13} = a_{13} \end{cases} \qquad (26.1) \qquad\qquad \begin{cases} \ell_{21} = a_{21}/u_{11} \\ \ell_{31} = a_{31}/u_{11} \end{cases}$$

$$\begin{cases} u_{22} = a_{22} - \ell_{21}u_{12} \\ u_{23} = a_{23} - \ell_{21}u_{13} \end{cases} \quad (26.2) \qquad\qquad \ell_{32} = (a_{32} - \ell_{31}u_{12})/u_{22}$$

$$u_{33} = a_{33} - \ell_{31}u_{13} - \ell_{32}u_{23}$$

Both sets of nine equations fall into blocks as indicated.   You will realise that, if the second set of nine equations is followed in the sequence u_{11}, u_{12}, u_{13}, ℓ_{21}, ℓ_{31}, u_{22}, u_{23}, ℓ_{32}, u_{33}, everything in each R.H.S. will be known by the time you need to use it.   Thus, for example, in (26.2) a_{23} is given and ℓ_{21}, u_{13} have already been found.

For bigger matrices a similar procedure can be followed.   Both sets of equations given in FRAME 26 will, of course, have to be extended to accommodate this situation.

See if you can fill in the blanks in the following:

$$\begin{pmatrix} 1 & 0 & 0 & 0 \\ & 1 & 0 & 0 \\ & & 1 & 0 \\ & & & 1 \end{pmatrix} \begin{pmatrix} & & & \\ 0 & & & \\ 0 & 0 & & \\ 0 & 0 & 0 & \end{pmatrix} = \begin{pmatrix} 2 & -1 & 0 & 0 \\ -3 & 5 & 1 & 0 \\ 0 & 2 & 4 & -1 \\ 0 & 0 & 7 & 10 \end{pmatrix}$$

****

$$\begin{pmatrix} 1 & 0 & 0 & 0 \\ -\frac{3}{2} & 1 & 0 & 0 \\ 0 & \frac{4}{7} & 1 & 0 \\ 0 & 0 & \frac{49}{24} & 1 \end{pmatrix} \begin{pmatrix} 2 & -1 & 0 & 0 \\ 0 & \frac{7}{2} & 1 & 0 \\ 0 & 0 & \frac{24}{7} & -1 \\ 0 & 0 & 0 & \frac{289}{24} \end{pmatrix} = \begin{pmatrix} 2 & -1 & 0 & 0 \\ -3 & 5 & 1 & 0 \\ 0 & 2 & 4 & -1 \\ 0 & 0 & 7 & 10 \end{pmatrix}$$

When using a computer, as you would do for large matrices, the equations for the ℓ's and the u's can easily be programmed and will produce simple loops.   If the original matrix is $n \times n$, then after (26.1) for example, there would be the additional equations

$$u_{14} = a_{14}, \quad u_{15} = a_{15}, \quad \ldots\ldots\ldots \quad u_{1n} = a_{1n} \quad \text{and after (26.2)}$$

$$u_{24} = a_{24} - \ell_{21}u_{14}, \quad \ldots\ldots\ldots\ldots \quad u_{2n} = a_{2n} - \ell_{21}u_{1n}$$

Thus, for example, the formulae for all the u's in the second row can be combined into the single equation $u_{2i} = a_{2i} - \ell_{21}u_{1i}$ where i takes the values 2, 3, 4, etc. up to n.

The diagram shown on page 133 is a flow chart for Choleski's factorisation process, for an $n \times n$ matrix.

One reason why this method is popular in computer programs is that storage space can be used very economically.   There is no need to store the zeros in either L or U, nor the 1's on the diagonal of L.   Those elements of L and U which have to be calculated can be fitted into an $n \times n$ array.   Furthermore, once an element of A has been used in the equations shown on page 133, it is never required again, so its place in the original $n \times n$ array can be used for the storage of the corresponding element of either L or U .   Thus, for example, for a 3×3 matrix A ,

$$\begin{pmatrix} a_{11} & a_{12} & a_{13} \\ a_{21} & a_{22} & a_{23} \\ a_{31} & a_{32} & a_{33} \end{pmatrix} \text{ would eventually be replaced by } \begin{pmatrix} u_{11} & u_{12} & u_{13} \\ \ell_{21} & u_{22} & u_{23} \\ \ell_{31} & \ell_{32} & u_{33} \end{pmatrix}$$

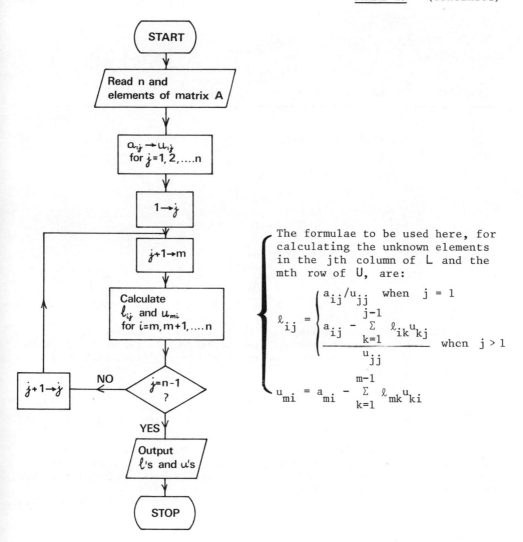

The formulae to be used here, for calculating the unknown elements in the jth column of L and the mth row of U, are:

$$\ell_{ij} = \begin{cases} a_{ij}/u_{jj} & \text{when } j = 1 \\ \dfrac{a_{ij} - \sum\limits_{k=1}^{j-1} \ell_{ik} u_{kj}}{u_{jj}} & \text{when } j > 1 \end{cases}$$

$$u_{mi} = a_{mi} - \sum_{k=1}^{m-1} \ell_{mk} u_{ki}$$

Returning now to 3×3 matrices, do you notice anything special about the following two triangular matrices and also about their product?

$$\begin{pmatrix} \ell_{11} & 0 & 0 \\ \ell_{21} & \ell_{22} & 0 \\ \ell_{31} & \ell_{32} & \ell_{33} \end{pmatrix} \begin{pmatrix} \ell_{11} & \ell_{21} & \ell_{31} \\ 0 & \ell_{22} & \ell_{32} \\ 0 & 0 & \ell_{33} \end{pmatrix}$$

****

Each is the transpose of the other.

Their product is
$$\begin{pmatrix} \ell_{11}^{\ 2} & \ell_{11}\ell_{21} & \ell_{11}\ell_{31} \\ \ell_{21}\ell_{11} & \ell_{21}^{\ 2} + \ell_{22}^{\ 2} & \ell_{21}\ell_{31} + \ell_{22}\ell_{32} \\ \ell_{31}\ell_{11} & \ell_{31}\ell_{21} + \ell_{32}\ell_{22} & \ell_{31}^{\ 2} + \ell_{32}^{\ 2} + \ell_{33}^{\ 2} \end{pmatrix}$$

and this is a symmetric matrix.

This suggests that if a matrix is symmetric it can be factorised into LU where $U = L'$.

As an example, let us take the matrix

$$\begin{pmatrix} 4 & -2 & 8 \\ -2 & 10 & -10 \\ 8 & -10 & 45 \end{pmatrix}$$

It is required to fill in the blanks in

$$\begin{pmatrix} & 0 & 0 \\ & & 0 \\ & 0 & 0 \end{pmatrix}\begin{pmatrix} & & \\ 0 & & \\ 0 & 0 & \end{pmatrix} = \begin{pmatrix} 4 & -2 & 8 \\ -2 & 10 & -10 \\ 8 & -10 & 45 \end{pmatrix}$$

Now in 30A, the element in position 11 (i.e. row 1, column 1) is $\ell_{11}^{\ 2}$ and so here $\ell_{11} = 2$, if the positive square root is taken. -2 is in the position 21 in the product and so $\ell_{11}\ell_{21} = -2$, i.e., $\ell_{21} = -1$. Similarly $\ell_{31} = 4$. We now have

$$\begin{pmatrix} 2 & 0 & 0 \\ -1 & & 0 \\ 4 & & \end{pmatrix}\begin{pmatrix} 2 & -1 & 4 \\ 0 & & \\ 0 & 0 & \end{pmatrix} = \begin{pmatrix} 4 & -2 & 8 \\ -2 & 10 & -10 \\ 8 & -10 & 45 \end{pmatrix}$$

Can you carry on and fill in the remainder of the blanks? Square roots can be taken either positive or negative but as there is no point in introducing this additional complication, take them all as being positive.
****

$$\begin{pmatrix} 2 & 0 & 0 \\ -1 & 3 & 0 \\ 4 & -2 & 5 \end{pmatrix}\begin{pmatrix} 2 & -1 & 4 \\ 0 & 3 & -2 \\ 0 & 0 & 5 \end{pmatrix}$$

The Application of the Factorisation Process to the Solution of Simultaneous Linear Equations

If the matrix A is resolved into factors LU then any equation such as $AX = B$ can be written as $LUX = B$. X is then found by first writing $UX = Y$ so that $LY = B$. Y can now be found, and after that X follows immediately. This may sound as if we are making life complicated, but it really works out quite simply as the examples will show.

To see how it works, the equations

$$\begin{aligned} 2x_1 \qquad + \quad x_3 &= 4 \\ -3x_1 + 4x_2 - 2x_3 &= -3 \\ x_1 + 7x_2 - 5x_3 &= 6 \end{aligned}$$

can be written as

$$\begin{pmatrix} 2 & 0 & 1 \\ -3 & 4 & -2 \\ 1 & 7 & -5 \end{pmatrix} \begin{pmatrix} x_1 \\ x_2 \\ x_3 \end{pmatrix} = \begin{pmatrix} 4 \\ -3 \\ 6 \end{pmatrix}$$

The square matrix has already been factorised in 25A.   So it follows that, using decimals instead of fractions, the equation $LUX = B$ becomes

$$\begin{pmatrix} 1 & 0 & 0 \\ -1 \cdot 5 & 1 & 0 \\ 0 \cdot 5 & 1 \cdot 75 & 1 \end{pmatrix} \begin{pmatrix} 2 & 0 & 1 \\ 0 & 4 & -0 \cdot 5 \\ 0 & 0 & -4 \cdot 625 \end{pmatrix} \begin{pmatrix} x_1 \\ x_2 \\ x_3 \end{pmatrix} = \begin{pmatrix} 4 \\ -3 \\ 6 \end{pmatrix}$$

If now $Y = \{y_1 \quad y_2 \quad y_3\}$, using $LY = B$ gives

$$\begin{pmatrix} 1 & 0 & 0 \\ -1 \cdot 5 & 1 & 0 \\ 0 \cdot 5 & 1 \cdot 75 & 1 \end{pmatrix} \begin{pmatrix} y_1 \\ y_2 \\ y_3 \end{pmatrix} = \begin{pmatrix} 4 \\ -3 \\ 6 \end{pmatrix}$$

The components of Y are found in the order y_1, y_2, y_3.   Multiplying the elements of R1 in L by the corresponding elements of Y and adding gives $y_1 = 4$. y_2 is now found by using R2 in L with Y.

What will be y_2 and then y_3?

****

32A

$y_2 = 3,$ $y_3 = -1 \cdot 25$

FRAME 33

Thus $Y = \{4 \quad 3 \quad -1 \cdot 25\}$ and so, as $UX = Y,$

$$\begin{pmatrix} 2 & 0 & 1 \\ 0 & 4 & -0 \cdot 5 \\ 0 & 0 & -4 \cdot 625 \end{pmatrix} \begin{pmatrix} x_1 \\ x_2 \\ x_3 \end{pmatrix} = \begin{pmatrix} 4 \\ 3 \\ -1 \cdot 25 \end{pmatrix}$$

The elements of X are now found in a similar manner to those of Y , but in the order x_3, x_2, x_1.   What will they be?

****

33A

$x_3 = 0 \cdot 270,$ $x_2 = 0 \cdot 784,$ $x_1 = 1 \cdot 865$
The residuals are $0 \cdot 000,$ $-0 \cdot 001,$ $0 \cdot 003$

FRAME 34

Use this method to solve the equations

$$\begin{aligned} 0 \cdot 7x_1 - 5 \cdot 4x_2 + 1 \cdot 0x_3 &= 4 \cdot 3 \\ 3 \cdot 5x_1 + 2 \cdot 2x_2 + 0 \cdot 8x_3 &= -7 \cdot 1 \\ 1 \cdot 0x_1 - 1 \cdot 5x_2 + 4 \cdot 3x_3 &= 3 \cdot 7 \end{aligned}$$

working to 2 decimal places.   (You have already solved these equations, or a similar set, by Gauss-Jordan in the miscellaneous examples of the previous programme.   When you have solved them by the present method you will be able to compare the working.)

****

34A

$$\begin{pmatrix} 0 \cdot 7 & -5 \cdot 4 & 1 \cdot 0 \\ 3 \cdot 5 & 2 \cdot 2 & 0 \cdot 8 \\ 1 \cdot 0 & -1 \cdot 5 & 4 \cdot 3 \end{pmatrix} = \begin{pmatrix} 1 & 0 & 0 \\ 5 & 1 & 0 \\ 1 \cdot 43 & 0 \cdot 21 & 1 \end{pmatrix} \begin{pmatrix} 0 \cdot 7 & -5 \cdot 4 & 1 \cdot 0 \\ 0 & 29 \cdot 2 & -4 \cdot 2 \\ 0 & 0 & 3 \cdot 75 \end{pmatrix}$$

$$Y = \{4 \cdot 3 \quad -28 \cdot 6 \quad 3 \cdot 56\}$$
$$X = \{-1 \cdot 69 \quad -0 \cdot 84 \quad 0 \cdot 95\}$$

Residuals are 0·003,   0·097,   −0·045

The Application of the Factorisation Process to Matrix Inversion

You have already seen that matrix inversion is equivalent to the solution of a number of sets of simultaneous equations, each set having the same left hand side, the right hand sides being the columns of the unit matrix. It follows then that the process we have just been studying can be extended to the finding of the inverse of a matrix.   The B in the equation AX = B will be in turn the columns of the unit matrix. However, instead of going through the process for each column of the unit matrix separately, the whole thing can be done at one go by combining each calculation in a manner similar to that in FRAME 6.   There the working for each of the three sets of equations (3.1), (4.2) and (5A.1) was combined into a single calculation.

To illustrate the method, the inverse of the coefficient matrix of the set of equations in FRAME 34 will be found, i.e., the inverse of

$$A = \begin{pmatrix} 0 \cdot 7 & -5 \cdot 4 & 1 \cdot 0 \\ 3 \cdot 5 & 2 \cdot 2 & 0 \cdot 8 \\ 1 \cdot 0 & -1 \cdot 5 & 4 \cdot 3 \end{pmatrix}$$

Now you have already found, in 34A, that this can be factorised into

$$\begin{pmatrix} 1 & 0 & 0 \\ 5 & 1 & 0 \\ 1 \cdot 43 & 0 \cdot 21 & 1 \end{pmatrix} \begin{pmatrix} 0 \cdot 7 & -5 \cdot 4 & 1 \cdot 0 \\ 0 & 29 \cdot 2 & -4 \cdot 2 \\ 0 & 0 & 3 \cdot 75 \end{pmatrix}$$

which is of the form LU .

This is now used to solve AX = B where $B = \begin{pmatrix} 1 & 0 & 0 \\ 0 & 1 & 0 \\ 0 & 0 & 1 \end{pmatrix}$

Once again, the equation AX = B is written as LUX = B or LY = B where UX = Y .

What will be the sizes of X and Y in this case?

****

35A

Each will be 3 × 3 .

Start by finding, by ordinary analytical methods, the eigenvalues and eigenvectors of the 2×2 matrix.

$$\begin{pmatrix} 5 & 2 \\ 2 & 8 \end{pmatrix}$$

****

43A

The characteristic equation is $\lambda^2 - 13\lambda + 36 = 0$.
$\lambda = 9, 4$.

Eigenvectors are $\{1 \;\; 2\}$ *for* $\lambda = 9$ *and* $\{2 \;\; -1\}$ *for* $\lambda = 4$.

FRAME 44

What we are going to do now may seem like a little bit of magic but don't worry about that for the moment.

If the matrix $\begin{pmatrix} 5 & 2 \\ 2 & 8 \end{pmatrix}$ is postmultiplied by $\begin{pmatrix} 1 \\ 0 \end{pmatrix}$, then

$\begin{pmatrix} 5 & 2 \\ 2 & 8 \end{pmatrix}\begin{pmatrix} 1 \\ 0 \end{pmatrix} = \begin{pmatrix} 5 \\ 2 \end{pmatrix}$ which can be written as $5\begin{pmatrix} 1 \\ 0 \cdot 4 \end{pmatrix}$.   Here 5 has been

taken outside the matrix so that what was the numerically largest element inside becomes <u>plus</u> 1.   Next

$$\begin{pmatrix} 5 & 2 \\ 2 & 8 \end{pmatrix}\begin{pmatrix} 1 \\ 0 \cdot 4 \end{pmatrix} = \begin{pmatrix} 5 \cdot 8 \\ 5 \cdot 2 \end{pmatrix} = 5 \cdot 8 \begin{pmatrix} 1 \\ 0 \cdot 90 \end{pmatrix}$$

Then $\begin{pmatrix} 5 & 2 \\ 2 & 8 \end{pmatrix}\begin{pmatrix} 1 \\ 0 \cdot 90 \end{pmatrix} = \begin{pmatrix} 6 \cdot 8 \\ 9 \cdot 2 \end{pmatrix} = 9 \cdot 2 \begin{pmatrix} 0 \cdot 74 \\ 1 \end{pmatrix}$

This time it is the lower element which is made +1 as $9 \cdot 2 > 6 \cdot 8$.

Continuing, $\begin{pmatrix} 5 & 2 \\ 2 & 8 \end{pmatrix}\begin{pmatrix} 0 \cdot 74 \\ 1 \end{pmatrix} = \begin{pmatrix} 5 \cdot 7 \\ 9 \cdot 48 \end{pmatrix} = 9 \cdot 48 \begin{pmatrix} 0 \cdot 60 \\ 1 \end{pmatrix}$

Now continue, doing the process 5 times more keeping the working to 2 decimal places for the first three times and then to 3 decimal places. Does anything particular appear to be happening?
****

44A

Successive R.H.S.'s are:

$9 \cdot 20 \begin{pmatrix} 0 \cdot 54 \\ 1 \end{pmatrix}, \quad 9 \cdot 08 \begin{pmatrix} 0 \cdot 52 \\ 1 \end{pmatrix}, \quad 9 \cdot 04 \begin{pmatrix} 0 \cdot 51 \\ 1 \end{pmatrix}, \quad 9 \cdot 020 \begin{pmatrix} 0 \cdot 504 \\ 1 \end{pmatrix}, \quad 9 \cdot 008 \begin{pmatrix} 0 \cdot 502 \\ 1 \end{pmatrix}$

The scalar multiple appears to be tending to 9, the larger value of λ
and the matrix to $\{0 \cdot 5 \;\; 1\}$ *which you found to be the eigenvector*
associated with the eigenvalue 9.   Remember, of course, that when an
eigenvector is quoted, only the ratios between its components are given
so that $\{1 \;\; 2\}$, $\{0 \cdot 5 \;\; 1\}$ *or* $\{-2 \;\; -4\}$, *etc., all represent the*
eigenvector associated with the eigenvalue 9.

If you continue in this way, further applications of the process give

$9 \cdot 004 \begin{pmatrix} 0 \cdot 501 \\ 1 \end{pmatrix}$, $9 \cdot 002 \begin{pmatrix} 0 \cdot 500 \\ 1 \end{pmatrix}$, $9 \cdot 000 \begin{pmatrix} 0 \cdot 500 \\ 1 \end{pmatrix}$, and after this no

further change takes place.

Now see if the same process works for the matrix $\begin{pmatrix} 4 & 2 \\ 1 & 5 \end{pmatrix}$

by (i)   finding the eigenvalues and eigenvectors analytically,
    (ii)   proceeding as above.   Once again start off with the vector
          {1 0}.   Keep your working to 2 decimal places.
          ****

Characteristic equation is $\lambda^2 - 9\lambda + 18 = 0$ *with roots* $\lambda = 3, 6.$
Eigenvectors are {-2 1} *for* $\lambda = 3$ *and* {1 1} *for* $\lambda = 6.$

$\begin{pmatrix} 4 & 2 \\ 1 & 5 \end{pmatrix} \begin{pmatrix} 1 \\ 0 \end{pmatrix} = \begin{pmatrix} 4 \\ 1 \end{pmatrix} = 4 \begin{pmatrix} 1 \\ 0 \cdot 25 \end{pmatrix}$, $\begin{pmatrix} 4 & 2 \\ 1 & 5 \end{pmatrix} \begin{pmatrix} 1 \\ 0 \cdot 25 \end{pmatrix} = 4 \cdot 5 \begin{pmatrix} 1 \\ 0 \cdot 50 \end{pmatrix}$

Further successive results are

$5 \begin{pmatrix} 1 \\ 0 \cdot 70 \end{pmatrix}$, $5 \cdot 4 \begin{pmatrix} 1 \\ 0 \cdot 83 \end{pmatrix}$, $5 \cdot 66 \begin{pmatrix} 1 \\ 0 \cdot 91 \end{pmatrix}$, $5 \cdot 82 \begin{pmatrix} 1 \\ 0 \cdot 95 \end{pmatrix}$, $5 \cdot 90 \begin{pmatrix} 1 \\ 0 \cdot 97 \end{pmatrix}$,

$5 \cdot 94 \begin{pmatrix} 1 \\ 0 \cdot 98 \end{pmatrix}$, $5 \cdot 96 \begin{pmatrix} 1 \\ 0 \cdot 99 \end{pmatrix}$, $5 \cdot 98 \begin{pmatrix} 1 \\ 0 \cdot 99 \end{pmatrix}.$

*A similar thing appears to be happening, i.e., the scalar multiple
tending to the eigenvalue 6 and the matrix to the eigenvector* {1 1}.
Closer results can only be obtained if more decimal places are now taken.

The matrix $\begin{pmatrix} 4 & 1 & -1 \\ 2 & 3 & -1 \\ -2 & 1 & 5 \end{pmatrix}$ has eigenvalues 2, 4 and 6, and

corresponding eigenvectors {1 -1 1}, {1 1 1} and {1 1 -1}.
Working to 2 decimal places and starting with the vector {1 0 0} see
whether a similar process to that just used will produce one of the
eigenvalues and one of the eigenvectors.
          ****

$$\begin{pmatrix} 4 & 1 & -1 \\ 2 & 3 & -1 \\ -2 & 1 & 5 \end{pmatrix} \begin{pmatrix} 1 \\ 0 \\ 0 \end{pmatrix} = \begin{pmatrix} 4 \\ 2 \\ -2 \end{pmatrix} = 4 \begin{pmatrix} 1 \\ 0 \cdot 5 \\ -0 \cdot 5 \end{pmatrix}$$

$$\begin{pmatrix} 4 & 1 & -1 \\ 2 & 3 & -1 \\ -2 & 1 & 5 \end{pmatrix} \begin{pmatrix} 1 \\ 0 \cdot 5 \\ -0 \cdot 5 \end{pmatrix} = \begin{pmatrix} 5 \\ 4 \\ -4 \end{pmatrix} = 5 \begin{pmatrix} 1 \\ 0 \cdot 8 \\ -0 \cdot 8 \end{pmatrix}$$

Further successive results are $5 \cdot 6 \begin{pmatrix} 1 \\ 0 \cdot 93 \\ -0 \cdot 93 \end{pmatrix}$, $5 \cdot 86 \begin{pmatrix} 1 \\ 0 \cdot 98 \\ -0 \cdot 98 \end{pmatrix}$,

$5 \cdot 96 \begin{pmatrix} 1 \\ 0 \cdot 99 \\ -0 \cdot 99 \end{pmatrix}$, $5 \cdot 98 \begin{pmatrix} 1 \\ 1 \\ -1 \end{pmatrix}$, $6 \begin{pmatrix} 1 \\ 1 \\ -1 \end{pmatrix}$ *and once again the process works.*

FRAME 47

You will have noticed that, when starting the process just described, we have always commenced with {1 0} or {1 0 0}.   There is nothing special about these vectors and, in place of the former, for example, {0 1}, {1 1}, {1 -1} or any other vector of the correct size could be used.   You will remember that all iterative processes require an initial estimate of the quantity sought to be taken.   In the absence of any other information {1 0} is as reasonable a starting vector as any other.

Another thing you will have noticed is that, in the case of each example we have looked at so far, the numerically largest or dominant eigenvalue and its associated eigenvector have been obtained.   Is this to be expected with other matrices, or is it just a fluke?   The answer is that it will usually happen but we shall not justify it here.   If, however, you are interested, you will find a proof of it in the appendix at the end of this programme.

FRAME 48

A flow diagram for finding the dominant eigenvalue and associated eigenvector is shown on page 142.

FRAME 49

Eigenvalues and Eigenvectors - The Numerically Least Eigenvalue

So far we have concentrated on the numerically largest eigenvalue, that being the easiest to obtain.   Assuming that all the eigenvalues are of the same sign, which will be taken as positive for the sake of simplicity, the next one we will look at is the smallest.

If λ is an eigenvalue and $\underline{X}$ the associated eigenvector of a matrix A, then these satisfy the equation $A\underline{X} = \lambda\underline{X}$.

Subtracting $p\underline{X}$ from each side of this equation gives

$$A\underline{X} - p\underline{X} = \lambda\underline{X} - p\underline{X} \qquad\qquad (A - pI)\underline{X} = (\lambda - p)\underline{X}$$

This equation can be interpreted as saying that $\lambda - p$ is an eigenvalue of the matrix A pI and $\underline{X}$ is the associated eigenvector.   In other words, subtracting p from each element on the leading diagonal of a matrix has the effect of subtracting p from each eigenvalue but leaves the eigenvectors unaltered.

Now suppose that p is taken equal to or greater than the dominant eigenvalue of A .   What can you say about the dominant eigenvalue of $A - pI$ in relation to the eigenvalues of A ?
****

Flow diagram for FRAME 48.

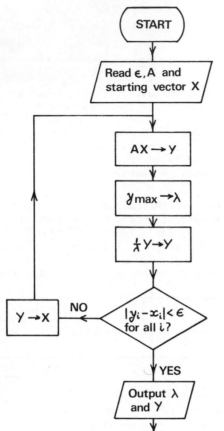

If A is an $n \times n$ matrix then $\underline{X}$ is an $n \times 1$ column matrix $\{x_1, x_2, \ldots x_n\}$ with, say, $x_1 = 1$ and $x_2 = x_3 = \ldots = x_n = 0$.

y_{max} is the numerically largest element of $\underline{Y}$.

This criterion uses agreement between two successive approximations to the eigenvector. Agreement between successive values of λ could also be used.

$\Big($ Programs for finding the numerically greatest eigenvalue and corresponding eigenvector by iteration can be found in references (2), (5), (8) and (9).

A program for finding the numerically smallest eigenvalue is given in reference (4). $\Big)$

49A

It will be p less than the numerically smallest eigenvalue of A.

Thus, if the dominant eigenvalue of $A - pI$ is found, adding p to it will give the numerically smallest eigenvalue of A. The corresponding eigenvector will be the same as that found for the dominant eigenvalue of $A - pI$. (When using a computer, p would be taken equal to the dominant eigenvalue. The only reason for taking it greater is that the arithmetic can sometimes be eased, when working manually.)

Taking as an example the matrix $\begin{pmatrix} 5 & 2 \\ 2 & 8 \end{pmatrix}$ (see FRAMES 43–45), the dominant eigenvalue was found to be 9. Thus if $A = \begin{pmatrix} 5 & 2 \\ 2 & 8 \end{pmatrix}$, $A - 9I = \begin{pmatrix} -4 & 2 \\ 2 & -1 \end{pmatrix}$.

Use the technique just developed to find the dominant eigenvalue of $A - 9I$ and the associated eigenvector.
****

50A

Starting from $\{1 \quad 0\}$ *leads very quickly to* $-5 \begin{pmatrix} 1 \\ -0 \cdot 5 \end{pmatrix}$.

Remember (from FRAME 44) that the unit element must always be arranged to be +1. The dominant eigenvalue is −5 and the associated eigenvector $\{1 \quad -0 \cdot 5\}$.

The corresponding eigenvalue of A is therefore $-5 + 9 = 4$ and the associated eigenvector $\{1 \quad -0 \cdot 5\}$. 4 is its lesser eigenvalue.

Now use this method to find the smallest eigenvalue and associated eigenvector of the matrix in FRAME 46.
****

51A

Subtracting 6 (the dominant eigenvalue) from each element on the leading diagonal gives

$$\begin{pmatrix} -2 & 1 & -1 \\ 2 & -3 & -1 \\ -2 & 1 & -1 \end{pmatrix}$$

The dominant eigenvalue of this matrix is −4 and the corresponding eigenvector is $\{1 \quad -1 \quad 1\}$.

The smallest eigenvalue of the original matrix is $-4 + 6 = 2$ *and the corresponding eigenvector is* $\{1 \quad -1 \quad 1\}$. *This agrees with the result quoted in FRAME 46.*

When considering the buckling of a strut, it is found that the smallest eigenvalue of a certain matrix is useful in finding the critical buckling load. (This problem was considered in more detail on pages 8:11 to 8:14 in Vol. 2 of our book "Mathematics for Engineers and Scientists".) Now the largest eigenvalue in this situation is of no particular interest, but, in order to find numerically the smallest eigenvalue, we have had, so far, first to find this largest eigenvalue. It is natural to question, then, whether the smallest can be found without the necessity first of finding the largest.   To answer this question, start by finding $A^{-1}\underline{X}$ from the equation $A\underline{X} = \lambda\underline{X}$.

****

52A

$A\underline{X} = \lambda\underline{X}$, $\underline{X} = A^{-1}\lambda\underline{X} = \lambda A^{-1}\underline{X}$ *(as λ is a scalar)*, $\lambda^{-1}\underline{X} = A^{-1}\underline{X}$

FRAME 53

Does this result suggest a way of finding the smallest eigenvalue?

****

53A

The equation $A^{-1}\underline{X} = \lambda^{-1}\underline{X}$ shows that $1/\lambda$ is an eigenvalue of A^{-1}.   If the iteration process used a moment ago is applied to A^{-1}, the largest value of $1/\lambda$ will be obtained and hence the smallest value of λ.

FRAME 54

The inverse of $\begin{pmatrix} 5 & 2 \\ 2 & 8 \end{pmatrix}$ is $\dfrac{1}{36}\begin{pmatrix} 8 & -2 \\ -2 & 5 \end{pmatrix}$,   i.e., $\begin{pmatrix} 0\cdot2222 & -0\cdot0556 \\ -0\cdot0556 & 0\cdot1389 \end{pmatrix}$

Find the largest eigenvalue for this matrix, using the iteration technique.   **

54A

The largest eigenvalue is $0\cdot25$.   The corresponding eigenvector is $\{1 \quad -0\cdot5\}$.

FRAME 55

The smallest eigenvalue of the original matrix is therefore 4, which agrees with the value found in 43A.   Furthermore 52A shows that if $\underline{X}$ is the eigenvector of A corresponding to eigenvalue λ, then $\underline{X}$ is the eigenvector of A^{-1} corresponding to eigenvalue λ^{-1}.   This means that $\{1 \quad -0\cdot5\}$ is the eigenvector of the original matrix corresponding to the eigenvalue 4.

Now find the smallest eigenvalue and corresponding eigenvector of the matrix you were working with in FRAME 46.

The inverse of this matrix is $\begin{pmatrix} 0\cdot333 & -0\cdot125 & 0\cdot042 \\ -0\cdot167 & 0\cdot375 & 0\cdot042 \\ 0\cdot167 & -0\cdot125 & 0\cdot208 \end{pmatrix}$

****

55A

The largest eigenvalue of the inverse is $0\cdot5$ and the associated eigenvector $\{1 \quad -1 \quad 1\}$.   The smallest eigenvalue of the given matrix is thus 2 and the associated eigenvector $\{1 \quad -1 \quad 1\}$.

FRAME 56

We thus have two methods for finding the smallest eigenvalue.   The first requires the largest eigenvalue to be found before the smallest but the second requires the inverse of the original matrix to be found.   Both

(continued)

methods therefore require two major calculations and so one cannot be
recommended particularly in preference to the other.

FRAME 57

If the eigenvalues are all negative, both of the processes just described
will lead to the numerically smallest one.   p, however (when using the
method described first) will be negative and should be numerically equal
to or larger than the numerical value of the dominant eigenvalue.   Thus
if the dominant eigenvalue is $-7 \cdot 3$,   p can be taken as $-7 \cdot 3$ or as -8,
say.

In practical problems giving rise to eigenvalues, the eigenvalues are
usually of the same sign (or zero).   If you meet a case where some are
positive and others negative, the method using the inverse matrix will
still lead to the numerically smallest eigenvalue.   Can you see what will
happen if the $A - pI$ method is applied to a matrix having eigenvalues
$6 \cdot 7$, $3 \cdot 6$, $0 \cdot 7$ and $-3 \cdot 1$,   say?

****

57A

The dominant eigenvalue of $A - 7I$, *say, will be* $-10 \cdot 1$ *and it is this
eigenvalue that the method will produce.   The algebraically least
eigenvalue of* A *(i.e.,* $-3 \cdot 1$ *) will thus emerge instead of the numerically
least (* $0 \cdot 7$ *).*

FRAME 58

The Remaining Eigenvalues and Eigenvectors

In the case of a 3×3 matrix, the third eigenvalue follows immediately
as the sum of the eigenvalues is equal to the trace (the sum of the
elements on the leading diagonal).   Note that this result will also give
us the second eigenvalue of a 2×2 matrix and so could have been used
previously.   The purpose of using the $A - pI$ and A^{-1} methods on a 2×2
matrix was just to have simple examples on which to demonstrate these
techniques.   If the matrix is larger, the intermediate eigenvalues can
be found by a method called DEFLATION.   We shall illustrate it by
applying it to just one example.   However, if your course does not
include this topic, you can omit the next few frames and proceed directly
to the miscellaneous examples in FRAME 72.

FRAME 59

If five flywheels, equal in all respects, are attached to a central shaft
which is free to rotate in its bearings, the following equations, giving
the angular displacements, hold:

$$I\ddot{\theta}_1 = -C(\theta_1 - \theta_2)$$
$$I\ddot{\theta}_2 = C(\theta_1 - \theta_2) - C(\theta_2 - \theta_3)$$
$$I\ddot{\theta}_3 = C(\theta_2 - \theta_3) - C(\theta_3 - \theta_4)$$
$$I\ddot{\theta}_4 = C(\theta_3 - \theta_4) - C(\theta_4 - \theta_5)$$
$$I\ddot{\theta}_5 = C(\theta_4 - \theta_5)$$

I is the M. of I. of each flywheel about its axis,   C the torsional
stiffness of the shaft and θ_1, θ_2, etc., the angular displacements of the
first, second, flywheel, etc.

Assuming oscillatory motion, so that each θ is of the form

$$\theta = \theta_{max} \sin(\omega t + \alpha)$$

and putting $\dfrac{I\omega^2}{C} = \lambda$, leads to the matrix equation

$$\begin{pmatrix} 1 & -1 & 0 & 0 & 0 \\ -1 & 2 & -1 & 0 & 0 \\ 0 & -1 & 2 & -1 & 0 \\ 0 & 0 & -1 & 2 & -1 \\ 0 & 0 & 0 & -1 & 1 \end{pmatrix} \begin{pmatrix} \theta_1 \\ \theta_2 \\ \theta_3 \\ \theta_4 \\ \theta_5 \end{pmatrix} = \lambda \begin{pmatrix} \theta_1 \\ \theta_2 \\ \theta_3 \\ \theta_4 \\ \theta_5 \end{pmatrix}$$

A non-zero solution will involve the eigenvalues of the square matrix, A, on the L.H.S. of this equation.   (This problem for a smaller system, and also some other problems leading to a similar matrix equation, is given on pages 8:3 to 8:5 in Vol. 2 of our "Mathematics for Engineers and Scientists".)   Incidentally, although what follows does not depend upon this, A is an example of what is called a tridiagonal matrix.   All non-zero elements in such a matrix lie on the leading diagonal and on two lines adjacent and parallel to this diagonal.

Now you already know how to find the largest and smallest eigenvalues of the above 5×5 matrix.   Start by obtaining the largest eigenvalue and the corresponding eigenvector, working eventually to 4 decimal places.

HINT: The symmetry of the matrix about both diagonals means some, at least, of the eigenvectors may also be symmetrical.   Therefore, instead of starting from $\{1 \ \ 0 \ \ 0 \ \ 0 \ \ 0\}$, we suggest you start from $\{1 \ \ 0 \ \ 0 \ \ 0 \ \ 1\}$, say.

****

59A

$3\cdot6180 \qquad \{0\cdot3090 \quad -0\cdot8090 \quad 1 \quad -0\cdot8090 \quad 0\cdot3090\}$

Subtracting $4I$ from the original matrix (in order to find the least eigenvalue) gives

$$\begin{pmatrix} -3 & -1 & 0 & 0 & 0 \\ -1 & -2 & -1 & 0 & 0 \\ 0 & -1 & -2 & -1 & 0 \\ 0 & 0 & -1 & -2 & -1 \\ 0 & 0 & 0 & -1 & -3 \end{pmatrix} = A_1 \quad \text{(say)}$$

This time, if a start is made with $\{1 \ \ 0 \ \ 0 \ \ 0 \ \ 1\}$, it is found that -4 is an eigenvalue and $\{1 \ \ 1 \ \ 1 \ \ 1 \ \ 1\}$ is the associated eigenvector. Both the eigenvalue and eigenvector here are exact.

---

0 is therefore an eigenvalue of the original matrix A in FRAME 59 and $\{1 \ \ 1 \ \ 1 \ \ 1 \ \ 1\}$ the associated eigenvector.   There are three more eigenvalues to be found.   To find another, which, as it happens, will turn out to be the next largest, the deflation process is applied as follows.

FRAME 61   (continued)

Let λ_1 be the largest eigenvalue of A and $\underline{x}_1$ the corresponding eigenvector with its first element made equal to 1.

What will $\underline{x}_1$ be in this example?

****

61A

$\{1 \quad -2 \cdot 6181 \quad 3 \cdot 2362 \quad -2 \cdot 6181 \quad 1\}$

To obtain this, each element in the vector in 59A is divided by 0·3090.

FRAME 62

Now let $\underline{a}_1$ be the first row of A.   A new matrix, B , is formed according to the formula $B = A - \underline{x}_1\underline{a}_1$.

What will this matrix be here?

****

62A

$$B = \begin{pmatrix} 1 & -1 & 0 & 0 & 0 \\ -1 & 2 & -1 & 0 & 0 \\ 0 & -1 & 2 & -1 & 0 \\ 0 & 0 & -1 & 2 & -1 \\ 0 & 0 & 0 & -1 & 1 \end{pmatrix} - \begin{pmatrix} 1 \\ -2 \cdot 6181 \\ 3 \cdot 2362 \\ -2 \cdot 6181 \\ 1 \end{pmatrix} (1 \quad -1 \quad 0 \quad 0 \quad 0)$$

$$= \begin{pmatrix} 0 & 0 & 0 & 0 & 0 \\ 1 \cdot 6181 & -0 \cdot 6181 & -1 & 0 & 0 \\ -3 \cdot 2362 & 2 \cdot 2362 & 2 & -1 & 0 \\ 2 \cdot 6181 & -2 \cdot 6181 & -1 & 2 & -1 \\ -1 & 1 & 0 & -1 & 1 \end{pmatrix}$$

FRAME 63

Now let λ_2 be another eigenvalue (other than zero) of A, and $\underline{x}_2$ the associated eigenvector which has 1 as its first element.   These are, at present, both unknown.

Then
$$\begin{aligned} B(\underline{x}_1 - \underline{x}_2) &= (A - \underline{x}_1\underline{a}_1)(\underline{x}_1 - \underline{x}_2) \\ &= A(\underline{x}_1 - \underline{x}_2) - \underline{x}_1\underline{a}_1(\underline{x}_1 - \underline{x}_2) \\ &= A\underline{x}_1 - A\underline{x}_2 - \underline{x}_1\underline{a}_1\underline{x}_1 + \underline{x}_1\underline{a}_1\underline{x}_2 \end{aligned} \qquad (63.1)$$

As λ_1 is an eigenvalue of A with associated eigenvector $\underline{x}_1$, $A\underline{x}_1 = \lambda_1\underline{x}_1$. For a similar reason, $A\underline{x}_2 = \lambda_2\underline{x}_2$.   Also the sum of the elements of the first row of A each multiplied by the corresponding elements of $\underline{x}_1$ give the first element of $\lambda_1\underline{x}_1$.   This is equal to λ_1 as the first element of $\underline{x}_1$ has been arranged to be 1.   If this is not immediately obvious, look and see how the starred 3·6181 is obtained.

$$\begin{pmatrix} \cdots 1\cdots & -1\cdots & 0\cdots & 0\cdots & 0 \\ -1 & 2 & -1 & 0 & 0 \\ 0 & -1 & 2 & -1 & 0 \\ 0 & 0 & -1 & 2 & -1 \\ 0 & 0 & 0 & -1 & 1 \end{pmatrix} \begin{pmatrix} 1 \\ -2 \cdot 6181 \\ 3 \cdot 2362 \\ -2 \cdot 6181 \\ 1 \end{pmatrix} = \begin{pmatrix} 3 \cdot 6181* \\ -9 \cdot 4724 \\ 11 \cdot 7084 \\ -9 \cdot 4724 \\ 3 \cdot 6181 \end{pmatrix}$$

Apart from a discrepancy of 1 in the last place of decimals due to round-off, this is the value of λ_1.   This means that $\underline{a_1 x_1} = \lambda_1$.   In a similar way $\underline{a_1 x_2} = \lambda_2$.

Thus (63.1) gives

$$B(\underline{x_1} - \underline{x_2}) = \lambda_1 \underline{x_1} - \lambda_2 \underline{x_2} - \underline{x_1}\lambda_1 + \underline{x_1}\lambda_2$$
$$= \lambda_2 (\underline{x_1} - \underline{x_2}) \text{ as } \underline{x_1}\lambda_1 = \lambda_1\underline{x_1}, \ \lambda_1 \text{ being a scalar.}$$

This equation gives us information about an eigenvalue of B and the associated eigenvector.   Can you say what this information is?
****

63A

λ_2 is an eigenvalue and $\underline{x_1} - \underline{x_2}$ the associated eigenvector.

FRAME 64

Similarly, if λ_r is another eigenvalue of A with associated eigenvector $\underline{x_r}$, λr and $\underline{x_1} - \underline{x_r}$ will be an eigenvalue and eigenvector respectively of B.   This means that all the eigenvalues of B are also eigenvalues of A.

The characteristic equation of B is

$$\begin{vmatrix} -\lambda & 0 & 0 & 0 & 0 \\ 1\cdot6181 & -0\cdot6181 - \lambda & -1 & 0 & 0 \\ -3\cdot2362 & 2\cdot2362 & 2 - \lambda & -1 & 0 \\ 2\cdot6181 & -2\cdot6181 & -1 & 2 - \lambda & -1 \\ -1 & 1 & 0 & -1 & 1 - \lambda \end{vmatrix} = 0$$

i.e. $$-\lambda \begin{vmatrix} -0\cdot6181 - \lambda & -1 & 0 & 0 \\ 2\cdot2362 & 2 - \lambda & -1 & 0 \\ -2\cdot6181 & -1 & 2 - \lambda & -1 \\ 1 & 0 & -1 & 1 - \lambda \end{vmatrix} = 0 \quad (64.1)$$

This shows that 0 is one eigenvalue of B and that the other eigenvalues are the same as those given by the fourth order determinant in (64.1).

Turning now to the eigenvector $\underline{x_1} - \underline{x_2}$, let this be represented by $\{y_1 \ \ y_2 \ \ y_3 \ \ y_4 \ \ y_5\}$.   Can you say anything about any of the y's?
****

64A

$y_1 = 0$ as the first element in both $\underline{x_1}$ and $\underline{x_2}$ is unity.

FRAME 65

This means that $B(\underline{x_1} - \underline{x_2}) = \lambda_2(\underline{x_1} - \underline{x_2})$ can be written as

$$\begin{pmatrix} 0 & 0 & 0 & 0 & 0 \\ 1\cdot6181 & -0\cdot6181 & -1 & 0 & 0 \\ -3\cdot2362 & 2\cdot2362 & 2 & -1 & 0 \\ 2\cdot6181 & -2\cdot6181 & -1 & 2 & -1 \\ -1 & 1 & 0 & -1 & 1 \end{pmatrix} \begin{pmatrix} 0 \\ y_2 \\ y_3 \\ y_4 \\ y_5 \end{pmatrix} = \lambda_2 \begin{pmatrix} 0 \\ y_2 \\ y_3 \\ y_4 \\ y_5 \end{pmatrix}$$

Now this matrix equation represents five simultaneous linear equations. The first of these equations is simply $0 = 0$ and the other four are the same as those given by the matrix equation

$$\begin{pmatrix} -0\cdot6181 & -1 & 0 & 0 \\ 2\cdot2362 & 2 & -1 & 0 \\ -2\cdot6181 & -1 & 2 & -1 \\ 1 & 0 & -1 & 1 \end{pmatrix} \begin{pmatrix} y_2 \\ y_3 \\ y_4 \\ y_5 \end{pmatrix} = \lambda_2 \begin{pmatrix} y_2 \\ y_3 \\ y_4 \\ y_5 \end{pmatrix} \qquad (65.1)$$

$\Big($Compare the square matrix in this equation with the determinant in (64.1).$\Big)$

We now proceed to find an eigenvalue of the 4×4 matrix in (65.1) and the corresponding eigenvector.   You should be able to do this for yourself.   You will, of course, obtain its dominant eigenvalue.
****

65A

$2\cdot6180$ $\{0\cdot2361 \quad -0\cdot7640 \quad 1 \quad -0\cdot4721\}$

FRAME 66

Thus $\lambda_2 = 2\cdot6180$ and an eigenvector of B is $\{0 \ 0\cdot2361 \ -0\cdot7640 \ 1 \ -0\cdot4721\}$.   The word 'an' has been stressed as any multiple of this is also an eigenvector.   But $\underline{x}_1 - \underline{x}_2$ is also an eigenvector and so it can be assumed that

$$\underline{x}_1 - \underline{x}_2 = k\{0 \quad 0\cdot2361 \quad -0\cdot7640 \quad 1 \quad -0\cdot4721\} = k\underline{Y} \quad \text{say.}$$

$\underline{x}_2$, the eigenvector we are looking for, is then given by

$$\underline{x}_2 = \underline{x}_1 - k\underline{Y} \qquad (66.1)$$

in which the only unknown is k.

To find k, premultiply (66.1) by $\underline{a}_1$.   Thus $\underline{a}_1\underline{x}_2 = \underline{a}_1\underline{x}_1 - \underline{a}_1k\underline{Y}$.

But in FRAME 63 it was seen that $\underline{a}_1\underline{x}_2 = \lambda_2$ and $\underline{a}_1\underline{x}_1 = \lambda_1$

$\therefore \ \lambda_2 = \lambda_1 - k\underline{a}_1\underline{Y}$ \qquad or \qquad $k = \dfrac{\lambda_1 - \lambda_2}{\underline{a}_1\underline{Y}}$

What will k be in the present situation?
****

66A

$$\frac{3\cdot6180 - 2\cdot6180}{(1 \ -1 \ 0 \ 0 \ 0)\begin{pmatrix} 0 \\ 0\cdot2361 \\ -0\cdot7640 \\ 1 \\ -0\cdot4721 \end{pmatrix}} = \frac{1}{-0\cdot2361} = -4\cdot2355$$

FRAME 67

Then $\underline{x}_2 = \begin{pmatrix} 1 \\ -2\cdot6181 \\ 3\cdot2362 \\ -2\cdot6181 \\ 1 \end{pmatrix} + 4\cdot2355 \begin{pmatrix} 0 \\ 0\cdot2361 \\ -0\cdot7640 \\ 1 \\ -0\cdot4721 \end{pmatrix}$

$= \{1 \quad -1\cdot6181 \quad 0\cdot0003 \quad 1\cdot6174 \quad -0\cdot9996\}$

Three eigenvalues have now been obtained, the two largest and the smallest. The next to the smallest can be found by treating the matrix A_1 of FRAME 60, i.e.,

$$\begin{pmatrix} -3 & -1 & 0 & 0 & 0 \\ -1 & -2 & -1 & 0 & 0 \\ 0 & -1 & -2 & -1 & 0 \\ 0 & 0 & -1 & -2 & -1 \\ 0 & 0 & 0 & -1 & -3 \end{pmatrix} \quad (68.1)$$

in the same way as we have just done A.

What will be (i) B_1, the matrix corresponding to B in 62A,
(ii) the square matrix corresponding to that in (65.1),
in this case? ***

(i)

$$B_1 = \begin{pmatrix} -3 & -1 & 0 & 0 & 0 \\ -1 & -2 & -1 & 0 & 0 \\ 0 & -1 & -2 & -1 & 0 \\ 0 & 0 & -1 & -2 & -1 \\ 0 & 0 & 0 & -1 & -3 \end{pmatrix} - \begin{pmatrix} 1 \\ 1 \\ 1 \\ 1 \\ 1 \end{pmatrix} (-3 \quad -1 \quad 0 \quad 0 \quad 0)$$

$$= \begin{pmatrix} 0 & 0 & 0 & 0 & 0 \\ 2 & -1 & -1 & 0 & 0 \\ 3 & 0 & -2 & -1 & 0 \\ 3 & 1 & -1 & -2 & -1 \\ 3 & 1 & 0 & -1 & -3 \end{pmatrix}$$

(ii)

$$\begin{pmatrix} -1 & -1 & 0 & 0 \\ 0 & -2 & -1 & 0 \\ 1 & -1 & -2 & -1 \\ 1 & 0 & -1 & -3 \end{pmatrix}$$

Applying the standard process to this matrix B_1 gives eigenvalue $-3 \cdot 6181$ and eigenvector $\{0 \quad 0 \cdot 1909 \quad 0 \cdot 4999 \quad 0 \cdot 8090 \quad 1\}$.

What will these give for an eigenvalue and eigenvector of the original matrix A? ***

The matrix (68.1) will have eigenvalue $-3 \cdot 6181$.
The value of k here will be

$$\frac{-4 - (-3 \cdot 6181)}{(-3 \quad -1 \quad 0 \quad 0 \quad 0) \begin{pmatrix} 0 \\ 0 \cdot 1909 \\ 0 \cdot 4999 \\ 0 \cdot 8090 \\ 1 \end{pmatrix}} = 2 \cdot 0005$$

and the relevant eigenvector of (68.1) will be

$$
\begin{pmatrix} 1 \\ 1 \\ 1 \\ 1 \\ 1 \end{pmatrix} - 2 \cdot 0005 \begin{pmatrix} 0 \\ 0 \cdot 1909 \\ 0 \cdot 4999 \\ 0 \cdot 8090 \\ 1 \end{pmatrix} = \begin{pmatrix} 1 \\ 0 \cdot 6181 \\ 0 \cdot 0000 \\ -0 \cdot 6184 \\ -1 \cdot 0005 \end{pmatrix}
$$

The eigenvalue of A *will be* $-3 \cdot 6181 + 4 = 0 \cdot 3819$ *and the associated
eigenvector* $\{1 \quad 0 \cdot 6181 \quad 0 \cdot 0000 \quad -0 \cdot 6184 \quad -1 \cdot 0005\}$.

FRAME 70

Finally the last eigenvalue can be found from the fact that the sum of the
eigenvalues is equal to the trace.   This gives
$8 - (3 \cdot 6180 + 2 \cdot 6180 + 0 \cdot 3819 + 0) = 1 \cdot 3821$ for the remaining eigenvalue.
The corresponding eigenvector is given by the equation

$$
\begin{pmatrix} 1 & -1 & 0 & 0 & 0 \\ -1 & 2 & -1 & 0 & 0 \\ 0 & -1 & 2 & -1 & 0 \\ 0 & 0 & -1 & 2 & -1 \\ 0 & 0 & 0 & -1 & 1 \end{pmatrix} \begin{pmatrix} z_1 \\ z_2 \\ z_3 \\ z_4 \\ z_5 \end{pmatrix} = 1 \cdot 3821 \begin{pmatrix} z_1 \\ z_2 \\ z_3 \\ z_4 \\ z_5 \end{pmatrix}
$$

from which the z-vector is $\{-0 \cdot 8086 \quad 0 \cdot 3090 \quad 1 \quad 0 \cdot 3090 \quad -0 \cdot 8086\}$.

FRAME 71

If the matrix A is larger still, then having found λ_2 and $\underline{x}_2$, the
deflation process can be repeated on B to give λ_3 and $\underline{x}_3$ and so on.
Unfortunately, each application of the deflation process renders the end
product less accurate.   Various modifications have been suggested to
overcome this loss of accuracy and also to improve the convergence of the
basic process for finding the highest eigenvalue.   These modifications
will not be pursued here but you should now have an idea of the basic
processes so that you can proceed further with the study of the subject
should you find this necessary.

FRAME 72

Miscellaneous Examples

In this frame a collection of miscellaneous examples is given for you to
try.   Answers are supplied in FRAME 73, together with such working as is
considered helpful.

1.   Express in the form LU the matrix
$$
\begin{pmatrix} 4 & 2 & 1 & 1 \\ 8 & 6 & 4 & 3 \\ 4 & -2 & 3 & 1 \\ 8 & 8 & 12 & 9 \end{pmatrix}
$$

where L is lower triangular with unit elements on its diagonal and U
is upper triangular.

Use this to solve the simultaneous equations

$$4x_1 + 2x_2 + x_3 + x_4 = 25$$
$$8x_1 + 6x_2 + 4x_3 + 3x_4 = 61$$
$$4x_1 - 2x_2 + 3x_3 + x_4 = 17$$
$$8x_1 + 8x_2 + 12x_3 + 9x_4 = 79$$

leaving your answer in fractional form.   (L.U.)

2. In the analysis of the transient behaviour of a five plate absorption tower, the matrix

$$\begin{pmatrix} 1 \cdot 173 & -0 \cdot 634 & 0 & 0 & 0 \\ -0 \cdot 539 & 1 \cdot 173 & -0 \cdot 634 & 0 & 0 \\ 0 & -0 \cdot 539 & 1 \cdot 173 & -0 \cdot 634 & 0 \\ 0 & 0 & -0 \cdot 539 & 1 \cdot 173 & -0 \cdot 634 \\ 0 & 0 & 0 & -0 \cdot 539 & 1 \cdot 173 \end{pmatrix}$$

occurred.   Find, working to 4 decimal places, the inverse matrix
(a) by the Gauss–Jordan process with pivoting,   (b) by the LU
factorisation process.   (If you do not have a machine available,
round-off all numbers in the matrix to 1 decimal place and work to
2 decimal places only.)

3. If you read FRAMES 59 – 71, find all the eigenvalues and corresponding eigenvectors of the matrix

$$A = \begin{pmatrix} 3 \cdot 6 & 2 \cdot 5 & -1 \cdot 1 & 1 \cdot 3 \\ 2 \cdot 3 & 1 \cdot 2 & 2 \cdot 1 & 3 \cdot 1 \\ -1 \cdot 3 & 1 \cdot 9 & 0 \cdot 9 & 2 \cdot 5 \\ 1 \cdot 0 & 2 \cdot 8 & 2 \cdot 2 & 2 \cdot 7 \end{pmatrix}$$

working throughout to 2 decimal places.   Otherwise find only two of them.

4. The iteration $\underline{x}^{(k+1)} = \underline{x}^{(k)} + A\underline{f}(\underline{x}^{(k)})$ where $\underline{x}$ is a vector, $\underline{f}$ is a vector of functions and A is a constant matrix, converges in certain circumstances to a root of $\underline{f}(\underline{x}) = 0$.

If $\underline{x}^{(o)} = \begin{pmatrix} 2 \\ 2 \end{pmatrix}$, $\underline{f}(\underline{x}) = \begin{pmatrix} x^2 + y^2 - 9 \\ x - y^2 + 1 \end{pmatrix}$ and $A = -\dfrac{1}{10}\begin{pmatrix} 2 & 2 \\ 1 & -2 \end{pmatrix}$

use the above algorithm to find a solution of $\underline{f}(\underline{x}) = 0$ near

$\underline{x}^{(o)}$ correct to two decimal places.   (C.E.I.)

Notes: (i) The notation $\underline{x}^{(n)}$ is used here to denote the nth iterate.

(ii) If you have read APPENDIX D of the second programme
(Solution of Non-Linear Equations) in this Unit, you have
seen how the Newton–Raphson method can be extended to a
pair of simultaneous equations.   The above question is a
slight modification of this process expressed in matrix
form.

5. Working to 1 decimal place find all the eigenvalues and eigenvectors of

$$\begin{pmatrix} 5 & 2 & 0 \\ 2 & 10 & 2 \\ 0 & 2 & 10 \end{pmatrix}$$

Answers to Miscellaneous Examples

1.
$$L = \begin{pmatrix} 1 & 0 & 0 & 0 \\ 2 & 1 & 0 & 0 \\ 1 & -2 & 1 & 0 \\ 2 & 2 & 1 & 1 \end{pmatrix} \qquad U = \begin{pmatrix} 4 & 2 & 1 & 1 \\ 0 & 2 & 2 & 1 \\ 0 & 0 & 6 & 2 \\ 0 & 0 & 0 & 3 \end{pmatrix}$$

$Y = \{25 \quad 11 \quad 14 \quad -7\}$ $\qquad$ $X = \{77/18 \quad 32/9 \quad 28/9 \quad -7/3\}$

2. (a)
$$\begin{pmatrix} 1 \cdot 4086 & 1 \cdot 2103 & 0 \cdot 9771 & 0 \cdot 7026 & 0 \cdot 3797 \\ 1 \cdot 0288 & 2 \cdot 2391 & 1 \cdot 8075 & 1 \cdot 2999 & 0 \cdot 7025 \\ 0 \cdot 7061 & 1 \cdot 5365 & 2 \cdot 5138 & 1 \cdot 8076 & 0 \cdot 9770 \\ 0 \cdot 4316 & 0 \cdot 9393 & 1 \cdot 5368 & 2 \cdot 2393 & 1 \cdot 2103 \\ 0 \cdot 1983 & 0 \cdot 4316 & 0 \cdot 7062 & 1 \cdot 0290 & 1 \cdot 4086 \end{pmatrix}$$

(b)
$$L = \begin{pmatrix} 1 & 0 & 0 & 0 & 0 \\ -0 \cdot 4595 & 1 & 0 & 0 & 0 \\ 0 & -0 \cdot 6113 & 1 & 0 & 0 \\ 0 & 0 & -0 \cdot 6863 & 1 & 0 \\ 0 & 0 & 0 & -0 \cdot 7305 & 1 \end{pmatrix}$$

$$U = \begin{pmatrix} 1 \cdot 173 & -0 \cdot 634 & 0 & 0 & 0 \\ 0 & 0 \cdot 8817 & -0 \cdot 634 & 0 & 0 \\ 0 & 0 & 0 \cdot 7854 & -0 \cdot 634 & 0 \\ 0 & 0 & 0 & 0 \cdot 7379 & -0 \cdot 634 \\ 0 & 0 & 0 & 0 & 0 \cdot 7099 \end{pmatrix}$$

$$Y = \begin{pmatrix} 1 & 0 & 0 & 0 & 0 \\ 0 \cdot 4595 & 1 & 0 & 0 & 0 \\ 0 \cdot 2809 & 0 \cdot 6113 & 1 & 0 & 0 \\ 0 \cdot 1928 & 0 \cdot 4195 & 0 \cdot 6863 & 1 & 0 \\ 0 \cdot 1408 & 0 \cdot 3064 & 0 \cdot 5013 & 0 \cdot 7305 & 1 \end{pmatrix}$$

Inverse is
$$\begin{pmatrix} 1 \cdot 4086 & 1 \cdot 2102 & 0 \cdot 9770 & 0 \cdot 7025 & 0 \cdot 3797 \\ 1 \cdot 0289 & 2 \cdot 2391 & 1 \cdot 8076 & 1 \cdot 2998 & 0 \cdot 7025 \\ 0 \cdot 7061 & 1 \cdot 5366 & 2 \cdot 5138 & 1 \cdot 8076 & 0 \cdot 9770 \\ 0 \cdot 4317 & 0 \cdot 9393 & 1 \cdot 5368 & 2 \cdot 2393 & 1 \cdot 2103 \\ 0 \cdot 1983 & 0 \cdot 4316 & 0 \cdot 7062 & 1 \cdot 0290 & 1 \cdot 4086 \end{pmatrix}$$

3. The dominant eigenvalue works out to be 7·38 with eigenvector $\{0 \cdot 84 \quad 0 \cdot 98 \quad 0 \cdot 50 \quad 1\}$.

In this example, two eigenvalues are negative and the least eigenvalue is not the numerically smallest. Forming $A - pI$ where p is taken as 7·5 leads to −9·60 as an eigenvalue of $A - 7 \cdot 5I$ with eigenvector $\{-0 \cdot 58 \quad 1 \quad -0 \cdot 82 \quad -0 \cdot 08\}$. The corresponding eigenvalue of A is −2·10, with the same eigenvector.

Deflating A gives $B = \begin{pmatrix} 0 & 0 & 0 & 0 \\ -1 \cdot 91 & -1 \cdot 72 & 3 \cdot 39 & 1 \cdot 58 \\ -3 \cdot 46 & 0 \cdot 40 & 1 \cdot 56 & 1 \cdot 72 \\ -3 \cdot 28 & -0 \cdot 18 & 3 \cdot 51 & 1 \cdot 15 \end{pmatrix}$

and the dominant eigenvalue of $\begin{pmatrix} -1 \cdot 72 & 3 \cdot 39 & 1 \cdot 58 \\ 0 \cdot 40 & 1 \cdot 56 & 1 \cdot 72 \\ -0 \cdot 18 & 3 \cdot 51 & 1 \cdot 15 \end{pmatrix}$ is $3 \cdot 96$

with eigenvector $\{0 \cdot 78 \quad 0 \cdot 84 \quad 1\}$. $k = 1 \cdot 47$ and this leads to an eigenvalue of A being $3 \cdot 96$ with eigenvector $\{1 \quad 0 \cdot 02 \quad -0 \cdot 63 \quad -0 \cdot 28\}$.

The trace is now used to give the fourth eigenvalue which is $-0 \cdot 86$. Using the definition $A\underline{X} = \lambda\underline{X}$ for this last eigenvector leads to $\{-0 \cdot 07 \quad -0 \cdot 71 \quad -0 \cdot 70 \quad 1\}$.   This value for $\underline{X}$ was found using the Gauss–Jordan process with pivoting on the set of equations

$$4 \cdot 46u_1 + 2 \cdot 5u_2 - 1 \cdot 1u_3 = -1 \cdot 3$$
$$2 \cdot 3u_1 + 2 \cdot 06u_2 + 2 \cdot 1u_3 = -3 \cdot 1$$
$$-1 \cdot 3u_1 + 1 \cdot 9u_2 + 1 \cdot 76u_3 = -2 \cdot 5$$

where $u_1 = x_1/x_4$, $u_2 = x_2/x_4$ and $u_3 = x_3/x_4$.

This example, particularly the finding of the dominant eigenvalue of $A - 7 \cdot 5I$, illustrates that the process can sometimes be very slowly convergent.   You will remember that we remarked on the rate of convergence in FRAME 71.

Alternatively, in this example, the inversion method can be used to find the numerically least eigenvalue and it is not then necessary to use the deflation method.

4. $\underline{x}^{(1)} = \begin{pmatrix} 2 \\ 2 \end{pmatrix} - \frac{1}{10}\begin{pmatrix} 2 & 2 \\ 1 & -2 \end{pmatrix}\begin{pmatrix} -1 \\ -1 \end{pmatrix} = \begin{pmatrix} 2 \cdot 4 \\ 1 \cdot 9 \end{pmatrix}$

Continuing the iterations leads to the value $\underline{x} = \{2 \cdot 37 \quad 1 \cdot 84\}$.

5. $12 \cdot 2$, $\{0 \cdot 3 \quad 1 \quad 0 \cdot 8\}$; $4 \cdot 2$, $\{1 \quad -0 \cdot 4 \quad 0 \cdot 1\}$; $8 \cdot 6$, $\{-0 \cdot 4 \quad -0 \cdot 7 \quad 1\}$.

APPENDIX

Justification of Process for Numerically Largest Eigenvalue

A square matrix of any size may be taken to do this. Generally one would take an $n \times n$ matrix, but we will take a specific value for n, say 4. A 4 × 4 matrix, A, will have four eigenvalues, which can be denoted by λ_1, λ_2, λ_3 and λ_4. There will be four corresponding eigenvectors (each of which will have four elements) and they can be denoted by v_1, v_2, v_3, v_4. Remembering that a column matrix is often regarded as a column vector, each of these eigenvectors can also be regarded as a 4 × 1 matrix.

Now we have seen that the process adopted in FRAMES 44 - 46 always gave the numerically greatest or dominant eigenvalue and the corresponding eigenvector. Let this eigenvalue be the one denoted by λ_1.

Returning to FRAMES 44 and 45, the starting vector taken was $\{1 \quad 0\}$ and in FRAME 46 it was $\{1 \quad 0 \quad 0\}$. Actually, as was pointed out in the main programme, any vector of the correct size could have been used instead. It's just a case of starting off with something and when you have no idea of the actual eigenvector itself, it's obviously sensible to start off with something as simple as possible. Here then $\{1 \quad 0 \quad 0 \quad 0\}$ could quite well be used but for purposes of working out the theory, we will assume that the arbitary vector V_0 is taken as a start. It will, of course, have 4 components as A is assumed to be a 4 × 4 matrix. It is now necessary to show that, as the process adopted in FRAMES 44 - 46 is repeated, $\lambda_1 v_1$ eventually appears.

The first step in the theory is to express V_0 in terms of the eigenvectors v_1, v_2, v_3, v_4. To get some idea as to how this can possibly be done let us go back to the simple matrix of FRAME 44. There, there were two eigenvectors, $\{1 \quad 2\}$ and $\{2 \quad -1\}$, which can be called v_1 and v_2, and the starting vector V_0 was $\{1 \quad 0\}$. Now, and don't worry where the 1/5 and 2/5 have come from for the moment,

$$\frac{1}{5}\begin{pmatrix} 1 \\ 2 \end{pmatrix} + \frac{2}{5}\begin{pmatrix} 2 \\ -1 \end{pmatrix} = \begin{pmatrix} 1/5 \\ 2/5 \end{pmatrix} + \begin{pmatrix} 4/5 \\ -2/5 \end{pmatrix} = \begin{pmatrix} 1 \\ 0 \end{pmatrix}$$

This shows that, in this particular case, we can write

$$\begin{pmatrix} 1 \\ 0 \end{pmatrix} = k_1 \begin{pmatrix} 1 \\ 2 \end{pmatrix} + k_2 \begin{pmatrix} 2 \\ -1 \end{pmatrix} \qquad (A3.1)$$

i.e., $\quad V_0 = k_1 v_1 + k_2 v_2$

provided k_1 and k_2 are given suitable values.

To find k_1 and k_2, notice that (A3.1) is equivalent to

$$\begin{pmatrix} 1 \\ 0 \end{pmatrix} = \begin{pmatrix} k_1 + 2k_2 \\ 2k_1 - k_2 \end{pmatrix}$$

155

MATRIX ALGEBRA, EIGENVALUES AND EIGENVECTORS

Equating corresponding elements gives

$$1 = k_1 + 2k_2 \qquad\qquad 0 = 2k_1 - k_2$$

i.e., two simultaneous equations for k_1 and k_2. You can easily verify that these give 1/5 and 2/5 respectively for k_1 and k_2.

What will be the values of the k's for the starting vector $\{1 \quad 0\}$ to be written as $k_1\underline{v}_1 + k_2\underline{v}_2$ in the case of the matrix $\begin{pmatrix} 4 & 2 \\ 1 & 5 \end{pmatrix}$ in FRAME 45? Take $\underline{v}_1$ as $\{1 \quad 1\}$ and $\underline{v}_2$ as $\{-2 \quad 1\}$.

****

$$\begin{pmatrix} 1 \\ 0 \end{pmatrix} = k_1 \begin{pmatrix} 1 \\ 1 \end{pmatrix} + k_2 \begin{pmatrix} -2 \\ 1 \end{pmatrix} \qquad\qquad k_1 = 1/3, \quad k_2 = -1/3$$

(In each of these two examples the eigenvector linked with k_1 has been that corresponding to the dominant eigenvalue. This doesn't matter here, but fits in with what has been done later.)

In the case of the 3×3 matrix in FRAME 46, it is possible to write the starting vector $\underline{V}_0$ as $k_1\underline{v}_1 + k_2\underline{v}_2 + k_3\underline{v}_3$. Thus, in this case,

$$\begin{pmatrix} 1 \\ 0 \\ 0 \end{pmatrix} = k_1 \begin{pmatrix} 1 \\ 1 \\ -1 \end{pmatrix} + k_2 \begin{pmatrix} 1 \\ 1 \\ 1 \end{pmatrix} + k_3 \begin{pmatrix} 1 \\ -1 \\ 1 \end{pmatrix}$$

(The eigenvector linked with k_1 is again that corresponding to the dominant eigenvalue.)

What will be the required values of the k's here?
****

$$1 = k_1 + k_2 + k_3, \qquad 0 = k_1 + k_2 - k_3, \qquad 0 = -k_1 + k_2 + k_3,$$
from which $k_1 = 1/2,$ $\quad k_2 = 0,$ $\quad k_3 = 1/2.$

These examples suggest that, in the case of the 4×4 matrix A, $\underline{V}_0$ can be written as $k_1\underline{v}_1 + k_2\underline{v}_2 + k_3\underline{v}_3 + k_4\underline{v}_4$.

Having decided on a starting vector $\underline{V}_0$, the next step was to pre-multiply it by A, thus forming $A\underline{V}_0$. As we now have

$$\underline{V}_0 = k_1\underline{v}_1 + k_2\underline{v}_2 + k_3\underline{v}_3 + k_4\underline{v}_4, \quad A\underline{V}_0 = k_1 A\underline{v}_1 + k_2 A\underline{v}_2 + k_3 A\underline{v}_3 + k_4 A\underline{v}_4.$$

Now, as $\underline{v}_1$ is the eigenvector corresponding to eigenvalue λ_1,

$$A\underline{v}_1 = \lambda_1\underline{v}_1$$

Similarly $A\underline{v}_2 = \lambda_2\underline{v}_2$ etc., and so

$$A\underline{V}_0 = k_1\lambda_1\underline{v}_1 + k_2\lambda_2\underline{v}_2 + k_3\lambda_3\underline{v}_3 + k_4\lambda_4\underline{v}_4$$

The next step was to arrange the largest element in $A\underline{V}_0$ to be 1. Suppose this was done by dividing by b_0. (In FRAME 44, b_0 was 5, in 45A it was 4 and in 46A it was also 4.) If the next vector to be pre-multiplied by A be denoted by $\underline{V}_1$, then

$$\underline{V_1} = \frac{1}{b_0} (k_1 \lambda_1 \underline{v_1} + k_2 \lambda_2 \underline{v_2} + k_3 \lambda_3 \underline{v_3} + k_4 \lambda_4 \underline{v_4})$$

which can be written as

$$\underline{V_1} = \frac{\lambda_1}{b_0} \left(k_1 \underline{v_1} + k_2 \frac{\lambda_2}{\lambda_1} \underline{v_2} + k_3 \frac{\lambda_3}{\lambda_1} \underline{v_3} + k_4 \frac{\lambda_4}{\lambda_1} \underline{v_4} \right)$$

The next step is to pre-multiply $\underline{V_1}$ by A.   Doing this gives

$$A\underline{V_1} = \frac{\lambda_1}{b_0} A \left(k_1 \underline{v_1} + k_2 \frac{\lambda_2}{\lambda_1} \underline{v_2} + k_3 \frac{\lambda_3}{\lambda_1} \underline{v_3} + k_4 \frac{\lambda_4}{\lambda_1} \underline{v_4} \right)$$

But as $A\underline{v_1} = \lambda_1 \underline{v_1}$ etc., this becomes

$$A\underline{V_1} = \frac{\lambda_1}{b_0} \left(k_1 \lambda_1 \underline{v_1} + k_2 \frac{\lambda_2^2}{\lambda_1} \underline{v_2} + k_3 \frac{\lambda_3^2}{\lambda_1} \underline{v_3} + k_4 \frac{\lambda_4^2}{\lambda_1} \underline{v_4} \right)$$

Once again, the largest element on the R.H.S. is made unity by dividing by a suitable numerical factor, say b_1.   Then the next vector to be pre-multiplied by A is

$$\underline{V_2} = \frac{\lambda_1}{b_0 b_1} \left(k_1 \lambda_1 \underline{v_1} + k_2 \frac{\lambda_2^2}{\lambda_1} \underline{v_2} + k_3 \frac{\lambda_3^2}{\lambda_1} \underline{v_3} + k_4 \frac{\lambda_4^2}{\lambda_1} \underline{v_4} \right)$$

$$= \frac{\lambda_1^2}{b_0 b_1} \left[k_1 \underline{v_1} + k_2 \left(\frac{\lambda_2}{\lambda_1} \right)^2 \underline{v_2} + k_3 \left(\frac{\lambda_3}{\lambda_1} \right)^2 \underline{v_3} + k_4 \left(\frac{\lambda_4}{\lambda_1} \right)^2 \underline{v_4} \right]$$

You should now be able to see that a pattern is being established. Proceeding in the same way, and using dividing factors b_2, b_3,   etc., as necessary, what will be   (a) $\underline{V_3}$ and   (b) $\underline{V_m}$?

****

$$\underline{V_3} = \frac{\lambda_1^3}{b_0 b_1 b_2} \left[k_1 \underline{v_1} + k_2 \left(\frac{\lambda_2}{\lambda_1} \right)^3 \underline{v_2} + k_3 \left(\frac{\lambda_3}{\lambda_1} \right)^3 \underline{v_3} + k_4 \left(\frac{\lambda_4}{\lambda_1} \right)^3 \underline{v_4} \right]$$

$$\underline{V_m} = \frac{\lambda_1^m}{b_0 b_1 \ldots b_{m-1}} \left[k_1 \underline{v_1} + k_2 \left(\frac{\lambda_2}{\lambda_1} \right)^m \underline{v_2} + k_3 \left(\frac{\lambda_3}{\lambda_1} \right)^m \underline{v_3} + k_4 \left(\frac{\lambda_4}{\lambda_1} \right)^m \underline{v_4} \right]$$

Now λ_1 is the numerically greatest λ and so all of $\left(\frac{\lambda_2}{\lambda_1} \right)^m$, $\left(\frac{\lambda_3}{\lambda_1} \right)^m$, ...

will tend to zero as m is increased.   When m is sufficiently large for them to be negligible,

$$\underline{V_m} = \frac{\lambda_1^m}{b_0 b_1 \ldots b_{m-1}} k_1 \underline{v_1} \qquad (A8.1)$$

Pre-multiplying again by A ,

$$A\underline{V_m} = \frac{\lambda_1^m}{b_0 b_1 \ldots b_{m-1}} k_1 A\underline{v_1}$$

157

$$= \lambda_1 \; \frac{\lambda_1^m}{b_0 b_1 \; \cdots \; b_{m-1}} \; k_1 \underline{v_1}$$

$$\therefore \quad A\underline{v_m} = \lambda_1 \underline{v_m} \qquad\qquad\qquad\qquad \text{(A8.2)}$$

But this is exactly the equation that must be satisfied by λ_1 and $\underline{v_m}$ for λ_1 to be an eigenvalue and $\underline{v_m}$ the corresponding eigenvector. Note that λ_1 would also be the value of the next b, i.e. b_m.   The link between $\underline{v_m}$ and $\underline{v_1}$, as is shown by (A8.1), is that the elements of one are proportional to the elements of the other and you will remember that an eigenvector is not unique in that if $\{1 \;\; -2 \;\; 1 \;\; 3\}$ is an eigenvector then so also are $\{-1 \;\; 2 \;\; -1 \;\; -3\}$, $\{4 \;\; -8 \;\; 4 \;\; 12\}$, $\{-3 \;\; 6 \;\; -3 \;\; -9\}$, etc. (A8.2) is therefore equivalent to $A\underline{v_1} = \lambda_1\underline{v_1}$.

The result of this analysis means that whatever vector is taken for $\underline{V_0}$ (apart, of course, from one whose components are all zero), the process adopted will lead to the numerically largest eigenvalue and the corresponding eigenvector.   The rapidity with which it will be obtained depends on the relative magnitudes of the λ's.   The smaller are $\frac{\lambda_2}{\lambda_1}$, $\frac{\lambda_3}{\lambda_1}$ etc. the quicker will their powers tend to zero.

Special treatment is necessary if the matrix has two numerically equal, but opposite in sign, dominant eigenvalues, for example, $\lambda_1 = 10$, $\lambda_2 = -10$, $\lambda_3 = 7$, $\lambda_4 = 2$.   We shall not consider this special case here.

The theory can, of course, be easily adapted to square matrices of other sizes.

Throughout the process b_0, b_1, b_2, ... have been chosen so that the numerically largest element in the vector has each time been reduced to +1.   Various other alternatives to this are possible without affecting the final outcome.   As an example, one such alternative is to choose the b's so that the first element in the vector becomes +1 each time.

1. In a certain moments problem in structural analysis, the equations

$$\begin{pmatrix} 3 & 1 & 2 \cdot 5 \\ 1 & 5 & 0 \cdot 4 \\ -40 & 50 & -204 \cdot 8 \end{pmatrix} \begin{pmatrix} \theta_b \\ \theta_c \\ \theta_a \end{pmatrix} = \begin{pmatrix} 0 \\ 0 \\ -600 \end{pmatrix}$$

occurred. Solve these equations for θ_a.

2.

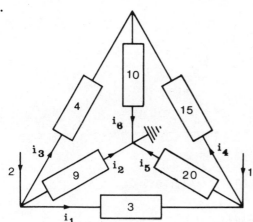

The following set of equations give the currents for the network shown:

$$i_1 + i_2 + i_3 = 2$$
$$i_3 + i_4 - i_6 = 0$$
$$-i_1 + i_4 + i_5 = 1$$
$$3i_1 - 9i_2 + 20i_5 = 0$$
$$15i_4 - 20i_5 + 10i_6 = 0$$
$$9i_2 - 4i_3 - 10i_6 = 0$$

All resistances are in ohms and all currents in amperes. Solve these equations for the currents by the Gauss–Jordan method, with partial pivoting, working to 4 decimal places.

3. The equation $\tanh x + \tan x = 0$ occurs in vibration theory. Find the smallest positive root.

4. Solve the equations

$$10x + 3y - 2z + 3w = 75, \qquad 2x - 10y + 25z + 5w = 60,$$
$$x + 40y + 5z - 8w = 50, \qquad 3x + 2y + 5z + 50w = 70$$

5. Find the value of x for which $x^4 - 0 \cdot 41x^3 + 1 \cdot 632x^2 - 9 \cdot 146x + 7 \cdot 260$ has a turning point. Then, if you read the appendix on Equal and Nearly Equal Roots, find the roots of the equation $x^4 - 0 \cdot 41x^3 + 1 \cdot 632x^2 - 9 \cdot 146x + 7 \cdot 260 = 0$.

6. Find all the eigenvalues and eigenvectors of $\begin{pmatrix} 1 & 0 & 2 \\ 0 & 1 & 3 \\ 1 & 1 & 1 \end{pmatrix}$.

7. p_1 is calculated from the formula $p_1 v_1^{1 \cdot 4} = p_2 v_2^{1 \cdot 4}$. Find the maximum relative error in p_1 if those in v_1, v_2 and p_2 are $0 \cdot 1$, $0 \cdot 15$ and $0 \cdot 2\%$ respectively.

8. The Gauss-Seidel iterative process can sometimes be adapted to solve non-linear equations. By rearranging the equations

$$x - 0 \cdot 1y^2 + 0 \cdot 06z^2 = 1 \cdot 2$$
$$y + 0 \cdot 3x^2 - 0 \cdot 01xz = 0 \cdot 6$$
$$z + 0 \cdot 2y^2 + 0 \cdot 3xy = 0 \cdot 8$$

so that only x, y and z respectively appear on the left hand sides, find the values of these variables to 3 decimal places.

9. Find, to 3 decimal places, the value of K for which $y = Ke^{x/10} - \ln x$ has the x-axis as a tangent.

10. Use the Choleski factorisation process to find the inverse of

$$\begin{pmatrix} 1 & 2 & 1 & 1 \\ 2 & 5 & 0 & -2 \\ 1 & 0 & 2 & 8 \\ 1 & -2 & 8 & 30 \end{pmatrix}$$

ANSWERS

1. $3 \cdot 62$.

2. $-0 \cdot 130$, $1 \cdot 394$, $0 \cdot 737$, $0 \cdot 223$, $0 \cdot 647$, $0 \cdot 960$.

3. $2 \cdot 365$.

4. $7 \cdot 4097$, $0 \cdot 9517$, $2 \cdot 0453$, $0 \cdot 7128$.

5. $1 \cdot 205$; $1 \cdot 210$, $1 \cdot 200$.

6. $3 \cdot 236\{0 \cdot 667 \quad 1 \quad 0 \cdot 745\}$; $1\{1 \quad -1 \quad 0\}$; $-1 \cdot 236\{0 \cdot 667 \quad 1 \quad -0 \cdot 745\}$.

7. $0 \cdot 55$ (Relative error in $x^{1 \cdot 4}$ is approximately $1 \cdot 4$ times relative error in x.)

8. $1 \cdot 172$, $0 \cdot 196$, $0 \cdot 723$.

9. $0 \cdot 984$.

10.
$$\begin{pmatrix} 0 \cdot 700 & -0 \cdot 500 & 1 \cdot 850 & -0 \cdot 550 \\ -0 \cdot 500 & 0 \cdot 500 & -0 \cdot 750 & 0 \cdot 250 \\ 1 \cdot 850 & -0 \cdot 750 & -0 \cdot 325 & -0 \cdot 025 \\ -0 \cdot 550 & 0 \cdot 250 & -0 \cdot 025 & 0 \cdot 075 \end{pmatrix}$$

UNIT 2
FINITE DIFFERENCES and
THEIR APPLICATIONS

This Unit comprises five programmes:

- (a) Least Squares
- (b) Finite Differences
- (c) Interpolation
- (d) Numerical Differentiation
- (e) Numerical Integration

Before reading these programmes, it is necessary that you are familiar with the following

Prerequisites

For (a) Maxima and minima.   Partial differentiation.   Solution of linear simultaneous equations by Gauss–Jordan as in Programme (c) of Unit 1.

For (b) Definition of a derivative.

For (c) The contents of (b).

For (d) The contents of (c).

For (e) Newton–Gregory and Everett interpolation formulae as in (c). Estimation of effect of round-off errors in calculations (as in FRAMES 16 – 32, pages 7 – 17).

Least Squares

Introduction

If you and your friends each measure the length of a line, it is quite likely that you will not all get exactly the same result.   For example, using a metre rule, the following figures for a particular length might be obtained by 10 different people: 5·127, 5·130, 5·125, 5·128, 5·126, 5·128, 5·129, 5·130, 5·126, 5·127 metres.   In these circumstances, what is the best value to take for the length of the line?   The answer you would probably give to this question is "The average of the 10 values which is 5·1276 m".   However, you would not be dogmatic about it and say that the length of the line is exactly 5·1276 m.

We are now going to look at this problem from a slightly different viewpoint.   This may seem to complicate the situation somewhat, but the ideas involved are important where the problem is not quite so simple.

Suppose the various measurements obtained are denoted by x_1, x_2, ... x_{10}, and the best estimate of the length is x.   The DEVIATIONS of the various readings from x are $x_1 - x$, $x_2 - x$, $x_3 - x$, etc.   Let S be the sum of the squares of these deviations, then

$$S = (x_1 - x)^2 + (x_2 - x)^2 + ... (x_{10} - x)^2 \quad (2.1)$$

As S is the sum of a number of squares, it can only be positive or zero, and it is extremely unlikely that it is zero.   It certainly cannot be zero in the example that has been taken here.   If x is chosen poorly, it is possible to make S quite large.   For example, if x is taken as 5·1, then S = 0·007 644 and, if as 5, then S = 0·162 844.   As a poor value of x makes S large, suppose x is chosen so that S becomes as small as possible.

What value of x will make S, as given in (2.1), as small as possible?
****

For S to be least, we must have $\frac{dS}{dx} = 0$.   This leads to the value of x as being

$$(x_1 + x_2 + ... + x_{10})/10$$

which, of course, is the average value, usually denoted by $\bar{x}$. $\frac{d^2S}{dx^2} = 20$ and this confirms that S is a minimum.

For this x, S = 0·000 026 4, and by comparison with this value, 0·007 644 and 0·162 844 are certainly large, however small they might have seemed to be when you were reading the last frame.

The process of making S as small as possible is called the METHOD OF LEAST SQUARES.   As you have seen, when only one variable is involved, it leads to the mean value.   If the number of x's is n instead of 10, then

$$S = (x_1 - x)^2 + (x_2 - x)^2 + \ldots + (x_n - x)^2$$

and putting $\dfrac{dS}{dx} = 0$ leads to

$$\bar{x} = (x_1 + x_2 + x_3 + \ldots\ldots\ldots\ldots + x_n)/n$$

Alternatively, the Σ notation can be used and then

$$S = \sum_{i=1}^{n} (x_i - x)^2 \qquad \text{and} \qquad \bar{x} = \frac{1}{n} \sum_{i=1}^{n} x_i$$

Can you recall another place in this book where we have taken the smallness of the sum of a number of squares as a criterion in deciding which of two solutions to a problem is better?

****

3A

In the solution of linear simultaneous equations. See FRAME 18, page 91.

Fitting the 'Best' Straight Line to a Set of Points

There are many cases in practice where, if two variables are involved, they are connected by a relation of the form $y = a + bx$. One of the best known is the elongation y produced in a wire when it is subjected to a load x. This, of course, is only of the form $y = a + bx$ provided that the elastic limit of the wire has not been reached. Another example is the length of a rod which is heated to various temperatures.

If the constants a and b are known in a particular case, there is no problem. However, sometimes the only information that you may be given, or can obtain by experiment, is a number of corresponding values of x and y. If this happens, and there is justification for the assumption that the law connecting x and y is linear, the question arises as to what values should be given to a and b to get the straight line best fitting the points (x, y) when these are plotted on a graph.

You might be wondering how a knowledge of this line can help us. Firstly, taking the load-extension situation, it can be used to estimate the length of the wire for any other load, provided that load is within the range of values used in the original experiment. Secondly it can be used for calibration purposes.

One way in which this 'best' straight line can be found is to guess its position by means of a taut thread or transparent ruler placed over a plot of the points. But, just as measurements of the length of a line will vary when made by different people, so also will guesses as to the best straight line through a number of points. It is therefore desirable to place the definition of the best straight line on to a more mathematical basis.

A criterion that is often used for the best straight line through a number of points is based on the idea of minimising the sum of a number of squares. It will be assumed that the x measurements are correct and only the y

FRAME 5   (continued)

measurements are subject to experimental error.   Thus it will be assumed
that the load hung on the wire or the temperature of the rod are known
exactly and only the measurements of the lengths are subject to error.   In
practice this will not be quite the case but it is hoped that any errors in
the load and temperature are very small (so that they can be ignored) in
comparison with those in the lengths.

FRAME 6

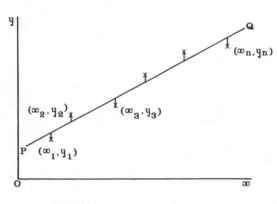

Suppose the points shown have
been obtained, the x values
being assumed correct, and that
a line PQ has been drawn
between them.   The points may
lie more nearly on the line
than is apparent from the
diagram but if they are shown
very close to the line it is
difficult to see what is
happening.   In any case, the
appearance is only relative as
it can be considerably altered
simply by changing the scale on, say, the y axis.

Now, a small deviation d from the line can be associated with each point.
These deviations are shown by the short vertical lines.   Each d can be
defined as

observed value of y at the point - value of y as given by the line for
the same x.

The actual length shown in the figure for each point is $|d|$.   It will be
obvious from the figure that some d's are positive and others negative.
If the equation of the line is y = a + bx then, for the observed point
(x_1, y_1), $d_1 = y_1 - (a + bx_1)$ and similarly for all the other points.

FRAME 7

Now, in FRAME 5, it was suggested that a criterion for the best straight
line is based on the minimising of the sum of a number of squares and here
it is Σd_i^2 that is used for this purpose.   You might, however, think that
something simpler than this could be used.   The simplest sum, you would
probably say, is Σd_i and wonder whether the use of this could lead
anywhere.   The trouble with this is that, as was previously remarked,
some d's are positive, while others are negative.   The tendency, then,
would be for the various d's to cancel each other out in Σd_i.   This
might suggest that we should try to make Σd_i zero.   If this is done, it
is found that the only information obtained is that the line must pass
through the centroid of the points, and this is insufficient to fix it
completely.

Taking the squares of the deviations eliminates the effect of positive and
negative d's.   Another way in which this could be achieved would be to

take $\Sigma|d_i|$.   The trouble with this, however, is that it is rather more difficult to deal with the mathematics of a process where moduli are involved.

As these two possibilities both have snags, it seems reasonable to take what you will probably agree is the next simplest sum, i.e. Σd_i^2, and see whether this will lead to any definite conclusion.

Denoting Σd_i^2 by S, we then have

$$S = (y_1 - a - bx_1)^2 + (y_2 - a - bx_2)^2 + \ldots + (y_n - a - bx_n)^2 \quad (7.1)$$

$$\text{or} \quad S = \sum_{i=1}^{n} (y_i - a - bx_i)^2 \quad (7.2)$$

In this expression all the x's and y's are fixed, being the coordinates of given points.   a and b are regarded as the variables as it is by changing these that the position of the line is varied.   Thus, if b is altered the slope of the line changes so that it effectively rotates, and if a is altered the line moves bodily upwards or downwards.

Now if u is a function of one variable t, it is necessary for $\frac{du}{dt}$ to be zero if u is to have a minimum value.   The corresponding requirement for S, a function of the two variables a and b, to have a minimum value is that $\frac{\partial S}{\partial a}$ and $\frac{\partial S}{\partial b}$ should be zero simultaneously.   Theoretically this will not distinguish between a maximum and a minimum just as $\frac{du}{dt} = 0$ doesn't either, but in practice this doesn't matter here as it is very easy to make Σd_i^2 as large as you like.

What equations will result from either (7.1) or (7.2) when you put $\frac{\partial S}{\partial a} = 0$ and $\frac{\partial S}{\partial b} = 0$?

****

7A

i)   Using (7.1) these give

$$-2(y_1 - a - bx_1) - 2(y_2 - a - bx_2) - \ldots -2(y_n - a - bx_n) = 0$$
$$\text{and } -2x_1(y_1 - a - bx_1) - 2x_2(y_2 - a - bx_2) - \ldots -2x_n(y_n - a - bx_n) = 0$$

i.e., $(y_1 - a - bx_1) + (y_2 - a - bx_2) + \ldots + (y_n - a - bx_n) = 0 \quad (7A.1)$
and $x_1(y_1 - a - bx_1) + x_2(y_2 - a - bx_2) + \ldots + x_n(y_n - a - bx_n) = 0$

ii)   Using (7.2), these give

$$\sum_{i=1}^{n} \{(-2)(y_i - a - bx_i)\} = 0 \quad \text{and} \quad \sum_{i=1}^{n} \{(-2x_i)(y_i - a - bx_i)\} = 0$$

i.e., $\sum_{i=1}^{n} (y_i - a - bx_i) = 0 \quad (7A.2) \quad$ and $\quad \sum_{i=1}^{n} x_i(y_i - a - bx_i) = 0 \quad (7A.3)$

FRAME 8

As you will realise, the pairs of results given in 7A are simply two slightly different ways of stating the same things.   The second of them

165

LEAST SQUARES

is more compact and then:

From (7A.2), $\sum\limits_{i=1}^{n} y_i - na - b \sum\limits_{i=1}^{n} x_i = 0$ or $na + b \sum\limits_{i=1}^{n} x_i = \sum\limits_{i=1}^{n} y_i$ (8.1)

From (7A.3), $\sum\limits_{i=1}^{n} x_i y_i - a \sum\limits_{i=1}^{n} x_i - b \sum\limits_{i=1}^{n} x_i^2 = 0$

or $a \sum\limits_{i=1}^{n} x_i + b \sum\limits_{i=1}^{n} x_i^2 = \sum\limits_{i=1}^{n} x_i y_i$ (8.2)

If you had difficulty in seeing where na comes from in (8.1), have a look back at (7A.1) which is the expanded form of (7A.2).   In it there are n brackets, each containing a.

(8.1) and (8.2) are two simultaneous equations for a and b.   They are called the NORMAL EQUATIONS.

If (8.1.) is divided by n it gives information about one of the points through which the line y = a + bx passes.   Can you spot what this is?

****

8A

$a + b\left(\dfrac{1}{n} \sum\limits_{i=1}^{n} x_i\right) = \dfrac{1}{n} \sum\limits_{i=1}^{n} y_i$

$\therefore$ $a + b\bar{x} = \bar{y}$ as $\dfrac{1}{n} \sum\limits_{i=1}^{n} x_i$ is the mean of the x values, i.e., $\bar{x}$ and

similarly $\dfrac{1}{n} \sum\limits_{i=1}^{n} y_i = \bar{y}$. The straight line found in this way therefore

passes through $(\bar{x}, \bar{y})$, i.e., through the centroid of the observed points.

As an example, let us take the case of a rod that is heated to various temperatures, its length being measured at intervals of 10°C from 10°C to 70°C.   The following table shows the results obtained, T, assumed free from error, being the temperature in °C and ℓ the length in mm:

T	10	20	30	40	50	60	70
ℓ	962·3	962·5	962·6	962·9	963·0	963·2	963·4

It is known that the law connecting T and ℓ is linear.   If it is of the form $\ell = a + bT$,   what are the best values for a and b?

What will the normal equations be for this example?   Give them in the forms corresponding to (8.1) and (8.2), i.e., without substituting the numerical values.

****

9A

$na + b \sum\limits_{i=1}^{n} T_i = \sum\limits_{i=1}^{n} \ell_i$, $a \sum\limits_{i=1}^{n} T_i + b \sum\limits_{i=1}^{n} T_i^2 = \sum\limits_{i=1}^{n} T_i \ell_i$

Here n = 7, ΣT = 280, ΣT^2 = 14 000, $\Sigma \ell$ = 6739·9 and $\Sigma T\ell$ = 269 647.

(The abbreviated notation ΣT for $\sum\limits_{i=1}^{n} T_i$ has now been used.)

The normal equations are thus

$$7a + 280b = 6739\cdot9 \qquad 280a + 14\,000b = 269\,647$$

from which a = 962·1, b = 0·018 214, and so, to three significant
figures, ℓ = 962 + 0·0182T   (10.1)

There are two points of interest to notice about this example. The first
concerns the arithmetic involved.

If a calculating machine is available, all or some of the sums ΣT^2, $\Sigma T\ell$,
ΣT and $\Sigma \ell$ can be formed simultaneously on the machine without writing down
any intermediate figures. If such a machine is not available, it is
necessary to record the individual values of T^2 and $T\ell$. In this case,
it is best to extend the table given in FRAME 9 as follows:

T	ℓ	T^2	$T\ell$
10	962·3	100	9623
20	962·5	400	19 250
30	962·6	900	28 878
40	962·9	1600	38 516
50	963·0	2500	48 150
60	963·2	3600	57 792
70	963·4	4900	67 438
280	6739·9	14 000	269 647

The second point concerns the Physics of the problem. If you have studied
this subject, you will know that the equation for the linear expansion of a
rod is usually given in the form

$$\ell = \ell_0(1 + \alpha T)$$

ℓ_0 being the length of the rod at zero temperature and α the coefficient
of expansion. Putting equation (10.1) in this form gives

$$\ell = 962(1 + 0\cdot000\,0189T)$$

Comparing this value of α with a table of values of α for different
materials suggests that the rod could possibly have been made of brass.

Now try the following example:

In a wind tunnel test, a set of helicoidal propellers of different pitch
ratios gave, at constant thrust, torque ratios as in the following table:

167

LEAST SQUARES

Pitch ratio	0·3	0·5	0·7	0·9	1·2	1·5	1·8	2·2	2·6
Torque ratio	0·316	0·533	0·753	0·979	1·310	1·650	1·980	2·436	2·875

Find the best line of the form $y = a + bx$ to fit these values, where x
denotes the pitch ratio (assumed free from error) and y the torque ratio.
Compare the values given for y by the formula you obtain with those in the
table.   ***************************************

13A

$\Sigma x = 11 \cdot 7$, $\Sigma x^2 = 20 \cdot 17$, $\Sigma y = 12 \cdot 832$, $\Sigma xy = 22 \cdot 2147$

Normal equations are

$$9a + 11 \cdot 7b = 12 \cdot 832 \qquad 11 \cdot 7a + 20 \cdot 17b = 22 \cdot 2147$$

giving

$$a = -0 \cdot 024\,43, \qquad b = 1 \cdot 115\,54, \qquad y = -0 \cdot 0244 + 1 \cdot 116x$$

The values of y, as given by this equation, are

$$0 \cdot 310, \quad 0 \cdot 534, \quad 0 \cdot 757, \quad 0 \cdot 980, \quad 1 \cdot 315, \quad 1 \cdot 650, \quad 1 \cdot 984, \quad 2 \cdot 431, \quad 2 \cdot 877.$$

When calculating the equation of the best straight line, it is advisable,
within reason, not to round off individual calculations as these are
performed.   (An exception to this rule however is in division, where you
can't usually help it.)   Rather, it is better to wait right till the end
and only round off the final values.   Obviously, errors have occurred in
the measurements of y and these are going to affect the result.   As you
saw in the first programme in Unit 1, rounding off can seriously affect the
results of certain arithmetical operations and it is better not to
introduce the possibility of this happening, thus making your result even
more inaccurate.

Extension of the Least Squares Process to Laws reducible to a Linear Form

Although some physical laws are of the straight line type, there are many
that are not.   In such cases the best straight line calculation can be
performed but the result of doing so can be an extremely bad fit.   As an
example take the points (-2,4), (-1,1), (0,0), (1,1), (2,4), (3,9) and
find the 'best' straight line passing through them.   Then plot the points
on a graph and also the line you have found.
          ****

15A

$$\Sigma x = 3, \quad \Sigma x^2 = 19,$$
$$\Sigma y = 19, \quad \Sigma xy = 27$$

Normal equations are

$$6a + 3b = 19$$
$$3a + 19b = 27$$

giving $a = 8/3, \quad b = 1$
$$y = \frac{8}{3} + x$$

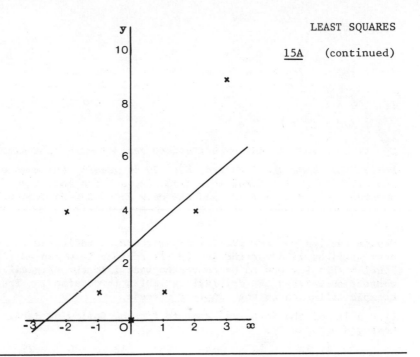

The points given in FRAME 15 lie exactly on the parabola $y = x^2$.   It is obvious from your graph that the line you obtained is by no means a good representation of the plotted points.   Fortunately, however, in practical cases, theory often provides us with a clue as to the type of curve on which a series of experimental points should lie.   For example, if a gas is compressed isothermally, theory states that the pressure should be inversely proportional to the volume.   On the other hand, if it is compressed adiabatically, the pressure and volume should be connected by a law of the form $pv^{\gamma} = c$.

By making a suitable change of one or both variables it is sometimes possible to reduce the law to one which is of the straight line form.   For example, if $pv = c$, then $p = \dfrac{c}{v}$ and if the substitution $\dfrac{1}{v} = V$ is made, $p = cV$ and the graph of p against V will be linear.   In the case of $pv^{\gamma} = c$, $\log p + \gamma \log v = \log c$ and if you put $\log p = P$ and $\log v = V$,   then $P + \gamma V = \log c$ and this is again linear.

Suggest alternative forms and/or suitable substitutions which would make the following laws linear:

i)   Variables i and p: law $p = ki^{n}$
ii)   Variables x and y: law $y = ke^{cx}$

iii) Variables w and I: law $w = \dfrac{a}{I} + b$

iv)   Variables A and d: law $A = ad^2 + bd$

$$**************************************$$

169

i) $\log p = \log k + n \log i$: $\log p = P$ and $\log i = I$

ii) $\log y = \log k + cx \log e$: $\log y = Y$ and $c \log e = C$

iii) $\frac{1}{I} = i$

iv) $A/d = ad + b$: $\frac{A}{d} = Y$

The letters used in any substitutions are, of course, a matter of choice.

Note: When logs are involved, base 10 is usually the most convenient choice. In (ii) natural logs might be used instead, then the equation becomes $\ln y = \ln k + cx$ *and the only substitution necessary is* $\ln y = Y$.

Having carried out the preliminary algebra as indicated in FRAME 17, the best straight line for the law in the revised form can easily be found. Finally this law has to be converted back into the original form, so that a connection between the original variables is obtained. The following example illustrates the complete process.

Find a law of the form $pv^\gamma = c$ to fit the following data. Assume v is measured accurately.

v	10	12	14	16	18	20	22	24
p	39·7	30·9	24·9	20·5	17·5	15·1	13·1	11·6

As was seen in FRAME 17, the linear law corresponding to $pv^\gamma = c$ is of the form $P + \gamma V = \log c$ or $P = C - \gamma V$ where $P = \log p$, $V = \log v$, $C = \log c$. The base of the logs is immaterial here but 10 is probably the most convenient.

Form the table showing corresponding values of P and V and find the best straight line law $P = C - \gamma V$ or $P = C + \beta V$ where $\beta = -\gamma$.

        ****

V	1·0000	1·0792	1·1461	1·2041	1·2553	1·3010	1·3424	1·3802
P	1·5988	1·4900	1·3962	1·3118	1·2430	1·1790	1·1173	1·0645

Here 4 figure logs have been used.

$\Sigma V = 9·7083$, $\Sigma V^2 = 11·9034$, $\Sigma P = 10·4006$, $\Sigma VP = 12·4498$

Normal equations are $8C + 9·7083\beta = 10·4006$

                    $9·7083C + 11·9034\beta = 12·4498$

from which $C = 3·008$ *and* $\beta = -1·407$. *Thus* $\gamma = 1·407$ *and* $c = 1019$ *and the law is* $pv^{1·407} = 1019$.

The actual working in this example has not been carried through to the full number of decimal places, as each V^2 and VP would be to 8 decimal places. This is a case where the "within reason" qualification of FRAME 14 would be used.

Extension of Least Squares Process to Polynomial Laws

There are some non-linear laws, such as $y = a + bx + cx^2$, for example, where the method of least squares can be applied directly. Indeed, in the case of a law such as this, it would be impossible to find an associated law of a straight line form. The reason for this is that here there are three unknown constants whereas, in the equation of a straight line, only two constants occur. Now as the normal equations are solved simultaneously to find the constants in the equation of the straight line or curve being found, it follows that in the case of the parabola $y = a + bx + cx^2$ there must be three normal equations. It is necessary then to see how these are formed.

As before, suppose n points (x_1, y_1), (x_2, y_2), ..., (x_n, y_n) have been found by experiment, all the x's being assumed free from error. To each point (x_i, y_i) there will be an associated deviation $y_i - (a + bx_i + cx_i^2)$, this being the difference between the observed value of y_i and the value of y_i as calculated from $y = a + bx + cx^2$ when $x = x_i$. The sum of the squares of the deviations, i.e.,

$$S = \sum_{i=1}^{n} (y_i - a - bx_i - cx_i^2)^2$$

is now minimised as before. However, this time S is regarded as a function of three independent variables a, b and c. The values of these will serve to fix the parabola.

For S to be a minimum when it is a function of a, b and c, it is necessary that the conditions $\frac{\partial S}{\partial a} = 0$, $\frac{\partial S}{\partial b} = 0$, $\frac{\partial S}{\partial c} = 0$ are satisfied simultaneously. Three normal equations will thus be obtained for a, b and c. Find these equations, expressing them in forms similar to (8.1) and (8.2).

****

21A

$$\frac{\partial S}{\partial a} = \sum_{i=1}^{n} \left\{ -2(y_i - a - bx_i - cx_i^2) \right\}$$

$$\frac{\partial S}{\partial b} = \sum_{i=1}^{n} \left\{ -2x_i(y_i - a - bx_i - cx_i^2) \right\}$$

$$\frac{\partial S}{\partial c} = \sum_{i=1}^{n} \left\{ -2x_i^2(y_i - a - bx_i - cx_i^2) \right\}$$

Putting each of these equal to zero leads to

$$na + b\Sigma x_i + c\Sigma x_i^2 = \Sigma y_i \qquad (21A.1)$$
$$a\Sigma x_i + b\Sigma x_i^2 + c\Sigma x_i^3 = \Sigma x_i y_i \qquad (21A.2)$$
$$a\Sigma x_i^2 + b\Sigma x_i^3 + c\Sigma x_i^4 = \Sigma x_i^2 y_i \qquad (21A.3)$$

The summation limits are here understood to be from $i = 1$ to n.

LEAST SQUARES

Unlike the straight line case, the best parabola does not pass through the centroid of the points. How can you demonstrate this from equation (21A.1)?

****

22A

Dividing by n gives $a + b\bar{x} + c(\Sigma x_i^2)/n = \bar{y}$
and this equation shows that $(\bar{x}, \bar{y})$ *does not satisfy*

$$a + bx + cx^2 = y \qquad as \qquad (\Sigma x_i^2)/n \neq \bar{x}^2$$

The following table gives, at time t, the displacement s of an object moving under the action of a constant force. What will be the best law relating s and t? Here it is to be assumed that the values of t are correct.

t	0	1	2	3	4	5
s	5	11	24	49	78	114

As you probably know, if an object moves under constant acceleration (implied by the action of a constant force), the equation of motion is of the form $s = ut + \frac{1}{2}ft^2$ if $s = 0$ when $t = 0$. In our problem, $s \neq 0$ when $t = 0$ and the correspondingly modified equation can be written as $s = a + bt + ct^2$.

What will be the three normal equations corresponding to (21A.1), (21A.2) and (21A.3)? Don't insert the numerical values yet.

****

23A

$$na + b\Sigma t_i + c\Sigma t_i^2 = \Sigma s_i, \qquad a\Sigma t_i + b\Sigma t_i^2 + c\Sigma t_i^3 = \Sigma t_i s_i$$
$$a\Sigma t_i^2 + b\Sigma t_i^3 + c\Sigma t_i^4 = \Sigma t_i^2 s_i$$

What will these equations become when the numerical values are inserted?

****

24A

$$6a + 15b + 55c = 281$$
$$15a + 55b + 225c = 1088$$
$$55a + 225b + 979c = 4646$$

These three equations now have to be solved and any appropriate method can be used. Try solving them by the Gauss–Jordan process with pivoting, working to 2 decimal places.

****

25A

$$a = 3 \cdot 98 \qquad b = 3 \cdot 49 \qquad c = 3 \cdot 72$$

The required law is thus $s = 3 \cdot 98 + 3 \cdot 49t + 3 \cdot 72t^2$.

If the normal equations for this example are written in matrix form, they are $\begin{pmatrix} 6 & 15 & 55 \\ 15 & 55 & 225 \\ 55 & 225 & 979 \end{pmatrix} \begin{pmatrix} a \\ b \\ c \end{pmatrix} = \begin{pmatrix} 281 \\ 1088 \\ 4646 \end{pmatrix}$

In the coefficient matrix, you will notice that the elements on the trailing diagonal (from bottom left to top right) are equal to each other (55). Also the elements on any line parallel to this are equal to each other (15 on one and 225 on the other). If you examine the equations (21A.1), (21A.2) and (21A.3) you will see that this must always occur.

FRAME 27

The method of extension of this technique to higher degree polynomials follows immediately.

Can you say what the normal equations would be if it is desired to fit the curve
$$y = a_0 + a_1x + a_2x^2 + a_3x^3$$
to a set of points (x_i, y_i)? If not, see if you can work them out.
****

27A

$na_0 + a_1\Sigma x_i + a_2\Sigma x_i^2 + a_3\Sigma x_i^3 = \Sigma y_i$

$a_0\Sigma x_i + a_1\Sigma x_i^2 + a_2\Sigma x_i^3 + a_3\Sigma x_i^4 = \Sigma x_i y_i$

$a_0\Sigma x_i^2 + a_1\Sigma x_i^3 + a_2\Sigma x_i^4 + a_3\Sigma x_i^5 = \Sigma x_i^2 y_i$

$a_0\Sigma x_i^3 + a_1\Sigma x_i^4 + a_2\Sigma x_i^5 + a_3\Sigma x_i^6 = \Sigma x_i^3 y_i$

FRAME 28

The pattern in these equations should now be obvious, as also should be the property mentioned in FRAME 26 for the example there. It is now very easy to see what the normal equations will be for even higher degree polynomials, and a flow diagram for these is given on page 174, taking the case of a polynomial of degree m and n(> m) data points. Unfortunately, however, there can be some difficulty with the arithmetic, including a tendency towards ill-conditioning, when the degree of the polynomial is increased to more than about four.

FRAME 29

Use of a False Origin or Coding

Sometimes, when doing an example manually, it is helpful to subtract a number from all the x readings, from all the y readings, or numbers from both sets. This is simply to reduce the amount of arithmetic involved.

Returning to the example in FRAME 9, we might decide to subtract 40 from all the T's and 963 from all the ℓ's. 40 is the mean of the T's and, when the mean is a simple figure, this is the best number to use. If $X = T - 40$ and $Y = \ell - 963$, then a table showing values of X and Y is

X	-30	-20	-10	0	10	20	30
Y	-0·7	-0·5	-0·4	-0·1	0·0	0·2	0·4

Find the best straight line through these points.
****

Flow diagram for FRAME 28.

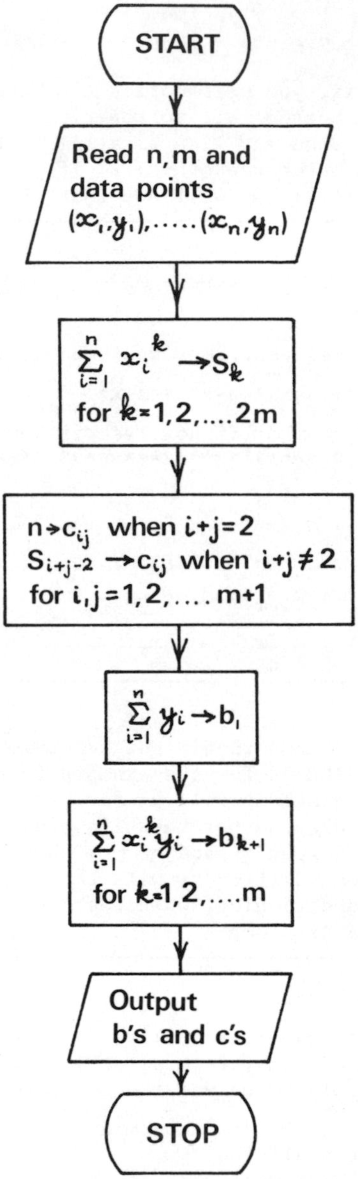

START

Read n, m and
data points
$(x_1, y_1), \ldots (x_n, y_n)$

$$\sum_{i=1}^{n} x_i^{\,k} \rightarrow S_k$$
for $k = 1, 2, \ldots 2m$

$n \rightarrow c_{ij}$ when $i+j=2$
$S_{i+j-2} \rightarrow c_{ij}$ when $i+j \neq 2$
for $i, j = 1, 2, \ldots m+1$

$$\sum_{i=1}^{n} y_i \rightarrow b_1$$

$$\sum_{i=1}^{n} x_i^{\,k} y_i \rightarrow b_{k+1}$$
for $k = 1, 2, \ldots m$

Output
b's and c's

STOP

(Programs using the method of least squares can be found in references
(3), (7) and (8).)

The normal equations for a line of the form $Y = A + BX$ *will be*

$$nA + B\Sigma X_i = \Sigma Y_i$$
$$A\Sigma X_i + B\Sigma X_i{}^2 = \Sigma X_i Y_i$$

i.e.
$$7A = -1\cdot 1$$
$$2800B = 51$$

$\therefore$ $A = -0\cdot 1571,$ $B = 0\cdot 018\ 214,$ $Y = -0\cdot 1571 + 0\cdot 018\ 214X$

FRAME 30

This equation must now be put into terms of T and ℓ.   As $X = T - 40$ and $Y = \ell - 963$,

$\ell - 963 = -0\cdot 1571 + 0\cdot 018\ 214(T - 40)$ i.e. $\ell = 962\cdot 11 + 0\cdot 018\ 214T$

or, to three significant figures

$$\ell = 962 + 0\cdot 0182T$$

You will notice the arithmetic here was much simpler.   However, when using a computer, there is really nothing to be gained.

FRAME 31

The Best Straight Line when Both Variables are subject to Error

Throughout this programme, it has been assumed that the independent variable is not subject to error, or, if it is, then the error is negligible.   Sometimes it may happen that both variables are subject to error.   If this is the case, the situation becomes somewhat more difficult and we shall not consider it here.

FRAME 32

Miscellaneous Examples

In this frame a collection of miscellaneous examples is given for you to try.   Answers are provided in FRAME 33, together with such working as is considered helpful.

1. If a gas is heated at constant volume the pressure at $t^{\circ}C$, p_t, in terms of p_0, its pressure at $0^{\circ}C$ is given by $p_t = p_0(1 + \alpha_v t)$, α_v being the coefficient of increase in pressure at constant volume. Find the best law of this form to fit the data, assuming values of t to be correct.

t	5	10	15	20	25	30	35	40
p_t	10·182	10·364	10·545	10·727	10·909	11·091	11·273	11·455

2. In statistics the best straight line as specified in FRAMES 4 – 8 is known as the line of regression of y on x.   In a similar way the line of regression of x on y is that line for which the sum of the squares of the <u>horizontal</u> distances from the points (x_i, y_i) to the line PQ is minimised.   In this case the equation of the line is assumed to be $x = a' + b'y$.   By adopting a similar process to that in FRAMES 6 – 8, find the normal equations for this regression line.

3. The following table gives index numbers of food retail prices of two
    countries for a number of years.

Year	1960	1962	1964	1966	1968	1970	1972
Country A	182	176	169	186	200	227	232
Country B	157	161	163	172	197	201	212

    Denoting readings for Country A by x and those for B by y, find the
    two regression lines.

4. The conductive heating of steel pipes at a given temperature was
    tested and gave the following readings for the power p and the
    current i:

i	25	20	15	10	5
p	0·432	0·305	0·191	0·103	0·035

    Find the best law of the form $p = ci^k$ to fit this data, assuming
    the readings of i to be correct.

5. The following table gives observations of the stopping distance of a
    vehicle travelling at speed v

v (km/h)	30	40	50	60	70
s (m)	90	138	206	292	396

    Write down the normal equations for a least-squares fit of form

    (a) $s = a + bv + cv^2$,   (b) $s = bv + cv^2$

    Select one of these formulae, giving a reason for your choice and
    hence determine the relevant parameters of the regression.   (C.E.I.)

    (Note: The parameters of the regression are a, b and c.   Readings
    of v are assumed to be free of error.)

6. In certain cases, y might be a function of two independent variables,
    x and t, say.   If the relationship between the variables is
    $y = a + bt + cx$,   find the normal equations by minimising
    $S = \Sigma(y - a - bt - cx)^2$.   Assume that x and t are without error.

7. The yield of a chemical process was measured at three temperatures,
    each with two concentrations of a particular reactant, as recorded
    below:

Temperature, $t^{\circ}C$	40	40	50	50	60	60
Concentration, x	0·2	0·4	0·2	0·4	0·2	0·4
Yield, y	38	42	41	46	46	49

    Use the method of least squares to find the best values of the
    coefficients a, b, c in the equation $y = a + bt + cx$, and from your
    equation estimate the yield at $70^{\circ}C$ with concentration 0·5.   (L.U.)

Answers to Miscellaneous Examples

1. Writing the law in the form $p_t = p_o + \beta t$ leads to $p_t = 10 + 0·036\ 37t$
    which gives $p_t = 10(1 + 0·003\ 637t)$.

    If coding is used in this example, it would be reasonable to subtract

20, say, from all the values of t and 10 or 11 from all the values of p_t. The mean of the t's, which is $22\frac{1}{2}$, is not such a simple figure to use as 20.

2. $S = \sum_{i=1}^{n} (x_i - a' - b'y_i)^2$

Normal equations are

$$na' + b'\Sigma y_i = \Sigma x_i \quad (33.1) \qquad a'\Sigma y_i + b'\Sigma y_i^2 = \Sigma y_i x_i$$

You will notice that these equations are similar to (8.1) and (8.2) but with the x's and y's interchanged. As $\Sigma y_i = n\bar{y}$ and $\Sigma x_i = n\bar{x}$, (33.1) is equivalent to $a' + b'\bar{y} = \bar{x}$ and so this regression line also passes through the centroid of the points.

3.   n = 7, $\Sigma x = 1372$ $\Sigma x^2 = 272\ 610$
$\Sigma y = 1263$, $\Sigma y^2 = 230\ 877$ $\Sigma xy = 250\ 660$

Line of regression of y on x is $y = 15 \cdot 488 + 0 \cdot 8415x$
Line of regression of x on y is $x = 8 \cdot 568 + 1 \cdot 0388y$

If coding is used one could subtract, say, 200 from x and 180 from y.   Then, if X = x - 200, Y = y - 180,

X	-18	-24	-31	-14	0	27	32
Y	-23	-19	-17	-8	17	21	32

n = 7, $\Sigma X = -28$, $\Sigma X^2 = 3810$
$\Sigma Y = 3$, $\Sigma Y^2 = 2997$ $\Sigma XY = 3100$

These lead to the same results.   For manual working these numbers are much easier to deal with.

4. $\log p = \log c + k \log i$ or $P = C + kI$

I	1·3979	1·3010	1·1761	1·0000	0·6990
P	-0·3645	-0·5157	-0·7190	-0·9872	-1·4559

$P = -2 \cdot 5481 + 1 \cdot 560\ 53I$
$p = 0 \cdot 002\ 83i^{1 \cdot 56}$, quoting to 3 significant figures.

5. The normal equations for (a) are

$$na + b\Sigma v + c\Sigma v^2 = \Sigma s$$
$$a\Sigma v + b\Sigma v^2 + c\Sigma v^3 = \Sigma vs$$
$$a\Sigma v^2 + b\Sigma v^3 + c\Sigma v^4 = \Sigma v^2 s$$

and for (b)

$$b\Sigma v^2 + c\Sigma v^3 = \Sigma vs$$
$$b\Sigma v^3 + c\Sigma v^4 = \Sigma v^2 s$$

In each case $\sum_{i=1}^{n} v_i$, etc., have been abbreviated to Σv, etc.

As s = 0 obviously corresponds to v = 0, the relationship between them should not contain a and so it would appear possible to use the simpler (b) equation.   However the point (0,0) is outside the range of the figures and the best parabola through the given points may not pass through the origin.   The calculation given here is for (a).

To simplify the normal equations, take $V = v - 50$.   The table is then

V	-20	-10	0	10	20
s	90	138	206	292	396

and, if we now assume $s = A + BV + CV^2$, the normal equations are

$$nA \quad + C\Sigma V^2 \quad = \Sigma s$$
$$B\Sigma V^2 \qquad\qquad = \Sigma Vs$$
$$A\Sigma V^2 \quad + C\Sigma V^4 \quad = \Sigma V^2 s$$

as $\Sigma V = 0$ and $\Sigma V^3 = 0$

Here $n = 5$, $\Sigma V^2 = 1000$ $\Sigma V^4 = 340\ 000$
       $\Sigma s = 1122$, $\Sigma Vs = 7660$ $\Sigma V^2 s = 237\ 400$

$\therefore$

$$5A \qquad + 1000C = 1122$$
$$1000B \qquad\qquad = 7660$$
$$1000A \qquad + 340\ 000C = 237\ 400$$

$\therefore$ $A = 205 \cdot 83$, $B = 7 \cdot 66$, $C = 0 \cdot 092\ 86$

$$s = 205 \cdot 83 + 7 \cdot 66V + 0 \cdot 092\ 86V^2$$
$$= 205 \cdot 83 + 7 \cdot 66(v - 50) + 0 \cdot 092\ 86(v - 50)^2$$
$$= 54 \cdot 98 - 1 \cdot 626v + 0 \cdot 092\ 86v^2$$

6. $\frac{\partial S}{\partial a} = 0$ yields $na + b\Sigma t + c\Sigma x = \Sigma y$

   $\frac{\partial S}{\partial b} = 0$ yields $a\Sigma t + b\Sigma t^2 + c\Sigma xt = \Sigma yt$

   $\frac{\partial S}{\partial c} = 0$ yields $a\Sigma x + b\Sigma tx + c\Sigma x^2 = \Sigma yx$

7. The calculation will be simplified if we put $T = t - \bar{t} = t - 50$ and
   $X = x - \bar{x} = x - 0 \cdot 3$.   New table is:

T	-10	-10	0	0	10	10
X	-0·1	0·1	-0·1	0·1	-0·1	0·1
y	38	42	41	46	46	49

$n = 6$, $\Sigma T = 0$, $\Sigma X = 0$, $\Sigma y = 262$, $\Sigma T^2 = 400$, $\Sigma XT = 0$,
$\Sigma X^2 = 0 \cdot 06$, $\Sigma yT = 150$, $\Sigma yX = 1 \cdot 2$.

Assuming $y = A + BT + CX$, normal equations become $6A = 262$,
$400B = 150$, $0 \cdot 06C = 1 \cdot 2$.

$\therefore$ $A = 43 \cdot 667$, $B = 0 \cdot 375$, $C = 20$

$$y = 43 \cdot 667 + 0 \cdot 375T + 20X$$
$$= 43 \cdot 667 + 0 \cdot 375(t - 50) + 20(x - 0 \cdot 3)$$
$$= 18 \cdot 917 + 0 \cdot 375t + 20x$$

When $t = 70$, $x = 0 \cdot 5$, $y = 55 \cdot 2$.

Finite Differences

FRAME 1

Introduction

At the beginning of the first programme in Unit 1 of this book, you saw
some of the difficulties that you can meet if your knowledge of mathematics
is limited to analytical methods. In the other programmes in that first
Unit we saw how some problems, for example non-linear equations,
simultaneous linear equations and certain properties of matrices, can be
tackled by numerical methods.

In the first programme in this Unit, the problem of finding an analytical
expression to fit as best as possible given numerical data was considered.
Even so the method used there was based on a knowledge of the <u>form</u> of the
law that was to be expected. The result was a straight line, or curve,
which did not usually pass through any of the given points, when these
were plotted. Having found this straight line, or curve, it could then be
used to estimate the value of the function at intermediate points and also,
as it was an analytical formula, the values of derivatives and integrals.

FRAME 2

In this and the following programmes, we shall be looking at these
problems from a different viewpoint. Now given a set of points, there are
obviously many curves which do actually pass through all of them and our
object this time will effectively be to find the simplest curve which does
just this. By the term 'simplest curve' is meant the curve with the
simplest equation. Having found this curve it can be used for estimating
the value of the function at other points, and for differentiating and
integrating. However, there is one major difference in technique between
what we shall be doing now and least squares.

When finding a least squares law, once the form of the law has been
decided, the tabulated values are used to give us that law and are then
effectively forgotten. Now the various formulae that we shall get will be
expressed in terms of the tabular values and these values (or their
differences) will be used <u>only</u> after the formulae have been obtained. If
you have met Simpson's rule for the area under a curve, you are already
familiar with one particular integration formula which works in this way.
In the next Unit, the ideas involved in integration will be extended to
differential equations.

Many of the methods used will be based on what are known as finite
differences and in this programme we shall look at these and related ideas.

FRAME 3

Finite Differences

You have already been introduced to the idea of a difference table in the
first Unit in this book. There the table shown on page 180 was given in
which each number in any column to the right of the first is formed by
subtracting the two adjacent numbers in the column immediately to its left.

Thus the number enclosed in $\bigcirc$ is obtained by subtracting the number

in $\square$ from that in $\diamondsuit$. Apart from the entries in the left hand

column [which are the values of $x^3/10$ for $x = 0(1)10$, i.e., x
increasing by steps of 1 from 0 to 10], each subsequent entry is a
FINITE DIFFERENCE.   The second column of the table is the first column of
differences, known as the FIRST DIFFERENCES, and so on.   In this
particular case, the analytical formula for the entries in the left hand
column is known.   Usually this will not be so.

0·0				
	0·1			
0·1		0·6		
	0·7		0·6	
0·8		1·2		0
	1·9		0·6	
2·7		1·8		0
	3·7		0·6	
6·4		2·4		0
	6·1		0·6	
12·5		3·0		0
	9·1		0·6	
21·6		3·6		0
	12·7		0·6	
34·3		4·2		0
	16·9		0·6	
51·2		4·8		0
	21·7		0·6	
72·9		5·4		
	27·1			
100·0				

Now form a difference table, following the same layout as above, for the
values 437 166 47 8 1 2 11 52 173 446 up to the fifth
differences.   **************************************

3A

437					
	−271				
166		152			
	−119		−72		
47		80		24	
	−39		−48		0
8		32		24	
	−7		−24		0
1		8		24	
	1		0		0
2		8		24	
	9		24		0
11		32		24	
	41		48		0
52		80		24	
	121		72		
173		152			
	273				
446					

The values in the left hand column of the table are those of $y = f(x)$ where $f(x) = x^4 - 2x^3 + 3x^2 - x + 1$ and x takes, in turn, the values $-4(1)5$. You have found, in this example, that the 4th differences are all the same (24) and the 5th (and consequently the higher) differences are all zero. Similarly, you will notice that for the cubic in the first example, the 3rd differences are constant while the 4th and higher differences are zero. In like manner, if y is an nth degree polynomial, the nth differences will all be the same and all higher differences will be zero provided that the values of y are calculated at equally spaced values of x. As the values of x are important they are usually also shown in a difference table, as follows:

x	f(x)	1st diffs	2nd diffs	3rd diffs	4th diffs	5th diffs
-4	437					
		-271				
-3	166		152			
		-119		-72		
-2	47		80		24	
		-39		-48		0
-1	8		32		24	
		-7		-24		
0	1		8			
		1				
1	2					

This is a part of the table in 3A and here the various columns have been labelled.

The Link between Differencing and Differentiation

As you know, one way in which the formula for the derivative can be written is

$$f'(x) = \lim_{h \to 0} \frac{f(x + h) - f(x)}{h}$$

and at the point x_0

$$f'(x_0) = \lim_{h \to 0} \frac{f(x_0 + h) - f(x_0)}{h}$$

If you think of the process involved when first differences are being found, you will realise that all you are doing is finding the numerator of such an expression as that above, i.e., $f(x_0 + h) - f(x_0)$ where h is the difference between consecutive values of x. Differencing is thus the beginning of the differentiation process and finds the actual change in y between consecutive values of x instead of the limit of the rate of change. Also h remains finite instead of becoming infinitesimal and so we now have the CALCULUS OF FINITE DIFFERENCES as against the infinitesimal calculus.

If $f(x) = x^n$, where n is a positive integer, then

$$f(x + h) - f(x) = (x + h)^n - x^n$$
$$= nx^{n-1}h + \text{terms in lower powers of } x \qquad (5.1)$$

Because of this, if $f(x)$ is a polynomial of degree n in x, then the first

181

differences are the values of a polynomial of degree n - 1.   Similarly the second differences are the values of a polynomial of degree n - 2. Continuing in this way the nth differences are the values of a polynomial of degree zero, i.e., are all the same.   The next and subsequent differences are thus zero.

Making use of (5.1) and the paragraph that follows it, find the values of the nth differences of

$$a_0 x^n + a_1 x^{n-1} + \ldots\ldots\ldots + a_{n-1}x + a_n \qquad (5.2)$$

in terms of n and h.   Check that the difference tables you have already met in this programme agree with your result.

****

5A

As differencing takes place n times, the contributions of all terms other than the first will be zero.

The leading term in the polynomial giving the 1st differences is $a_0 n x^{n-1}h$.   The formula giving the differences of this is $a_0 n (x + h)^{n-1}h - a_0 n x^{n-1}h$, and the leading term of this polynomial is $a_0 n(n - 1)x^{n-2}h^2$.   (Note that this result can alternatively be obtained by applying $a_0 n x^{n-1}h$ to itself, so to speak.) This is the leading term in the formula for the 2nd differences of (5.2).   Similarly the leading term in the formula for the third differences is $a_0 n(n - 1)(n - 2)x^{n-3}h^3$ and so on.   The leading term (the only term in fact) in the formula for the nth differences is

$$a_0 n(n - 1)(n - 2) \ldots 1\, x^0 h^n = a_0\, n!\, h^n$$

The fact that the nth differences of a polynomial of degree n are constant enables us to extend a difference table to find further values of the polynomial, simply by additions instead of a series of additions and multiplications.   To illustrate the process, let us find the next entry after 446 in the table in 3A.   Repeating the bottom part only of this table, we have

11		32		24	
	41		48		
52		80		24	
	121		72		
173		152			
	273				
446					

The next entry in the last column will again be 24.   Then, as this is the difference of two entries in the preceding column, the next entry in that column is the sum of 72 and 24.   The calculation then proceeds as shown on the next page.

Now find (i) the next entry after 967 in the table on the next page, and (ii) the next two entries in the list -205 -35 7 17 assuming that these are values of a polynomial of degree 3 calculated at equal intervals of x .

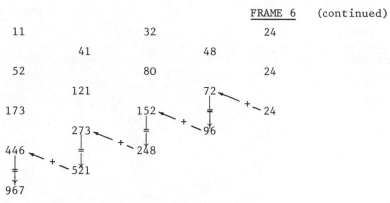

****

6A

(i) 1856;   (ii) 91, 325

$\left(\begin{array}{l}\textit{The figures }-205,\ -35,\ 7,\ 17\ \textit{are the values of}\ \ 2x^3\ -\ 4x^2\ +\ 5x\ +\ 7\\ \textit{at}\ \ x\ =\ -4(2)2.\end{array}\right.$

FRAME 7

Decimals in a Difference Table

Construct a difference table for $f(x) = 2x^3 - 4x^2 + 5x + 7$, $x = 0(0 \cdot 1)0 \cdot 6$.

****

7A

x	$f(x)$	1st diffs	2nd diffs	3rd diffs
0	*7·000*			
		0·462		
0·1	*7·462*		*−0·068*	
		0·394		*0·012**
0·2	*7·856*		*−0·056*	
		0·338		*0·012*
0·3	*8·194*		*−0·044*	
		0·294		*0·012*
0·4	*8·488*		*−0·032*	
		0·262		*0·012*
0·5	*8·750*		*−0·020*	
		0·242		
0·6	*8·992*			

As f(x) is a polynomial of degree 3, the third differences are constant. The portion above the dashed line must be found in the usual way.   That below the line can be found by the method of FRAME 6.   The figure starred should be checked against the formula $a_0 n! h^n$ before using it for further calculations.

FRAME 8

When a difference table contains decimals, the difference columns themselves are often written as whole numbers.   If you do this, you must

FRAME 8 (continued)

remember that these numbers are actually decimals and must be treated as such in any subsequent calculation with them. Using this idea, a slightly extended version of the table in FRAME 7 would appear as:

0	7•000			
		462		
0•1	7•462		-68	
		394		12
0•2	7•856		-56	
		338		12
0•3	8•194		-44	
		294		12
0•4	8•488		-32	
		262		12
0•5	8•750		-20	
		242		12
0•6	8•992		-8	
		234		12
0•7	9•226		4	
		238		
0•8	9•464			

FRAME 9

The Build-up of Errors in a Difference Table due to Errors in the Functional Values

Replace each of the values of f(x) in the table in FRAME 8 by that obtained after rounding to 2 decimal places. Then form the difference table for these rounded values, going as far as the 7th differences.

****

9A

x	*f(x)*	*1st diffs*	*2nd diffs*	*3rd diffs*	*4th diffs*	*5th diffs*	*6th diffs*	*7th diffs*
0	7•00							
		46						
0•1	7•46		-6					
		40		-1				
0•2	7•86		-7		5			
		33		4		-10		
0•3	8•19		-3		-5		18	
		30		-1		8		-29
0•4	8•49		-4		3		-11	
		26		2		-3		11
0•5	8•75		-2		0		0	
		24		2		-3		
0•6	8•99		0		-3			
		24		-1				
0•7	9•23		-1					
		23						
0•8	9•46							

The effect of rounding off the functional values to 2 decimal places is to introduce slight errors into them. Examine the difference table you have just formed and see how these errors have affected it. For a true cubic (as in FRAME 8) the first differences are relatively quite large, the second differences are smaller, the third differences are all the same and the fourth and higher differences are zero. But in 9A, although the 1st and 2nd differences behave reasonably well, the 3rd differences are certainly not constant and the rest are certainly not all zero. They do, in fact, start to increase and before long get quite large. Small errors are thus seen to mask completely the true values of the higher differences and consequently the columns on the right of such a difference table are extremely unreliable. Generally speaking, one would not use columns in a table like this after the magnitudes of the entries start increasing.

FRAME 11

All the difference tables so far constructed have been for polynomials. They can equally well be formed for other functions which may or, as is very often the case in practice, may not be known analytically. Construct the difference table as far as the 7th differences for sin x, $x = 0(0 \cdot 05)0 \cdot 5$, working to 4 decimal places. Again notice how the higher differences misbehave (i.e., increase in magnitude), due to presence of round-off errors in the functional values.

****

| | | | | *Differences* | | | | 11A |
x	$\sin x$	1st	2nd	3rd	4th	5th	6th	7th
0	0							
		500						
0·05	0·0500		−2					
		498		0				
0·10	0·0998		−2		−1			
		496		−1		−2		
0·15	0·1494		−3		−3		12	
		493		−4		10		−37
0·20	0·1987		−7		7		−25	
		486		3		−15		55
0·25	0·2473		−4		−8		30	
		482		−5		15		−57
0·30	0·2955		−9		7		−27	
		473		2		−12		45
0·35	0·3428		−7		−5		18	
		466		−3		6		
0·40	0·3894		−10		1			
		456		−2				
0·45	0·4350		−12					
		444						
0·50	0·4794							

FRAME 12

To examine further how small errors can affect differences let us make up a difference table of errors from functional values, just one of the functional values having an error of amount ε in it. As all other

185

readings are to be assumed correct, the error in each of these will be zero.

If the functional values are f_0, f_1, f_2, f_3 etc., say, and just one of them is in error, then the errors in the readings can be listed

0, 0, 0, 0, 0, 0, ε , 0, 0, 0, 0, 0, 0 ε being the one error.

Form a difference table of these errors, placing the errors in the functional values column and going as far as the 6th differences.
****

12A

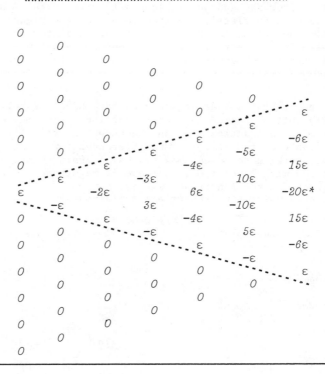

FRAME 13

From the table, you will see that

i)   the effect of a single error spreads out fanwise from that error, shown by the dashed lines.

ii)   each column of differences has one more entry affected by the error than the preceding column.

iii) individual readings in the column are affected by different amounts, those in the centre of the fan being affected more than those near the edge.   For example, the magnitude of the effect on the entry starred is 20 times the original error.

iv)   the multiples of ε in the nth difference column are the coefficients in the binomial expansion of $(1 - a)^n$.

It doesn't take much imagination to sense how much a difference table is affected when there are several readings containing errors.

If the only errors involved in a table are round-off errors, the worst situation that can arise, as far as a difference table is concerned, is when each reading is in error by $\pm\frac{1}{2}$ in the last decimal place, the signs alternating from term to term. Such an extreme situation is very unlikely to occur in practice, but in order to see the worst possible effect, form a difference table for the values $\frac{1}{2}$, $-\frac{1}{2}$, $\frac{1}{2}$, $-\frac{1}{2}$, $\frac{1}{2}$, $-\frac{1}{2}$, $\frac{1}{2}$, $-\frac{1}{2}$, $\frac{1}{2}$, $-\frac{1}{2}$, $\frac{1}{2}$ going as far as the sixth differences.

****

value	Δ	Δ²	Δ³	Δ⁴	Δ⁵	Δ⁶
$\frac{1}{2}$						
	-1					
$-\frac{1}{2}$		2				
	1		-4			
$\frac{1}{2}$		-2		8		
	-1		4		-16	
$-\frac{1}{2}$		2		-8		32
	1		-4		16	
$\frac{1}{2}$		-2		8		-32
	-1		4		-16	
$-\frac{1}{2}$		2		-8		32
	1		-4		16	
$\frac{1}{2}$		-2		8		-32
	-1		4		-16	
$-\frac{1}{2}$		2		-8		32
	1		-4		16	
$\frac{1}{2}$		-2		8		
	-1		4			
$-\frac{1}{2}$		2				
	1					
$\frac{1}{2}$						

You will see that, with this set-up, the error in, for example, the 6th differences is 64 times the error in each individual reading and is of magnitude 32 in the last decimal place. Generally, in the nth difference column, the magnitude of the error is 2^{n-1}.

So far, we have mainly concentrated on the effect of round-off errors in a difference table, these being unavoidable and a great nuisance. But other errors can occur and the build-up of these can be of assistance in that this build-up enables us to spot where such errors have occurred.

The entries in the left hand column of the following table (on page 188) have allegedly been formed from a cubic. If all entries are perfectly correct, then, as you know, there should be a constant column of differences and the presence of the figures 46, -114, 126, -34 in a column of sixes means that something has gone wrong here. Now the effect of an error spreads out fanwise from the reading that is in error and the construction of a fan backwards from the group 46 to -34 points to 2995 as being the guilty party.

Can you suggest what this reading should be and a probable cause of the error? Having changed this value accordingly, reconstruct the difference table and check it for reasonableness.

1111			
	353		
1464		68	
	421		6
1885		74	
	495		46
2380		120	
	615		−114
2995		6	
	621		126
3616		132	
	753		−34
4369		98	
	851		6
5220		104	
	955		6
6175		110	
	1065		6
7240		116	
	1181		
8421			

(dashed lines connect 2380 → 46 region and 2995 → 132 region)

****

15A

Changing 46 into 6 and working backwards replaces 2995 by 2955. This suggests that the wrong digit was repeated when copying. The revised difference table is as shown.

1111			
	353		
1464		68	
	421		6
1885		74	
	495		6
2380		80	
	575		6
2955		86	
	661		6
3616		92	
	753		6
4369		98	
	851		6
5220		104	
	955		6
6175		110	
	1065		
7240			

Some sets of figures may have more than one error in them.

Form a difference table for the figures   −48, −75, −132, −147, 0, 447, 1550 3333, 6288, 10 725, 17 052, 25 725, 37 248, 52 173   and suggest which figures are in error.

****

f	Δ	Δ^2	Δ^3	Δ^4
-48				
	-27			
-75		-30		
	-57		72	
-132		42		48
	-15		120	
-147		162		18
	147		138	
0		300		218
	447		356	
447		656		-332
	1103		24	
1550		680		468
	1783		492	
3333		1172		-182
	2955		310	
6288		1482		98
	4437		408	
10 725		1890		48
	6327		456	
17 052		2346		48
	8673		504	
25 725		2850		48
	11 523		552	
37 248		3402		
	14 925			
52 173				

The fan suggests that 447 and 1550 are both in error.

FRAME 17

Now suggest possible corrections. Then construct a fresh difference table to see if an improvement in the table results.

****

17A

The 48's in the last column give the clue.

447 should be 477 and 1550 should be 1500. Your difference table should then have a constant column of 48.

FRAME 18

A more awkward situation arises when a table contains round-off errors and one or more blunders into the bargain. Construct a difference table for the following set of values. Then examine it and see if you can make any suggestions as to a possible unintentional error.

0·0000, 0·0500, 0·1002, 0·1506, 0·2013, 0·2526, 0·3054, 0·3572, 0·4108, 0·4653, 0·5211.

****

The pattern of differences suggests an error in 0·3054. No difference column gives an exact amendment to be made to this figure, as, due to the presence of round-off, no one column follows a set, well-defined sequence. The sudden jump from 6 to 15 suggests an error of about 8 or 9 in the last decimal place. A difference of 9 would be obtained from a reversal of the last two digits. This would replace 0·3054 by 0·3045 and if this is done, the table becomes much smoother.

0·0000		
	500	
0·0500		2
	502	
0·1002		2
	504	
0·1506		3
	507	
0·2013		6
	513	
0·2526		15
	528	
0·3054		-10
	518	
0·3572		18
	536	
0·4108		9
	545	
0·4653		13
	558	
0·5211		

FRAME 19

Now try this example:

The following figures are rounded to five places of decimals and also contain two copying errors. Locate these two errors and suggest possible corrections.

0·000 00, 0·087 49, 0·167 33, 0·267 95, 0·363 97, 0·466 31, 0·577 35, 0·700 21, 0·839 10, 1·100 00, 1·191 75, 1·428 15, 1·732 05.

****

19A

A difference table locates the errors at 0·167 33, 1·100 00.

Changes in these to 0·176 33 and 1·000 00 produce a much more likely difference table.

FRAME 20

Finite Difference Notations

Due to the use made of the differential coefficient in various formulae, it is necessary to use a symbol to denote it. Similarly, a notation is necessary when formulae involving finite differences are used. But, as you might suspect, mathematicians are not content with just one notation. This is not due to awkwardness on their part, though, as it is found that while one notation is best for one particular problem, another is better in other circumstances.

The essential feature of any notation is that it should enable us to locate immediately each and every entry in a difference table. A start is made by choosing, at will, one entry in the x column and labelling it x_0. If h is the difference between consecutive values of x quoted in the table, then the other values of x will be

FRAME 20   (continued)

..., $x_0 - 3h$, $x_0 - 2h$, $x_0 - h$ before x_0 and $x_0 + h$, $x_0 + 2h$, $x_0 + 3h$,...,
after it.

The complete set in the table will then be

..., $x_0 - 3h$, $x_0 - 2h$, $x_0 - h$, x_0, $x_0 + h$, $x_0 + 2h$, $x_0 + 3h$, ...

For brevity, these values are usually denoted by

...., x_{-3}, x_{-2}, x_{-1}, x_0, x_1, x_2, x_3, ...

and so x_p is used for $x_0 + ph$.

Although the only values which will actually appear in a table are those
for which p is an integer, it is very valuable, as you will see later, to
extend this notation to other values of p.   Then, in general,

$$x_p = x_0 + ph$$

where p can take any value.

FRAME 21

In a similar way, the functional values are denoted by

....., f_{-3}, f_{-2}, f_{-1}, f_0, f_1, f_2, f_3, ...

f_0 being that value which corresponds to x_0, etc.   y or any other
convenient letter may, of course, be used instead of f.   Again, only the
values just listed will actually appear in a difference table, but the
notation is extended so that f_p indicates the value of the function when
$x = x_0 + ph$,   whatever the value of p.

FRAME 22

The Forward Difference Operator Δ

There are three notations in common use for the actual difference columns
themselves.   One of these involves what are known as FORWARD DIFFERENCES
and uses the symbol Δ as the FORWARD DIFFERENCE OPERATOR. Δf_n is then
defined by the equation

$$\Delta f_n = f_{n+1} - f_n$$

where here n is an integer.   Then

$$\Delta f_0 = f_1 - f_0, \qquad \Delta f_2 = f_3 - f_2, \qquad \Delta f_{-3} = f_{-2} - f_{-3}, \quad \text{etc.}$$

Using the symbols so far defined, the beginning of a difference table can
be expressed symbolically as, for example,

x_{-2}	f_{-2}	
		Δf_{-2}
x_{-1}	f_{-1}	
		Δf_{-1}
x_0	f_0	
		Δf_0
x_1	f_1	
		Δf_1
x_2	f_2	
		Δf_2
x_3	f_3	

You have already met the idea of an operator in mathematics. For example, D is sometimes used to denote the operation of differentiation and i is used in complex numbers to denote the operation of rotating a vector 90° anti-clockwise. Repeated differentiation is then denoted by D^2, D^3, etc., and repeated vector rotation by i^2, i^3, etc. The power of i need not be an integer as it is possible to rotate vectors through angles other than multiples of 90°, but powers of D must be integral as one cannot differentiate, say, $2\frac{1}{2}$ times.

In a similar way, $\Delta^2 f_n$ for example is used to denote $\Delta(\Delta f_n)$ and is an instruction to take the difference of a difference, which is a second difference. How this works can be seen by denoting Δf_n by u_n, say, then

$$\Delta(\Delta f_n) \; = \; \Delta u_n \; = \; u_{n+1} - u_n \; = \; \Delta f_{n+1} - \Delta f_n$$

Thus, for example,

$$\Delta^2 f_3 \; = \; \Delta f_4 - \Delta f_3, \qquad \Delta^2 f_{-1} \; = \; \Delta f_0 - \Delta f_{-1}$$

We can now go one stage further in the difference table in the last frame and write

x	f	Δf	$\Delta^2 f$
x_{-2}	f_{-2}		
		Δf_{-2}	
x_{-1}	f_{-1}		$\Delta^2 f_{-2}$
		Δf_{-1}	
x_0	f_0		$\Delta^2 f_{-1}$
		Δf_0	
x_1	f_1		$\Delta^2 f_0$
		Δf_1	
x_2	f_2		$\Delta^2 f_1$
		Δf_2	
x_3	f_3		

The extension of this will be obvious, Δ^3 being used for 3rd differences and so on. In general, $\Delta^r f_n$ is defined by the equation

$$\Delta^r f_n \; = \; \Delta^{r-1} f_{n+1} - \Delta^{r-1} f_n$$

and the table from x_{-2} to x_4 can be exhibited as

x	f	Δf	$\Delta^2 f$	$\Delta^3 f$	$\Delta^4 f$
x_{-2}	f_{-2}				
		Δf_{-2}			
x_{-1}	f_{-1}		$\Delta^2 f_{-2}$		
		Δf_{-1}		$\Delta^3 f_{-2}$	
x_0	f_0		$\Delta^2 f_{-1}$		$\Delta^4 f_{-2}$
		Δf_0		$\Delta^3 f_{-1}$	
x_1	f_1		$\Delta^2 f_0$		$\Delta^4 f_{-1}$
		Δf_1		$\Delta^3 f_0$	
x_2	f_2		$\Delta^2 f_1$		$\Delta^4 f_0$
		Δf_2		$\Delta^3 f_1$	
x_3	f_3		$\Delta^2 f_2$		
		Δf_3			
x_4	f_4				

and extended upwards, downwards and to the right as necessary.

You will notice that all the entries having a common power of Δ appear in the same vertical column and all entries having a common f suffix appear on a downwards sloping diagonal.

To relate these symbol forms with a specific set of figures, suppose we have the table

x	f(x)	Δ	Δ²	Δ³	Δ⁴	Δ⁵	Δ⁶
-6	0·250 00						
		-8333					
-4	0·166 67		4166				
		-4167		-2499			
-2	0·125 00		1667		1665		
		-2500		-834		-1187	
0	0·100 00		833		478		885
		-1667		-356		-302	
2	0·083 33		477		176		208
		-1190		-180		-94	
4	0·071 43		297		82		49
		-893		-98		-45	
6	0·062 50		199		37		32
		-694		-61		-13	
8	0·055 56		138		24		
		-556		-37			
10	0·050 00		101				
		-455					
12	0·045 45						

If 4 is labelled x_0 then $f_0 = 0 \cdot 071\,43$, $f_{-2} = 0 \cdot 100\,00$, Δf_3 indicates the entry -455, $\Delta^2 f_{-1}$, 297 and $\Delta^4 f_{-2}$, 82.   (Don't forget that -455 really means $-0 \cdot 004\,55$, etc.)

Now, for the above table:

1. What are the entries corresponding to

    i) $\Delta^2 f_2$, $\Delta^6 f_1$, $\Delta^5 f_4$, if $x_0 = -6$,
    ii) Δf_{-3}, $\Delta^3 f_2$, $\Delta^4 f_{-1}$ if $x_0 = 0$?

2. What are the symbols for

    i) -1190, 82, -356, if $x_0 = -4$,
    ii) 199, 885, -45 if $x_0 = 8$?

3. What is the value of h?
    ****

24A

1. i) 833, 208, -13   ii) -8333, -61, 176
2. i) Δf_3, $\Delta^4 f_2$, $\Delta^3 f_1$ ii) $\Delta^2 f_{-2}$, $\Delta^6 f_{-7}$, $\Delta^5 f_{-4}$
3. $h = 2$

FRAME 25

The following figure is a flow diagram for forming a difference table from n function values as far as the kth differences $(n > k)$.

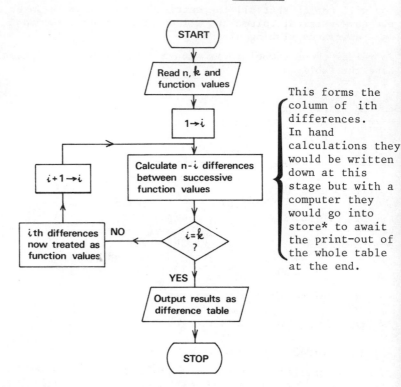

This forms the column of ith differences. In hand calculations they would be written down at this stage but with a computer they would go into store* to await the print-out of the whole table at the end.

This will happen automatically at the next step, when i is incremented, if the notation below is used.

*This can be achieved using a double suffix, e.g., f_{im} to denote $\Delta^i f_m$. The difference formula $\Delta^i f_m = \Delta^{i-1} f_{m+1} - \Delta^{i-1} f_m$ then becomes $f_{im} = f_{i-1,\, m+1} - f_{i-1,\, m}$ and the column of ith differences is obtained by applying this formula for $n - i$ consecutive values of m. Incrementing i by 1 when this is completed will make the differences just found into the values which have to be differenced the next time round.   The values of f_{im} will be stored as an array.

[A program for forming a difference table can be found in reference (5).]

Just as Δf_2, for example, can be expressed in terms of functional values, i.e., $f_3 - f_2$, so also can any of the higher differences.   Later, in for example, integration and the solution of differential equations, you will find that this is done quite frequently.

As an example,

$$\Delta^2 f_3 = \Delta f_4 - \Delta f_3 = (f_5 - f_4) - (f_4 - f_3) = f_5 - 2f_4 + f_3$$

Similarly, $\Delta^2 f_{-2} = f_0 - 2f_{-1} + f_{-2}$

By a similar method, other differences can be expressed in terms of functional values.   However, for the higher differences it is easier to use another operator – the shift operator E.

The Shift Operator E

The shift operator E simply has the effect of taking you forwards from one reading in a column to the next.   Thus

$$Ef_2 = f_3, \qquad E\Delta^2 f_{-1} = \Delta^2 f_0, \qquad E\Delta^5 f_0 = \Delta^5 f_1.$$

What do you think will be the effect of i) E^2, ii) E^5, iii) E^{-1}?
****

i)   To go forwards two readings in a column.
ii)   To go forwards five readings in a column.
iii)   To go backwards one reading in a column.   Remember the use of i^{-1}
for the rotation of a complex number vector through $90°$ clockwise.

Thus, $E^3 f_3 = f_6$, $E^4 \Delta^3 f_{-1} = \Delta^3 f_3$ and $E^{-2} \Delta^4 f_0 = \Delta^4 f_{-2}$.

Now $\Delta f_n = f_{n+1} - f_n = Ef_n - f_n = (E - 1)f_n$

and so Δ is symbolically equivalent to $E - 1$.   Alternatively, E is symbolically equivalent to $1 + \Delta$.

Can you suggest, without introducing Δ, an interpretation for $(E - 1)^2 f_3$ in terms of functional values?   Is your result the same as for $\Delta^2 f_3$ as found in FRAME 26?
****

$(E - 1)^2 f_3 \;=\; (E^2 - 2E + 1)f_3 \;=\; f_5 - 2f_4 + f_3 \qquad Yes$

Any higher difference can similarly be expressed in terms of functional values.   Thus, for example,

$$\Delta^4 f_{-2} \;=\; (E - 1)^4 f_{-2} \;=\; (E^4 - 4E^3 + 6E^2 - 4E + 1)f_{-2}$$
$$= \; f_2 - 4f_1 + 6f_0 - 4f_{-1} + f_{-2}$$

What will $\Delta^5 f_{-1}$ be when expressed in terms of functional values?
****

$f_4 - 5f_3 + 10f_2 - 10f_1 + 5f_0 - f_{-1}$

The Backward Difference Operator ∇

$f_{n+1} - f_n$ was expressed as Δf_n in FRAME 22, i.e., Δf_n was expressed in terms of f_n and the functional value one step forwards, that is, f_{n+1}. Sometimes it is desirable to express a difference using a functional value one step backwards instead of one step forwards.   To accomplish this the entry in the table in the same position as Δf_n is also written as ∇f_{n+1}. Then

$$\nabla f_{n+1} = f_{n+1} - f_n$$

and so $\nabla f_2 = f_2 - f_1$, $\nabla f_0 = f_0 - f_{-1}$, $\nabla f_n = f_n - f_{n-1}$.

FRAME 30   (continued)

It is very important to note that in a numerical table Δf_n and ∇f_{n+1} are exactly the same entry. Δf_n and ∇f_{n+1} are only two different symbols for the same thing.   Continuing, second differences follow in a manner similar to that obtaining when forward differences are used.   For example,

$$\nabla^2 f_n = \nabla f_n - \nabla f_{n-1} = (f_n - f_{n-1}) - (f_{n-1} - f_{n-2}) = f_n - 2f_{n-1} + f_{n-2}$$

Using backward differences, the symbolic table in FRAME 24 would appear as

x_{-2}	f_{-2}				
		∇f_{-1}			
x_{-1}	f_{-1}		$\nabla^2 f_0$		
		∇f_0		$\nabla^3 f_1$	
x_0	f_0		$\nabla^2 f_1$		$\nabla^4 f_2$
		∇f_1		$\nabla^3 f_2$	
x_1	f_1		$\nabla^2 f_2$		$\nabla^4 f_3$
		∇f_2		$\nabla^3 f_3$	
x_2	f_2		$\nabla^2 f_3$		$\nabla^4 f_4$
		∇f_3		$\nabla^3 f_4$	
x_3	f_3		$\nabla^2 f_4$		
		∇f_4			
x_4	f_4				

You will notice that this time all entries having the same f suffix appear on an upwards sloping diagonal.

Using the numerical table in FRAME 24, what will be

i)   the entries corresponding to ∇f_1, $\nabla^2 f_3$ and $\nabla^3 f_5$ if $x_0 = -6$,
ii)   the entries corresponding to $\nabla^4 f_{-2}$ and $\nabla^5 f_0$ if $x_0 = 6$,
iii)   the backward difference symbols for -1190, 82, -356 if $x_0 = -4$,
iv)   the backward difference symbols for 199, 885, -45 if $x_0 = 8$?
            ****

30A

i) -8333, 1667, -356 ii) 1665, -302
iii) ∇f_4, $\nabla^4 f_6$, $\nabla^3 f_4$ iv) $\nabla^2 f_0$, $\nabla^6 f_{-1}$, $\nabla^5 f_1$

FRAME 31

Can you now express ∇ in terms of E and show that $\nabla = E^{-1}\Delta$?
            ****

31A

$\nabla f_n = f_n - f_{n-1} = f_n - E^{-1}f_n = (1 - E^{-1})f_n$ $\therefore$ $\nabla = 1 - E^{-1}$

Also $1 - E^{-1} = E^{-1}(E - 1)$ $\therefore$ $\nabla = E^{-1}\Delta$

These two results can also be written as $E = (1 - \nabla)^{-1}$ and $\Delta = E\nabla$ respectively.

FRAME 32

In 24A, question No. 2, you expressed certain entries in the numerical table in FRAME 24 in terms of Δ.   Take these results and use $\Delta = E\nabla$ to express them in terms of ∇.   Then check that your answers agree with those you obtained in 30A.
            ****

$$\Delta f_3 = (E\nabla)f_3 = \nabla f_4 \qquad\qquad \Delta^4 f_2 = (E\nabla)^4 f_2 = \nabla^4 f_6$$

The others follow similarly.

The Central Difference Operator δ

In the case of the forward difference operator, all entries with the same f suffix appear on a downwards sloping diagonal. When using the backward difference operator they all appear on an upwards sloping diagonal. The CENTRAL DIFFERENCE OPERATOR δ is defined in such a way that entries with the same f suffix all appear on a horizontal line. This immediately leads to a snag - the odd differences are not placed on the same level as any of the functional values. Omitting these differences for a moment, the even differences can be relabelled as shown below - and remember, it is only a relabelling. The numerical values are the same as before.

x_{-2}	f_{-2}			
x_{-1}	f_{-1}		$\delta^2 f_{-1}$	
x_0	f_0		$\delta^2 f_0$	$\delta^4 f_0$
x_1	f_1	$\odot$	$\delta^2 f_1$	$\delta^4 f_1$
x_2	f_2		$\delta^2 f_2$	$\delta^4 f_2$
x_3	f_3		$\delta^2 f_3$	
x_4	f_4			

The dots indicate the awkward ones.

Now each entry shown by a dot is written on a level half way between two functional values. Thus, for example, the ringed dot is on a level half way between f_0 and f_1. It is therefore assumed to lie on the level of $f_{\frac{1}{2}}$, although this entry does not, of course, exist in the table. Using this idea the ringed entry is labelled $\delta f_{\frac{1}{2}}$. Extending this notation to the other dots, the table now appears as

x_{-2}	f_{-2}					
		$\delta f_{-1\frac{1}{2}}$				
x_{-1}	f_{-1}		$\delta^2 f_{-1}$			
		$\delta f_{-\frac{1}{2}}$		$\delta^3 f_{-\frac{1}{2}}$		
x_0	f_0		$\delta^2 f_0$		$\delta^4 f_0$	
		$\delta f_{\frac{1}{2}}$		$\delta^3 f_{\frac{1}{2}}$		
x_1	f_1		$\delta^2 f_1$		$\delta^4 f_1$	
		$\delta f_{1\frac{1}{2}}$		$\delta^3 f_{1\frac{1}{2}}$		
x_2	f_2		$\delta^2 f_2$		$\delta^4 f_2$	
		$\delta f_{2\frac{1}{2}}$		$\delta^3 f_{2\frac{1}{2}}$		
x_3	f_3		$\delta^2 f_3$			
		$\delta f_{3\frac{1}{2}}$				
x_4	f_4					

FRAME 34   (continued)

Notice that even powers of δ are always associated with integral suffixes of f and odd powers with fractional suffixes.

Returning now to the numerical table in FRAME 24, what will be

i)   the entries corresponding to $\delta^2 f_2$ and $\delta^5 f_{\frac{1}{2}}$ if $x_0 = -2$,

ii)   the entries corresponding to $\delta f_{1\frac{1}{2}}$ and $\delta^3 f_{-2\frac{1}{2}}$ if $x_0 = 4$,

iii)   the central difference symbols for -1190, 82, -356 if $x_0 = -4$,

iv)   the central difference symbols for 199, 885, -45 if $x_0 = 8$?

****

34A

i)   477, -1187   *ii) -694, -834*
iii) $\delta f_{3\frac{1}{2}}$, $\delta^4 f_4$, $\delta^3 f_{2\frac{1}{2}}$ *iv) $\delta^2 f_{-1}$, $\delta^6 f_{-4}$, $\delta^5 f_{-1\frac{1}{2}}$*

FRAME 35

The formula for any central difference can be expressed as

$$\delta^r f_n = \delta^{r-1} f_{n+\frac{1}{2}} - \delta^{r-1} f_{n-\frac{1}{2}}$$

provided n and r satisfy the association noted under the table in the last frame.

A few special cases of this are

$$\delta f_{1\frac{1}{2}} = f_2 - f_1, \quad \delta^4 f_2 = \delta^3 f_{2\frac{1}{2}} - \delta^3 f_{1\frac{1}{2}}, \quad \delta^3 f_{-\frac{1}{2}} = \delta^2 f_0 - \delta^2 f_1$$

FRAME 36

In order to express δ in terms of E, it is necessary to give a meaning to fractional powers of E.

In FRAME 20, we wrote x_p to denote the value of $x_0 + ph$, where p could take any value, and, in FRAME 21, we used f_p to denote the corresponding value of the function. The symbol E^p is used in a similar sense and indicates that one goes forwards p entries in the table, even if p is fractional, thus landing one on an "entry" that isn't really there. So f can also be written as $E^p f_0$. Using this notation, $f_{\frac{1}{2}}$ is written as $E^{\frac{1}{2}} f_0$, $f_{-\frac{1}{2}}$ as $E^{-\frac{1}{2}} f_0$ and, extending it, we can also write $E^{1\frac{1}{2}} f_1 = f_{2\frac{1}{2}}$, $E^{-2\frac{1}{2}} f_{-1\frac{1}{2}} = f_{-4}$, $E^{-\frac{1}{2}} f_2 = f_{1\frac{1}{2}}$, etc.

Now, as $\delta f_{\frac{1}{2}} = f_1 - f_0$ and as f_1 can be written as $E^{\frac{1}{2}} f_{\frac{1}{2}}$ and f_0 as $E^{-\frac{1}{2}} f_{\frac{1}{2}}$, $\delta f_{\frac{1}{2}} = E^{\frac{1}{2}} f_{\frac{1}{2}} - E^{-\frac{1}{2}} f_{\frac{1}{2}} = (E^{\frac{1}{2}} - E^{-\frac{1}{2}}) f_{\frac{1}{2}}$.

So here, δ is equivalent to $E^{\frac{1}{2}} - E^{-\frac{1}{2}}$. That it is true generally can be seen if the front δ is replaced by this in, say, $\delta(\delta^{r-1} f_n)$. Then

$$\delta(\delta^{r-1} f_n) = (E^{\frac{1}{2}} - E^{-\frac{1}{2}})(\delta^{r-1} f_n) = E^{\frac{1}{2}} \delta^{r-1} f_n - E^{-\frac{1}{2}} \delta^{r-1} f_n$$

$$= \delta^{r-1} E^{\frac{1}{2}} f_n - \delta^{r-1} E^{-\frac{1}{2}} f_n = \delta^{r-1} f_{n+\frac{1}{2}} - \delta^{r-1} f_{n-\frac{1}{2}}$$

But $\delta(\delta^{r-1}f_n) = \delta^r f_n$,   and, as was seen in FRAME 35, this is equal to
$\delta^{r-1}f_{n+\frac{1}{2}} - \delta^{r-1}f_{n-\frac{1}{2}}$.

In 31A you worked out the relation between ∇ and Δ, i.e., $\nabla = E^{-1}\Delta$.
See if you can now show that $\delta = E^{-\frac{1}{2}}\Delta$ and $\delta = E^{\frac{1}{2}}\nabla$.
****

$\delta = E^{\frac{1}{2}} - E^{-\frac{1}{2}} = E^{-\frac{1}{2}}(E - 1) = E^{-\frac{1}{2}}\Delta$
$\delta = E^{\frac{1}{2}} - E^{-\frac{1}{2}} = E^{\frac{1}{2}}(1 - E) \ = E^{\frac{1}{2}}\nabla$

You may have noticed, as you have been reading this programme that we have
sometimes reversed the order of two operators.   For example, in FRAME 36,
$E^{\frac{1}{2}}\delta^{r-1}$ was assumed to be equivalent to $\delta^{r-1}E^{\frac{1}{2}}$.   This sort of reversal
has not been justified but if you think about it you will see that it is
quite reasonable.   Thus, $E^2\Delta^4 f_0$ tells us to start from f_0, go four
columns diagonally downwards to the right and then two rows down
vertically. $\Delta^4 E^2 f_0$ tells us to go two rows down vertically and then
four columns diagonally downwards to the right.   The net effect is to
arrive at the same place.   Similarly, three columns to the right of f_{-1}
and then $4\frac{1}{2}$ rows down (i.e. $E^{4\frac{1}{2}}\delta^3 f_{-1}$) gets to the same place as $4\frac{1}{2}$ rows
down from f_{-1} and then three columns to the right $(\delta^3 E^{4\frac{1}{2}}f_{-1})$.

(You have met similar ideas in analytical work, for example, $\dfrac{\partial^2 u}{\partial x \partial y} = \dfrac{\partial^2 u}{\partial y \partial x}$
and $i^2 i^{1\frac{1}{2}} = i^{1\frac{1}{2}} i^2$.   The first of these is saying that the order of
differentiation doesn't matter and the second that a complex number vector
rotated through $135°$ and then $180°$ arrives in the same position as if it
is rotated first through $180°$ and then $135°$.)

The Averaging Operator μ

There is one further operator that will be required in later work.   This
is the MEAN or AVERAGING OPERATOR μ.   It is used to denote the mean or
average of two adjacent readings in a column.   Thus

$$\mu f_{\frac{1}{2}} = \tfrac{1}{2}(f_1 + f_0), \quad \mu\delta^2 f_{-\frac{1}{2}} = \tfrac{1}{2}(\delta^2 f_0 + \delta^2 f_{-1}), \quad \text{and so on.}$$

Be careful not to confuse $f_{\frac{1}{2}}$ with $\mu f_{\frac{1}{2}}$, etc.   The difference in meaning
is easily seen by means of a simple example.

If $f(x) = x^2$,   the following table of values can be formed:

x	0	0·5	1·0	1·5	2·0	2·5	3·0
f(x)	0	0·25	1·00	2·25	4·00	6·25	9·00

Now suppose x_0 is chosen to be $1·0$. $h = 0·5$ and so $x_1 = 1·5$,
$f_0 = 1·00$ and $f_1 = 2·25$.   Then $x_{\frac{1}{2}} = 1·25$ and so $f_{\frac{1}{2}} = 1·25^2 = 1·5625$,
but $\mu f_{\frac{1}{2}} = \tfrac{1}{2}(f_1 + f_0) = \tfrac{1}{2}(2·25 + 1·00) = 1·625$, which is not the same

FINITE DIFFERENCES

FRAME 39   (continued)

value.

Referring to the numerical table in FRAME 24, what will be the numbers corresponding to $\mu f_{1\frac{1}{2}}$, $\mu \delta^2 f_{-1\frac{1}{2}}$ and $\mu \delta f_2$ if $x_0 = 4$? Can you also find an expression for μ in terms of E?

****

39A

$0 \cdot 05903$, 655, -625.

$\mu = (E^{\frac{1}{2}} + E^{-\frac{1}{2}})/2$

FRAME 40

Having defined Δ, E, ∇ and δ, we have seen how to interpret operators such as Δ^2, E^3, δ^4, etc. Is it possible then, in a similar way, to attach a meaning to μ^2?

To investigate, let us see if we can interpret, say, $\mu^2 f_0$.

As $\Delta^2 f_0 = \Delta(\Delta f_0)$, it seems reasonable to assume at the outset that $\mu^2 f_0 = \mu(\mu f_0)$. Now μf_0 doesn't really mean anything from the way in which μ has been defined, but as $\mu = (E^{\frac{1}{2}} + E^{-\frac{1}{2}})/2$, let us agree that $\mu f_0 = \frac{1}{2}(E^{\frac{1}{2}} + E^{-\frac{1}{2}})f_0$, i.e., $\frac{1}{2}(f_{\frac{1}{2}} + f_{-\frac{1}{2}})$.

Then $\mu(\mu f_0) = \mu(f_{\frac{1}{2}} + f_{-\frac{1}{2}})/2 = (\mu f_{\frac{1}{2}} + \mu f_{-\frac{1}{2}})/2$ and both of these are defined.

Applying the requisite formulae for $\mu f_{\frac{1}{2}}$ and $\mu f_{-\frac{1}{2}}$, find an interpretation for $\mu^2 f_0$. **

40A

$\mu^2 f_0 = \{\frac{1}{2}(f_1 + f_0) + \frac{1}{2}(f_0 + f_{-1})\}/2 = (f_1 + 2f_0 + f_{-1})/4$

FRAME 41

Now, using $\mu = (E^{\frac{1}{2}} + E^{-\frac{1}{2}})/2$, express μ^2 in terms of E and see whether the result, when applied to f_0, agrees with what you have just found.

****

41A

$\mu^2 = \{(E^{\frac{1}{2}} + E^{-\frac{1}{2}})/2\}^2 = (E + 2 + E^{-1})/4$

$\mu^2 f_0 = \{(E + 2 + E^{-1})/4\}f_0 = (f_1 + 2f_0 + f_{-1})/4$

FRAME 42

Other Operational Formulae

There are many other formulae that have been developed for work in connection with finite differences. Apart from showing you a couple of examples by way of illustration, it is not proposed to go into these in detail. Any that are required for later work can be obtained when they are needed.

For the first of these examples, take $1 + \frac{1}{4}\delta^2$, express δ in terms of expand and simplify.

****

$$1 + \tfrac{1}{4}\delta^2 = 1 + \tfrac{1}{4}(E^{\frac{1}{2}} - E^{-\frac{1}{2}})^2 = 1 + (E - 2 + E^{-1})/4 = (E + 2 + E^{-1})/4$$

FRAME 43

Comparing this result with that obtained in 41A, you will see that $\mu^2 = 1 + \tfrac{1}{4}\delta^2$

For a second example, let us see if there is a result in this work analogous to the derivative of a product. Thus, $\Delta(f_n g_n)$ might be required.

By definition $\qquad \Delta(f_n g_n) = f_{n+1} g_{n+1} - f_n g_n \qquad\qquad (43.1)$

But $\quad \Delta g_n = g_{n+1} - g_n \quad$ and so $\quad g_{n+1} = g_n + \Delta g_n$

$\therefore \quad \Delta(f_n g_n) = f_{n+1}(g_n + \Delta g_n) - f_n g_n = (f_{n+1} - f_n)g_n + f_{n+1}\Delta g_n$

$$= g_n \Delta f_n + f_{n+1}\Delta g_n$$

Show that (43.1) can also be expressed as $f_n \Delta g_n + g_{n+1}\Delta f_n$.

****

43A

$\Delta f_n = f_{n+1} - f_n \qquad \therefore \quad f_{n+1} = f_n + \Delta f_n$

$\therefore \quad \Delta(f_n g_n) = (f_n + \Delta f_n)g_{n+1} - f_n g_n = f_n(g_{n+1} - g_n) + g_{n+1}\Delta f_n$

$$= f_n \Delta g_n + g_{n+1}\Delta f_n$$

FRAME 44

Some of the ideas developed in this programme may, at this stage, seem somewhat unusual. But when you get used to working with them, you will find that they are really no queerer than the ideas you came across in differentiation and integration, for example. The various techniques that you learn in mathematics were originally invented to perform certain functions and the methods of finite differences are just some of these techniques.

FRAME 45

Summary

In this frame, some of the definitions and formulae are listed for your convenience.

$\Delta f_n = f_{n+1} - f_n$

$\Delta^r f_n = \Delta^{r-1} f_{n+1} - \Delta^{r-1} f_n$

$\nabla f_n = f_n - f_{n-1}$

$\nabla^r f_n = \nabla^{r-1} f_n - \nabla^{r-1} f_{n-1}$

$\delta f_{n+\frac{1}{2}} = f_{n+1} - f_n$

$\delta^r f_n = \delta^{r-1} f_{n+\frac{1}{2}} - \delta^{r-1} f_{n-\frac{1}{2}}$

$E f_n = f_{n+1}$

$E^r f_n = f_{n+r}$

$\mu f_{n+\frac{1}{2}} = (f_{n+1} + f_n)/2$

$\Delta = E - 1$

$\nabla = 1 - E^{-1}$

$\delta = E^{\frac{1}{2}} - E^{-\frac{1}{2}}$

$E = 1 + \Delta$

$E = (1 - \nabla)^{-1}$

$\Delta = E\nabla$

$\nabla = E^{-1}\Delta$

$\delta = E^{-\frac{1}{2}}\Delta = E^{\frac{1}{2}}\nabla$

$\mu = (E^{\frac{1}{2}} + E^{-\frac{1}{2}})/2$

$\mu^2 = 1 + \frac{1}{4}\delta^2$

Miscellaneous Examples

In this frame a collection of miscellaneous examples is given for you to try.   Answers are provided in FRAME 47, together with such working as is considered helpful.

1.   -7, -6, -1, 2, 21   are five consecutive entries in the tabulation of a certain quartic.   Find the two entries preceding -7 and the two following 21.

2.   A polynomial f(x) of low degree is tabulated as follows:

x	-3	-2	-1	0	1	2	3	4	5	6
f(x)	-30	-12	0	3	4	5	9	19	38	69

Two errors in the values of f(x) are suspected.   Locate and correct them.

3.   There are some copying errors in the following table.   Find them and suggest probable corrections.

0·000 00,   0·036 59,   0·072 32,   0·107 23,   0·143 18,   0·174 80,
0·207 53,   0·239 61,   0·271 07,   0·301 93,   0·333 32,   0·361 98,
0·391 21,   0·419 95,   0·448 52,   0·476 03.

4.   Obtain the results

i) $\Delta\left(\dfrac{f_n}{g_n}\right) = \dfrac{g_n \, \Delta f_n - f_n \, \Delta g_n}{g_n g_{n+1}}$

ii) $\Delta\left(\dfrac{1}{g_n}\right) = \dfrac{-\Delta g_n}{g_n g_{n+1}}$

iii) $\Delta(\log f_n) = \log(f_{n+1}/f_n)$

5.   Show that

i) $\Delta - \nabla = \Delta\nabla$

ii) $\delta^2 = \Delta - \nabla$

iii) $\mu\delta = \frac{1}{2}(\Delta + \nabla)$

iv) $\mu + \frac{1}{2}\delta = E^{\frac{1}{2}}$

Answers to Miscellaneous Examples

1.   147,   26;   98,   299

2.   The last four 3rd differences suggest a polynomial of degree 3. Taking all the third differences as 3 leads to tabular entries of -1 and -1 instead of -12 and 0.

3. 0·000 00

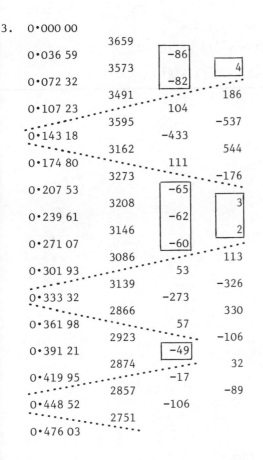

0·000 00			
	3659		
0·036 59		−86	
	3573		4
0·072 32		−82	
	3491		186
0·107 23		104	
	3595		−537
0·143 18		−433	
	3162		544
0·174 80		111	
	3273		−176
0·207 53		−65	
	3208		3
0·239 61		−62	
	3146		2
0·271 07		−60	
	3086		113
0·301 93		53	
	3139		−326
0·333 32		−273	
	2866		330
0·361 98		57	
	2923		−106
0·391 21		−49	
	2874		32
0·419 95		−17	
	2857		−89
0·448 52		−106	
	2751		
0·476 03			

The differences enclosed in rectangles seem to be following a reasonable pattern and fans would suggest errors in 0·143 18, 0·333 32 and 0·448 52.   There are insufficient differences possible at the end of the table to enable us to be reasonably sure at the moment, one way or the other, about 0·476 03.

The change from −82 to 104 in the 2nd differences column suggests that 0·143 18 is in error by something in the region of 180 and the reversal of the two digits 31 to 13 will make a change of this magnitude. 0·141 38 instead of 0·143 18 gives reasonable differences in the top fan.

The change from −60 to 53 in the 2nd differences column suggests that 0·333 32 is in error by about 110.   Changing it to 0·332 22 then produces reasonable differences in the second fan.

The change from −49 to −17 suggests an error of about 30 in 0·448 52.   0·448 25 and 0·448 22 are possible corrections and the latter gives better differences at the end of the table.   Finally, no change is indicated in 0·476 03.

4. i)
$$\Delta\left(\frac{f_n}{g_n}\right) = \frac{f_{n+1}}{g_{n+1}} - \frac{f_n}{g_n} = \frac{f_{n+1}g_n - f_n g_{n+1}}{g_n g_{n+1}}$$

$$= \frac{(f_{n+1} - f_n)g_n - f_n(g_{n+1} - g_n)}{g_n g_{n+1}} = \frac{g_n \Delta f_n - f_n \Delta g_n}{g_n g_{n+1}}$$

ii) Putting $f_n = 1$ and consequently $\Delta f_n = 0$ in (i) gives result.

iii) $\Delta(\log f_n) = \log f_{n+1} - \log f_n = \log(f_{n+1}/f_n)$

5. i) $\Delta - \nabla = (E - 1) - (1 - E^{-1}) = E - 2 + E^{-1}$

$\Delta\nabla = (E - 1)(1 - E^{-1}) = E - 2 + E^{-1}$

Hence result

ii) $\delta^2 = (E^{\frac{1}{2}} - E^{-\frac{1}{2}})^2 = E - 2 + E^{-1}$ and result follows from (i)

iii) $\mu\delta = \frac{1}{2}(E^{\frac{1}{2}} + E^{-\frac{1}{2}})(E^{\frac{1}{2}} - E^{-\frac{1}{2}}) = \frac{1}{2}(E - E^{-1})$

$\frac{1}{2}(\Delta + \nabla) = \frac{1}{2}\{(E - 1) + (1 - E^{-1})\} = \frac{1}{2}(E - E^{-1})$

iv) $\mu + \frac{1}{2}\delta = \frac{1}{2}(E^{\frac{1}{2}} + E^{-\frac{1}{2}}) + \frac{1}{2}(E^{\frac{1}{2}} - E^{-\frac{1}{2}}) = E^{\frac{1}{2}}$

Interpolation

Introduction and Linear Interpolation

You may already have met the idea of interpolation.   In case you haven't,
just a few words as to its meaning.   Suppose the length of a wire has
been measured for various loads suspended from it.   The following table
might result:

Load (kg)	0	1	2	3	4	5
Length (mm)	2027·1	2029·4	2031·8	2034·1	2036·5	2039·0

One could then ask – what would be the length of the wire if the load is
2·7, 4·1 or 6·3 kg?   The first two of these loads lie within the range
of the table.   Finding the length of the wire for either of these loads,
or for any other non-tabulated load between 0 and 5 kg, from the
information given is known as INTERPOLATION.   Finding the extension for
a load outside the range of the table is known as EXTRAPOLATION.
Unfortunately, extrapolation can be a dangerous process.   Do you know, or
if not, can you guess, why?

****

1A

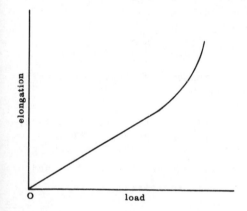

*The law connecting the variables
may only apply within certain limits.
For example, Hooke's law that the
elongation of a wire is proportional
to the load to which it is subjected
only applies provided the elastic
limit has not been reached.   After
that the material begins to flow and
the graph of elongation against load
curves sharply upwards, as is shown
in the diagram.*

When a set of numerical data is given as in the last frame, it may be
possible to find an analytical law fitting the data reasonably well – we
have already seen how to do this by least squares.   This law can then be
used to estimate the value of the dependent variable for a value of the
independent variable not given in the table.   However, in this programme
we shall be concerned with the problem of interpolating when the form of
the curve passing through a series of points is not known.   But, for
illustration purposes, some examples will be taken where the analytical
formula is known.   Doing this will enable us to compare the results
obtained by purely numerical methods with those given by analytical means.

In pre-computer days interpolation was also useful for filling in
intermediate values for tables of functions whose analytical forms were
known.   This use became of relatively little importance in early computer
days when memory facilities were very limited.   Then the values of

FRAME 2   (continued)

functions were more often calculated directly.   Now this trend is being reversed with the advent of larger and less expensive computer memories.

FRAME 3

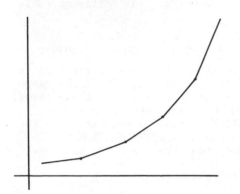

The simplest method of interpolation is that known as LINEAR INTERPOLATION. This simply joins any two consecutive points in a table by a straight line. If a series of points are given, then a series of straight lines are used, one such line joining each pair of consecutive points, as shown in the diagram.

The process can be illustrated by taking, for example, the figures given on page 1:2 in our Volume 1 of "Mathematics for Engineers and Scientists" for the cooling of a particular hot body, i.e.,

t	0	120	240	360	480	600	720
θ	100	86	74	64	56	49	44

where t, the time, is in seconds and θ, the temperature, in degrees Celsius.   From these figures, we can find the values of θ at, say, t = 144, 168, 192 and 216,   as follows:

During the 120 second interval from t = 120 to t = 240, θ falls by 12°.   In 24 seconds it therefore falls by 2·4 degrees.   The temperatures at the required times are thus 83·6, 81·2, 78·8 and 76·4.

Using this method, what will be the values of θ at t = 50, 75,   and at t = 400, 420?   ***

3A

Between t = 0 and t = 120, θ falls by 14°.   In 50 s it falls by $\frac{50}{120} \times 14 = 5\frac{5}{6}$ and in 75 s by $8\frac{3}{4}$.   Required values are $94\frac{1}{6}$, $91\frac{1}{4}$.

Similarly, when t = 400, $\theta = 61\frac{1}{3}$ and when t = 420, $\theta = 60$.

FRAME 4

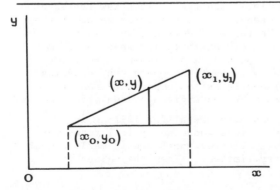

If the two given points are (x_0, y_0) and (x_1, y_1) it is a simple matter to express the connection between x and y by a mathematical formula. This connection is obviously the equation of the straight line joining (x_0, y_0) and (x_1, y_1),   i.e

$$\frac{y - y_0}{x - x_0} = \frac{y_1 - y_0}{x_1 - x_0} \quad \text{or} \quad y = y_0 + \frac{x - x_0}{x_1 - x_0}(y_1 - y_0)$$

FRAME 5

For our present purposes, it will be instructive to link up this formula with the notation that was used in the previous programme.   There, the general value of x was denoted by x_p and so, using y_p instead of y,

$$y_p = y_0 + \frac{x_p - x_0}{x_1 - x_0}(y_1 - y_0) \qquad (5.1)$$

Also we wrote $x_p = x_0 + ph$ where $h = x_1 - x_0$, and $y_1 - y_0 = \Delta y_0$.

With this notation (5.1) becomes

$$y_p = y_0 + p\Delta y_0 \qquad (5.2)$$

As, in FRAME 4, x was taken as lying between x_0 and x_1, p will lie between 0 and 1.

Now you remember that, when working with differences, you can choose any value of x in the table to be labelled x_0.   Returning now to the example you did at the end of FRAME 3, what values of t would you take to be t_0 and what would be the corresponding values of $\Delta\theta_0$ and p when finding θ for t = 50, 75, 400, 420?   Keep p to be within the range 0 to 1.
****

5A

Here, $t_p = t_0 + ph$ and $h = 120$.

For $t = 50$, take $t_0 = 0$ and then $\Delta\theta_0 = -14$, $p = \dfrac{t_p - t_0}{h} = \dfrac{5}{12}$.

For $t = 75$, take $t_0 = 0$ and then $\Delta\theta_0 = -14$, $p = \dfrac{5}{8}$.

For $t = 400$, take $t_0 = 360$ and then $\Delta\theta_0 = -8$, $p = \dfrac{1}{3}$.

For $t = 420$, take $t_0 = 360$ and then $\Delta\theta_0 = -8$, $p = \dfrac{1}{2}$.

FRAME 6

The Newton-Gregory Forward Difference Interpolation Formula

The method of interpolation adopted in the last few frames assumed that two adjacent points in a table were joined, when plotted, by a straight line.   In practice this is very seldom the case and to obtain a more accurate estimate of y for intermediate values of x, it is necessary to take into account the curvature of the graph between the two points. But this immediately raises the difficulty that we have, as a rule in this sort of work, no information about this curvature.   For example, suppose we are given a table of four pairs of values of x and y, which, when plotted, give Fig. (i).   Fig. (ii) shows just three of the many curves on which the four points might lie.   So, in order to proceed, it is necessary to guess which is the most likely form.   Fortunately, in practice, things usually behave fairly well unless some upset occurs.

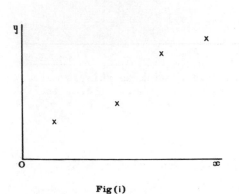

Fig (i)

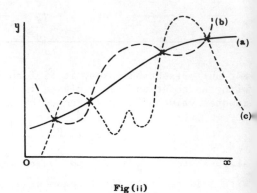

Fig (ii)

For example, in the case of the cooling of a hot body, cooling takes place continuously until the temperature of the surroundings is reached, and the rate of cooling decreases as room temperature is approached. This means that the temperature-time graph is of the form

and unless something else occurs, will not look like

or

We thus assume that, in the absence of any more information, (a) in Fig. (ii) is more likely than (b) or (c).

Many formulae which take into account the curvature of the graph have been evolved.   One of the simplest of these is the NEWTON-GREGORY (sometimes, alternatively, Gregory-Newton, or even just Newton) FORWARD DIFFERENCE FORMULA.   It makes no assumptions as to the actual form of the curve, but does assume that there is a polynomial function whose graph is not too far removed from the true curve.

In FRAME 36, page 198, it was seen that f_p, the value of a function f when x takes the value $x_0 + ph$, can be written as $E^p f_0$.   Also, in FRAME 28, page 195, E was found to be equivalent to $1 + \Delta$.

Thus $f_p = E^p f_0 = (1 + \Delta)^p f_0$.

The next step is to expand $(1 + \Delta)^p$ by the binomial theorem, giving

$$(1 + \Delta)^p = 1 + p\Delta + \frac{p(p-1)}{2!} \Delta^2 + \frac{p(p-1)(p-2)}{3!} \Delta^3 + \ldots\ldots\ldots\ldots$$

or, more shortly,

$$(1 + \Delta)^p = 1 + \binom{p}{1}\Delta + \binom{p}{2}\Delta^2 + \binom{p}{3}\Delta^3 + \ldots\ldots\ldots\ldots$$

Then, $f_p = \{1 + \binom{p}{1}\Delta + \binom{p}{2}\Delta^2 + \binom{p}{3}\Delta^3 + \ldots\ldots\ldots\ldots\}f_0$

i.e. $f_p = f_0 + \binom{p}{1}\Delta f_0 + \binom{p}{2}\Delta^2 f_0 + \binom{p}{3}\Delta^3 f_0 + \ldots\ldots\ldots\ldots$ (7.1)

and this is the required formula.   You will notice that the first two terms on the R.H.S. are equivalent to the linear interpolation formula (5.2).   Further, if the formula is truncated or, as may happen, automatically stops, at any other point, a polynomial in p results, e.g., if the last term included is $\binom{p}{3}\Delta^3 f_0$,   we have a polynomial of degree 3.

[If you haven't met the notation $\binom{p}{r}$ before, this simply stands for

$$\frac{p(p-1)(p-2)\ldots\ldots\ldots(p-r+1)}{r!}.$$ This is the same formula that

you have for ^pC_r.   But when dealing with permutations and combinations, ^pC_r only makes sense if p is a positive integer.   But, whatever the value of p, it is still possible to evaluate $\dfrac{p(p-1)(p-2)\ldots\ldots(p-r+1)}{r!}$

and it is convenient to have a symbol for this expression.   Note that, in it, the number of terms is the same in both numerator and denominator.]

As a very simple example of this formula, let us find $f(2\cdot31)$ from the table

x	0	1	2	3	4	5	6
f(x)	2	3	10	29	66	127	218

Start by suggesting what value for x_0 will make $0 \leqslant p < 1$,   stating the value of p and forming a difference table.

****

$x_0 = 2,$ $p = 0\cdot31$

x_{-2}	0	2				
			1			
x_{-1}	1	3		6		
			7		6	
x_0	2	<u>10</u>		12		0
			<u>19</u>		6	
x_1	3	29		<u>18</u>		0
			37		<u>6</u>	
x_2	4	66		24		<u>0</u>
			61		6	
x_3	5	127		30		
			91			
x_4	6	218				

(The reason for the underlining is given in the next frame.)

If you compare the table in 8A with the first one in FRAME 24, page 192, you will see that $f_0 = 10$, $\Delta f_0 = 19$, $\Delta^2 f_0 = 18$, $\Delta^3 f_0 = 6$, $\Delta^4 f_0 = 0$. The higher differences cannot be stated from the table, but you will recognise the table as that of a cubic, for which all these higher differences are zero.   The values just stated are the quantities underlined in 8A.   It is a good idea to indicate in this way which entries in your table you are going to use.   Now, substituting into (7.1) gives

$$f(2 \cdot 31) = 10 + \binom{0 \cdot 31}{1} 19 + \binom{0 \cdot 31}{2} 18 + \binom{0 \cdot 31}{3} 6$$

$$= 10 + \frac{0 \cdot 31}{1} \times 19 + \frac{0 \cdot 31 \times (-0 \cdot 69)}{2} \times 18 + \frac{0 \cdot 31 \times (-0 \cdot 69) \times (-1 \cdot 69)}{6} \times 6$$

$$= 10 + 5 \cdot 89 - 1 \cdot 9251 + 0 \cdot 361\ 491 \ = \ 14 \cdot 326\ 391$$

The actual cubic used here was $f(x) = x^3 + 2$ and, when $x = 2 \cdot 31$, this gives $14 \cdot 326\ 391$ and so, in this case, the value obtained from the formula (7.1) is exact.   This is because no term which does not give a non-zero contribution has been omitted.

Now, by the same method, find $f(1 \cdot 6)$ and check that your result is the same as $1 \cdot 6^3 + 2$.

****

Taking x_0 as 1 and $p = 0 \cdot 6$ gives

$$f(1 \cdot 6) = 3 + \binom{0 \cdot 6}{1} 7 + \binom{0 \cdot 6}{2} 12 + \binom{0 \cdot 6}{3} 6$$

$$= 3 + 0 \cdot 6 \times 7 + 0 \cdot 6 \times (-0 \cdot 4) \times 6 + 0 \cdot 6 \times (-0 \cdot 4) \times (-1 \cdot 4) = 6 \cdot 096$$

which agrees with $1 \cdot 6^3 + 2$.

Taking now $x_0 = 1$, what will be the expression giving f_p for any value of p?

****

$$3 + \binom{p}{1} 7 + \binom{p}{2} 12 + \binom{p}{3} 6$$

This can also be written as

$$3 + 7p + 6p(p - 1) + p(p - 1)(p - 2) \qquad (12.1)$$

and so is a cubic in p.   But $x_p = x_0 + ph$ and so, here, $p = x_p - 1$ as $x_0 = 1$ and $h = 1$.   (12.1) can thus be written as

$$3 + 7(x_p - 1) + 6(x_p - 1)(x_p - 2) + (x_p - 1)(x_p - 2)(x_p - 3) = x_p^3 + 2$$

which is the function from which the original table was formed.   This means that, for this example, the correct result will always be obtained whatever value is given to p.   It is thus not necessary in this case to restrict p to the range 0 to 1 although in practice this is usually don

If a difference table is formed from any other polynomial, it will be found that this interpolation formula, i.e., (7·1) is always exact provided that all the decimal places are kept. Furthermore, when a process similar to that in FRAME 12 is done, the original polynomial will always be recovered.

FRAME 14

More often than not, when a difference table is being formed, the values of f will not be those of a polynomial. The table in 11A, page 185, is an example. From the values in this table, find f(0·12) using the Newton-Gregory formula. In doing so, remember what was said in FRAME 10, page 185, about the use of the higher differences when there are round-off errors present. Give your answer to 4 decimal places.
****

14A

Taking $x_0 = 0\cdot10$, $p = \dfrac{0\cdot12 - 0\cdot10}{0\cdot05} = 0\cdot4$ *as* $h = 0\cdot05$.

$f(0\cdot12) = 0\cdot0998 + \begin{pmatrix} 0\cdot4 \\ 1 \end{pmatrix} \times 0\cdot0496 + \begin{pmatrix} 0\cdot4 \\ 2 \end{pmatrix} \times (-0\cdot0003) + \begin{pmatrix} 0\cdot4 \\ 3 \end{pmatrix} \times (-0\cdot0004) + \ldots$

$\approx 0\cdot0998 + 0\cdot019\,84 + 0\cdot000\,036 - 0\cdot000\,0256 = 0\cdot119\,6504 = 0\cdot1197$

Obviously one could take more terms in the interpolation formula but as the higher differences are unreliable, it is not only pointless to do so, but can also lead to an answer that is less accurate than that quoted.

FRAME 15

The differences used in the Newton-Gregory forward interpolation formula lie on a diagonal line sloping downwards to the right. Can you suggest when this would be inconvenient?
****

15A

If x_0 *is near the end of the table, the differences required will not be available.*

FRAME 16

The Newton-Gregory Backward Difference Formula

The difficulty in FRAME 15 would be overcome if it could be arranged that the differences required were on a diagonal line sloping <u>upwards</u> to the right. This is accomplished by the use of the NEWTON-GREGORY BACKWARD DIFFERENCE FORMULA. As before $f_p = E^p f_0$ but now E is eliminated by using the formula $E = (1 - \nabla)^{-1}$. (Derived in 31A, page 196.)

Thus $f_p = (1 - \nabla)^{-p} f_0$.

Expand the R.H.S. of this as far as the term in ∇^3.
****

16A

$\{1 + p\nabla + \dfrac{p(p + 1)}{2!}\nabla^2 + \dfrac{p(p + 1)(p + 2)}{3!}\nabla^3 + \ldots\ldots\ldots\ldots\}f_0$

$= f_0 + p\nabla f_0 + \dfrac{(p + 1)p}{2!}\nabla^2 f_0 + \dfrac{(p + 2)(p + 1)p}{3!}\nabla^3 f_0 + \ldots\ldots\ldots\ldots$

211

The formula can thus be written as

$$f_p = f_0 + \binom{p}{1}\nabla f_0 + \binom{p+1}{2}\nabla^2 f_0 + \binom{p+2}{3}\nabla^3 f_0 + \dots\dots$$

If only the first few terms in this are used, it, like the forward difference formula, becomes a polynomial in p.

As an example, let us find $f(4\cdot2)$ from the table in 8A. x_0 can be taken as 4, then $p = 0\cdot2$ and the entries used will be

$$6$$
$$18$$
$$37$$
$$66$$

(Note: backward differences of f_0 lie on a diagonal line sloping upwards to the right.)

and so $f(4\cdot2) = 66 + \binom{0\cdot2}{1}37 + \binom{1\cdot2}{2}18 + \binom{2\cdot2}{3}6$

$$= 66 + 7\cdot4 + 2\cdot16 + 0\cdot528 = 76\cdot088$$

which agrees with $(4\cdot2)^3 + 2$.

Now use this formula to find $f(0\cdot43)$ from the table of values in 11A, page 185. **

$h = 0\cdot05, \quad p = 0\cdot6$

$f(0\cdot43) = 0\cdot3894 + \binom{0\cdot6}{1} \times 0\cdot0466 + \binom{1\cdot6}{2} \times (-0\cdot0007) + \binom{2\cdot6}{3} \times 0\cdot0002$

$$= 0\cdot3894 + 0\cdot027\,96 - 0\cdot000\,34 + 0\cdot000\,08 \simeq 0\cdot4171$$

Other Finite Difference Interpolation Formulae

There are other formulae besides the two already dealt with. We shall not go into the derivation of these here but simply quote some of them and do some examples. Notice that, in each of them, if the formula is truncated, a polynomial in p results.

BESSEL'S FORMULA is

$$f_p = \mu f_{\frac{1}{2}} + (p - \tfrac{1}{2})\delta f_{\frac{1}{2}} + \frac{p(p-1)}{2!}\mu\delta^2 f_{\frac{1}{2}} + \frac{p(p-1)(p-\frac{1}{2})}{3!}\delta^3 f_{\frac{1}{2}}$$
$$+ \frac{(p+1)p(p-1)(p-2)}{4!}\mu\delta^4 f_{\frac{1}{2}} + \dots\dots\dots$$

STIRLING'S FORMULA is

$$f_p = f_0 + p\mu\delta f_0 + \frac{p^2}{2!}\delta^2 f_0 + \frac{p(p^2-1)}{3!}\mu\delta^3 f_0 + \frac{p^2(p^2-1)}{4!}\delta^4 f_0 + \dots\dots$$

EVERETT'S FORMULA is

$$f_p = \binom{q}{1}f_0 + \binom{q+1}{3}\delta^2 f_0 + \binom{q+2}{5}\delta^4 f_0 + \dots\dots\dots\dots$$
$$+ \binom{p}{1}f_1 + \binom{p+1}{3}\delta^2 f_1 + \binom{p+2}{5}\delta^4 f_1 + \dots\dots\dots\dots$$

where $q = 1 - p$. Also, as $\binom{q+1}{3} = \frac{(q+1)q(q-1)}{3!} = \frac{q(q^2-1)}{3!}$ and similarly $\begin{pmatrix} q + 1 \\ 5 \end{pmatrix} = \dfrac{q(q^2 - 1)(q^2 - 4)}{5!}$,   the formula can be expressed in the alternative form

$$f_p = qf_0 + \frac{q(q^2 - 1)}{3!} \delta^2 f_0 + \frac{q(q^2 - 1)(q^2 - 4)}{5!} \delta^4 f_0 + \ldots\ldots\ldots\ldots$$

$$+ pf_1 + \frac{p(p^2 - 1)}{3!} \delta^2 f_1 + \frac{p(p^2 - 1)(p^2 - 4)}{5!} \delta^4 f_1 + \ldots\ldots\ldots\ldots$$

which you may prefer.

All of these formulae are central difference formulae as they are expressed in terms of δ rather than Δ or ∇, although it is possible to use Δ or ∇.   We don't suggest that you try to learn these formulae. You will also find that they are not always quoted by different authors in exactly the same form.

The main point about using any of these formulae is that you should be able to interpret the various δ expressions in terms of the entries in your difference table.   For your convenience we repeat here the table in FRAME 34 on page 197.

x_{-2}	f_{-2}				
		$\delta f_{-1\frac{1}{2}}$			
x_{-1}	f_{-1}		$\delta^2 f_{-1}$		
		$\delta f_{-\frac{1}{2}}$		$\delta^3 f_{-\frac{1}{2}}$	
x_0	f_0		$\delta^2 f_0$		$\delta^4 f_0$
		$\delta f_{\frac{1}{2}}$		$\delta^3 f_{\frac{1}{2}}$	
x_1	f_1		$\delta^2 f_1$		$\delta^4 f_1$
		$\delta f_{1\frac{1}{2}}$		$\delta^3 f_{1\frac{1}{2}}$	
x_2	f_2		$\delta^2 f_2$		$\delta^4 f_2$
		$\delta f_{2\frac{1}{2}}$		$\delta^3 f_{2\frac{1}{2}}$	
x_3	f_3		$\delta^2 f_3$		
		$\delta f_{3\frac{1}{2}}$			
x_4	f_4				

As an example on the formulae in the last frame, let us find $f(2\cdot36)$ from the following table:

x	1·6	1·8	2·0	2·2	2·4	2·6	2·8	3·0
f(x)	0·0495	0·0605	0·0739	0·0903	0·1102	0·1346	0·1644	0·2009

Which value of x would you label x_0 and what will then be the values of p and q?   **

$x_0 = 2\cdot2,$ $p = 0\cdot8,$ $q = 0\cdot2$ (Note: $h = 0\cdot2$)

Now form the difference table for the values given in the last frame and then state the values of the various δ quantities required for each of the three formulae in FRAME 18.   State also which is the last column of differences you would use.

            ****

	x	f					
	1·6	0·0495					
			110				
	1·8	0·0605		24			
			134		6		
	2·0	0·0739		30		−1	
			164		5		6
x_0	2·2	0·0903		35		5	−12
			199		10		−6
x_1	2·4	0·1102		45		−1	11
			244		9		5
	2·6	0·1346		54		4	
			298		13		
	2·8	0·1644		67			
			365				
	3·0	0·2009					

Bessel: $\mu f_{\frac{1}{2}} = 0\cdot100\,25,\quad \delta f_{\frac{1}{2}} = 0\cdot0199,\quad \mu\delta^2 f_{\frac{1}{2}} = 0\cdot0040,\quad \delta^3 f_{\frac{1}{2}} = 0\cdot0010,$
$\mu\delta^4 f_{\frac{1}{2}} = 0\cdot0002$

Stirling: $f_0 = 0\cdot0903,\quad \mu\delta f_0 = 0\cdot018\,15,\quad \delta^2 f_0 = 0\cdot0035,\quad \mu\delta^3 f_0 = 0\cdot000\,75,$
$\delta^4 f_0 = 0\cdot0005$

Everett: $f_0 = 0\cdot0903,\quad \delta^2 f_0 = 0\cdot0035,\quad \delta^4 f_0 = 0\cdot0005$
$f_1 = 0\cdot1102,\quad \delta^2 f_1 = 0\cdot0045,\quad \delta^4 f_1 = -0\cdot0001$

(Remember that all the entries in the difference columns have to be multiplied by 10^{-4}.)

Up to the 4th differences (no more can be relied upon).

Bessel's formula now gives
$$0\cdot100\,25 + 0\cdot3 \times 0\cdot0199 + \frac{0\cdot8 \times (-0\cdot2)}{2} \times 0\cdot0040 + \frac{0\cdot8 \times (-0\cdot2) \times 0\cdot3}{6} \times 0\cdot0010$$

$$+ \frac{1\cdot8 \times 0\cdot8 \times (-0\cdot2) \times (-1\cdot2)}{24} \times 0\cdot0002 \simeq 0\cdot1059$$

Try working out the value obtained by either of the other two formulae.
****

Stirling's formula gives:
$$0\cdot0903 + 0\cdot8 \times 0\cdot018\,15 + \frac{0\cdot8^2}{2} \times 0\cdot0035$$

$$+ \frac{0\cdot8(0\cdot8^2 - 1)}{6} \times 0\cdot000\,75 + \frac{0\cdot8^2(0\cdot8^2 - 1)}{24} \times 0\cdot0005 \simeq 0\cdot1059$$

Everett's formula gives:
$$0\cdot2 \times 0\cdot0903 + \frac{1\cdot2 \times 0\cdot2 \times (-0\cdot8)}{6} \times 0\cdot0035$$

$$+ \frac{2\cdot2 \times 1\cdot2 \times 0\cdot2 \times (-0\cdot8) \times (-1\cdot8)}{120} \times 0\cdot0005$$

$$+ \; 0 \cdot 8 \times 0 \cdot 1102 \; + \; \frac{1 \cdot 8 \times 0 \cdot 8 \times (-0 \cdot 2)}{6} \times 0 \cdot 0045$$

$$+ \; \frac{2 \cdot 8 \times 1 \cdot 8 \times 0 \cdot 8 \times (-0 \cdot 2) \times (-1 \cdot 2)}{120} \times (-0 \cdot 0001) \; \simeq \; 0 \cdot 1059$$

FRAME 22

You have seen that several formulae are available for interpolation, all doing more or less the same job.   Which one is it best to use in any particular case?   In FRAME 15, you saw that if it is necessary to interpolate with x_0 near the end of a table, then the Newton-Gregory forward difference formula is not suitable as the required differences are not available.   Similarly the Newton-Gregory backward difference formula would not be used near the beginning of a table, nor would central difference formulae be used near the beginning or end of a table.

If the table is sufficiently long, there may be an overlap where more than one type of formula can be used.   In such cases it is usually better to use one of the central difference formulae as these are likely to converge more quickly than the others.   There are sometimes reasons for choosing one central difference formula in preference to another but we shall not go into these here.   Because of its symmetry, Everett's is sometimes a popular choice.   However, when $p = \frac{1}{2}$, one form is particularly suitable.   Can you suggest which one it is?

****

22A

Bessel's is the obvious choice as all terms involving a factor $p - \frac{1}{2}$ will become zero.

Everett's is another reasonable choice, for, if $p = \frac{1}{2}$, then $q = p$, and the coefficients in the two rows become the same.

FRAME 23

One other point in favour of central difference formulae is that as they tend to converge more quickly, their use does not involve such high order differences as forward or backward difference formulae.   Consequently they are not so susceptible to the effects of any round-off errors that may exist in the tabulated values.

FRAME 24

In working the example quoted in FRAME 19, only differences up to the fourth order were used, higher differences being too unreliable.   You also found that some of the terms in the formulae were so small when you calculated their values as to be negligible.   Generally, when using an interpolation formula, only the first few terms in it are necessary, the rest being either so small as to be negligible to the degree of accuracy to which you are working or unreliable because of round-off errors in the difference table.   It is because successive terms in a formula become smaller in magnitude and eventually negligible, i.e., the series in the formula converges, that one can use the binomial theorem in FRAME 7.

Unequal Intervals of Tabulation

All the work on interpolation done so far has depended on a function being tabulated at equal intervals of x.   But this might not happen and so the question immediately arises: Can anything be done in this case?

First of all, let us look at some of the places where our work will break down.   Right at the start, our table of finite differences will no longer behave nicely as it did before.   To see this try forming a difference table for

x	0	1	3	5	8	9	11	15	18	20
f(x)	-4	-3	5	21	60	77	117	221	320	396

going as far as the fourth differences.

****

```
0    -4    ★
                1
1    -3              7
                8          1
3     5              8           14
               16          15
5    21             23           -60
               39         -45
8    60            -22            90
               17          45
9    77             23           -4
               40          41
11   117            64          -110
              104         -69
15   221            -5            51
               99         -18
18   320           -23
               76
20   396
```

The equation connecting x and $f(x)$ was $f(x) = x^2 - 4$.   But the column of second differences is not now constant, neither are all higher differences zero.

Divided Differences

Another trouble that arises is that h is no longer constant, and this leads to difficulty with p.

Returning to the question of differences, you remember that forming a first difference is equivalent to the first part of the process of differentiation from first principles.   (See FRAME 5, page 181 if you have forgotten this.)   The next step in differentiation from first principles is to divide $f(x_0 + h) - f(x_0)$ by h.   The corresponding process here is to divide each first difference by the difference between the two corresponding values of x.   The result of this division is called a DIVIDED DIFFERENCE.   Thus, going down the column headed by ★ in 25A, 1 is divided by 1 - 0, 8 by 3 - 1, 16 by 5 - 3, 39 by 8 - and so on.   The results form the first column of a table of divided differences.   Just as the differences in the first column in an ordinary table are known as first differences, so these values are called FIRST DIVIDED DIFFERENCES.

Work out this column for the function f(x) in FRAME 25.
                 ****
                                                                        26A

0	-4	
		1
1	-3	
		4
3	5	
		8
5	21	
		13
8	60	
		17
9	77	
		20
11	117	
		26
15	221	
		33
18	320	
		38
20	396	

The SECOND DIVIDED DIFFERENCES are formed in a somewhat similar manner.

To find the entry labelled A, a fan is taken from A back to the tabulated values.   A is now the ordinary difference 4 - 1 divided by the difference of the two values of x corresponding to the front of the fan i.e.

3 - 0.   Thus $A = \dfrac{4 - 1}{3 - 0} = 1.$ Similarly

$B = \dfrac{8 - 4}{5 - 1} = 1$ and $C = \dfrac{13 - 8}{8 - 3} = 1.$ Continue

in this way and so find the remainder of the entries in the second divided differences column.
                 ****
                                                                        27A

All entries in this column are 1.

We now have the table shown on page 218.   Third and higher divided differences are found in a similar way.   Thus to find D a fan is constructed backwards to the functional values and then

$D = \dfrac{1 - 1}{5 - 0} = 0.$ Similarly $E = \dfrac{1 - 1}{8 - 1} = 0$ and so on.   It is easy to see

that in this example all the third and consequently higher divided differences are zero.

0	-4		
		1	
1	-3		1
		4	
3	5		1	D
		8	
5	21		1	E
		13	
8	60		1
		17	
9	77		1
		20	
11	117		1
		26	
15	221		1
		33	
18	320		1
		38	
20	396		

Now, form a divided differences table for

x	-2	-1	3	5	8	9	14	15	18	24
f(x)	-26	-14	34	226	994	1426	5446	6706	11 614	27 586

****

-2	-26			
		12		
-1	-14		0	
		12		2
3	34		14	
		96		2
5	226		32	
		256		2
8	994		44	
		432		2
9	1426		62	
		804		2
14	5446		76	
		1260		2
15	6706		94	
		1636		2
18	11 614		114	
		2662		
24	27 586			

and all higher divided differences are zero.

The table in the last frame was formed from the cubic $f(x) = 2x^3 - 2x - 1$
and you will have noticed that the <u>third</u> divided differences are constant.
In the quadratic in FRAME 25, the <u>second</u> divided differences were
constant.   In a similar way, the <u>fourth</u> divided differences of a quartic
are constant and so on.   All differences after a constant column will, of
course, be zero.

When forming divided differences, it is not even necessary for the x's to be in numerical order.   Take the order   x = 0, 3, 9, 5, 11, 18, 8, 20, 1, 15   and construct the new divided differences table for the function given by the table in FRAME 25.

****

30A

0	-4			
		3		
3	5		1	
		12		
9	77		1	
		14		
5	21		1	
		16		
11	117		1	
		29		
18	320		1	
		26		
8	60		1	
		28		
20	396		1	
		21		
1	-3		1	
		16		
5	221			

When we were considering ordinary differences, it was found useful to have a notation (actually three notations were used) for them.   In a similar way it is useful to have a notation for divided differences.   The following table shows one notation:

x_0 $f(x_0)$

$\qquad f(x_0, x_1)$

x_1 $f(x_1)$ $\qquad f(x_0, x_1, x_2)$

$\qquad f(x_1, x_2)$ $\qquad f(x_0, x_1, x_2, x_3)$

x_2 $f(x_2)$ $\qquad f(x_1, x_2, x_3)$ $\qquad f(x_0, x_1, x_2, x_3, x_4)$

$\qquad f(x_2, x_3)$ $\qquad f(x_1, x_2, x_3, x_4)$

x_3 $f(x_3)$ $\qquad f(x_2, x_3, x_4)$ $\qquad f(x_1, x_2, x_3, x_4, x_5)$

$\qquad f(x_3, x_4)$ $\qquad f(x_2, x_3, x_4, x_5)$

x_4 $f(x_4)$ $\qquad f(x_3, x_4, x_5)$ $\qquad f(x_2, x_3, x_4, x_5, x_6)$

$\qquad f(x_4, x_5)$ $\qquad f(x_3, x_4, x_5, x_6)$

x_5 $f(x_5)$ $\qquad f(x_4, x_5, x_6)$

$\qquad f(x_5, x_6)$

x_6 $f(x_6)$

As you know, ordinary differences can be expressed in terms of the functional values.   For example, $\Delta f_2 = f_3 - f_2$ and $\delta^2 f_0 = f_1 - 2f_0 + f_{-1}$. Divided differences can be expressed in terms of functional values and the values of x.   Thus, remembering the way in which divided differences are formed,

$$f(x_0, x_1) = \frac{f(x_1) - f(x_0)}{x_1 - x_0} \qquad (31.1)$$

$$f(x_1, x_2) = \frac{f(x_2) - f(x_1)}{x_2 - x_1} \qquad (31.2)$$

and so on for the other first divided differences.

What will be $f(x_3, x_4)$ and $f(x_5, x_6)$?

****

31A

$$f(x_3, x_4) = \frac{f(x_4) - f(x_3)}{x_4 - x_3} \qquad\qquad f(x_5, x_6) = \frac{f(x_6) - f(x_5)}{x_6 - x_5}$$

FRAME 32

A little later on, you will see a pattern emerging for the values of divided differences.   To fit in with this pattern, (31.1) can be written

as $f(x_0, x_1) = \dfrac{f(x_0)}{x_0 - x_1} + \dfrac{f(x_1)}{x_1 - x_0}$ and (31.2) as

$$f(x_1, x_2) = \frac{f(x_1)}{x_1 - x_2} + \frac{f(x_2)}{x_2 - x_1}$$

and similarly for the other first differences.

Turning now to second differences,

$$f(x_0, x_1, x_2) = \frac{f(x_1, x_2) - f(x_0, x_1)}{x_2 - x_0} \qquad (32.1)$$

$$f(x_1, x_2, x_3) = \frac{f(x_2, x_3) - f(x_1, x_2)}{x_3 - x_1} \qquad \text{etc.}$$

What will be the formula for $f(x_2, x_3, x_4)$?

****

32A

$$\frac{f(x_3, x_4) - f(x_2, x_3)}{x_4 - x_2}$$

FRAME 33

In (32.1), the formulae (31.2) and (31.1) can be used for the terms in the numerator.   Doing this gives

$$f(x_0, x_1, x_2) = \frac{\dfrac{f(x_2) - f(x_1)}{x_2 - x_1} - \dfrac{f(x_1) - f(x_0)}{x_1 - x_0}}{x_2 - x_0}$$

which can be arranged as

$$f(x_0, x_1, x_2) = \frac{f(x_0)}{(x_0 - x_1)(x_0 - x_2)} + \frac{f(x_1)}{(x_1 - x_0)(x_1 - x_2)} + \frac{f(x_2)}{(x_2 - x_0)(x_2 - x_1)}$$

$$(33.1)$$

Similarly,

$$f(x_1, x_2, x_3) = \frac{f(x_1)}{(x_1 - x_2)(x_1 - x_3)} + \frac{f(x_2)}{(x_2 - x_1)(x_2 - x_3)} + \frac{f(x_3)}{(x_3 - x_1)(x_3 - x_2)}$$

FRAME 34

As an example of a third divided difference,

$$f(x_0, x_1, x_2, x_3) = \frac{f(x_1, x_2, x_3) - f(x_0, x_1, x_2)}{x_3 - x_0}$$

Substitute for the expressions in the numerator and obtain this in a form corresponding to that of (33.1) for $f(x_0, x_1, x_2)$.

****

34A

$$f(x_0, x_1, x_2, x_3) = \frac{f(x_0)}{(x_0 - x_1)(x_0 - x_2)(x_0 - x_3)} + \frac{f(x_1)}{(x_1 - x_0)(x_1 - x_2)(x_1 - x_3)}$$

$$+ \frac{f(x_2)}{(x_2 - x_0)(x_2 - x_1)(x_2 - x_3)} + \frac{f(x_3)}{(x_3 - x_0)(x_3 - x_1)(x_3 - x_2)}$$

You should now be able to see the pattern that is emerging and how the first divided differences fit into it.

FRAME 35

What do you think will be the form for $f(x_0, x_1, x_2, \ldots\ldots, x_n)$?

****

35A

$$\frac{f(x_0)}{(x_0 - x_1)(x_0 - x_2) \ldots (x_0 - x_n)} + \frac{f(x_1)}{(x_1 - x_0)(x_1 - x_2) \ldots (x_1 - x_n)}$$

$$+ \frac{f(x_2)}{(x_2 - x_0)(x_2 - x_1)(x_2 - x_3) \ldots (x_2 - x_n)} + \ldots\ldots\ldots\ldots\ldots$$

$$+ \frac{f(x_n)}{(x_n - x_0)(x_n - x_1) \ldots (x_n - x_{n-1})} \qquad (35A.1)$$

FRAME 36

Lagrange's Interpolation Formula

It was pointed out in FRAME 6 that many curves can be drawn through a number of points.   It was also suggested that in practical situations, where the points represent the results of an experiment say, they are more likely to lie on a simple curve than one that jumps about all over the place.

INTERPOLATION

Now, if two points only are given, the algebraically simplest curve passing through them is a straight line, the equation of which is linear and contains two constants.   Taking the equation as $f(x) = a_0 + a_1 x$, these constants are a_0 and a_1.   What do you think will be the algebraically simplest curve passing through three given points, assuming that the three points are not collinear?

****

36A

A parabola, whose equation is quadratic and is of the form $f(x) = a_0 + a_1 x + a_2 x^2$, *containing three constants* a_0, a_1, a_2.

FRAME 37

Similarly, if four points are given which do not lie on either a straight line or parabola, the simplest curve passing through them is a cubic whose equation is of the form $f(x) = a_0 + a_1 x + a_2 x^2 + a_3 x^3$, containing four constants.   This argument can obviously be extended to n + 1 given points, in which case the simplest equation would be of the form $f(x) = a_0 + a_1 x + a_2 x^2 + \ldots\ldots\ldots + a_n x^n$, containing n + 1 constants. The Lagrange interpolation formula enables us to write down this equation immediately, although not quite in this form.

FRAME 38

In FRAMES 28-29 it was seen that the second divided differences of a quadratic are constant and the third are zero, also that for a cubic the third divided differences are constant and the fourth zero.   For a polynomial of degree n, the nth divided differences are constant and the (n + 1)th are zero.   Thus the (n + 1)th divided differences of $f(x) = a_0 + a_1 x + a_2 x^2 + \ldots\ldots\ldots + a_n x^n$ are zero.

What is the least number of functional values you must have in order to be able to find one (n + 1)th divided difference?

****

38A

n + 2

If you are given two functional values you can obtain one first difference if you are given three values you can obtain two first differences and one second difference, and so on.

FRAME 39

Now a polynomial of degree n can be made to pass through n + 1 points. But for an (n + 1)th difference we require n + 2 points.   For this extra point we take an unspecified value which we can call x and then f(x) which is unknown and is what we are seeking, is the corresponding functional value.   As the (n + 1)th divided difference is zero,

$$f(x_0, x_1, x_2, \ldots\ldots, x_{n-1}, x_n, x) = 0 \qquad (39.1)$$

where $x_0, x_1, x_2, \ldots\ldots, x_{n-1}, x_n$ are the x-values of the given n + 1 points.

What will (39.1) become when the L.H.S. is written in a form similar to (35A.1)?

****

$$\frac{f(x_0)}{(x_0 - x_1)(x_0 - x_2) \ldots (x_0 - x_n)(x_0 - x)}$$

$$+ \frac{f(x_1)}{(x_1 - x_0)(x_1 - x_2) \ldots (x_1 - x_n)(x_1 - x)} + \ldots \ldots$$

$$+ \frac{f(x_n)}{(x_n - x_0)(x_n - x_1) \ldots (x_n - x_{n-1})(x_n - x)}$$

$$+ \frac{f(x)}{(x - x_0)(x - x_1) \ldots (x - x_n)} = 0$$

This can be rewritten as

$$\frac{f(x)}{(x - x_0)(x - x_1) \ldots (x - x_n)} = \frac{f(x_0)}{(x - x_0)(x_0 - x_1)(x_0 - x_2) \ldots (x_0 - x_n)}$$

$$\frac{f(x_1)}{(x - x_1)(x_1 - x_0)(x_1 - x_2) \ldots (x_1 - x_n)} + \ldots \ldots$$

$$\frac{f(x_n)}{(x - x_n)(x_n - x_0)(x_n - x_1) \ldots (x_n - x_{n-1})}$$

and this is the LAGRANGE INTERPOLATION FORMULA.   x can now be given any
desired value.   Notice that although this formula has been derived to
deal with the case of a function tabulated at unequal intervals of x, it
can still be used even if the function is tabulated at equal intervals, as
the result depends only on the values of x and f(x) at the tabulated
points.

The formula can alternatively be written as

$$f(x) = \frac{(x - x_1)(x - x_2) \ldots \ldots (x - x_n)}{(x_0 - x_1)(x_0 - x_2) \ldots (x_0 - x_n)} f(x_0)$$

$$+ \frac{(x - x_0)(x - x_2) \ldots \ldots (x - x_n)}{(x_1 - x_0)(x_1 - x_2) \ldots (x_1 - x_n)} f(x_1) + \ldots \ldots$$

$$+ \frac{(x - x_0)(x - x_1) \ldots \ldots (x - x_{n-1})}{(x_n - x_0)(x_n - x_1) \ldots (x_n - x_{n-1})} f(x_n)$$

and in this form it is obvious that it is a polynomial of degree n.

As an example, let us find the polynomial that will fit (0, -4), (1, -3)
and (3, 5), three points which do not lie on a straight line.   As we are
given three points, n + 1 = 3, so n = 2 and the polynomial, which will
be of degree 2, is given by

FRAME 41   (continued)

$$\frac{f(x)}{(x - 0)(x - 1)(x - 3)} = \frac{-4}{(x - 0)(0 - 1)(0 - 3)} + \frac{-3}{(x - 1)(1 - 0)(1 - 3)}$$
$$+ \frac{5}{(x - 3)(3 - 0)(3 - 1)}$$

As you can verify, this reduces to $f(x) = x^2 - 4$.   The points were actually the first three from the table in FRAME 25.

Now find the polynomial of degree three that will fit the points (-1, -3), (0, -1),   (1, 1) and (3, 29).

****

41A

$$\frac{f(x)}{(x + 1)(x - 0)(x - 1)(x - 3)} = \frac{-3}{(x + 1)(-1 - 0)(-1 - 1)(-1 - 3)} +$$

$$\frac{-1}{(x - 0)(0 + 1)(0 - 1)(0 - 3)} + \frac{1}{(x - 1)(1 + 1)(1 - 0)(1 - 3)} +$$

$$\frac{29}{(x - 3)(3 + 1)(3 - 0)(3 - 1)}$$

which reduces to $f(x) = x^3 + x - 1.$

FRAME 42

If the value of f(x) is required when x = 2, say, this can now be found immediately.   If this <u>only</u> is required (i.e., the algebraic formula is not asked for first) then f(2) can be found directly from

$$\frac{f(2)}{(2 + 1)(2 - 0)(2 - 1)(2 - 3)} = \frac{-3}{(2 + 1)(-1 - 0)(-1 - 1)(-1 - 3)} +$$

$$\frac{-1}{(2 - 0)(0 + 1)(0 - 1)(0 - 3)} + \frac{1}{(2 - 1)(1 + 1)(1 - 0)(1 - 3)} +$$

$$\frac{29}{(2 - 3)(3 + 1)(3 - 0)(3 - 1)}$$

giving f(2) = 9.

FRAME 43

A flow diagram for using Lagrangian Interpolation to find the value of a function f(x) at a given value of x is shown on page 225.

FRAME 44

One other thing remains to be said about Lagrange's formula.   Suppose the whole of the table in FRAME 25 were given.   As there are 10 points, it is possible that the polynomial sought is of degree 9, although actually we know that it is only of degree 2.   If you wish, you can proceed directly, using the formula with all 10 points, but this will entail an unnecessary amount of work.   As there is always the possibility that a set of n points may lie on a curve of degree less than n - 1, it may be as well to test first to see if this is so.   The method of testing is to construct a divided differences table.   If this gives a constant column, then the order of this column is the degree of the polynomial sought. For example, the table in 28A indicates that the relevant polynomial is of

Flow diagram for FRAME 43.

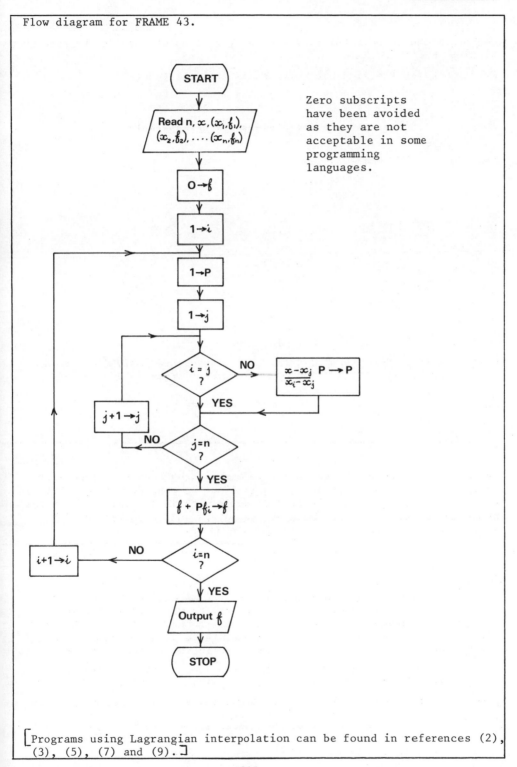

Zero subscripts
have been avoided
as they are not
acceptable in some
programming
languages.

[Programs using Lagrangian interpolation can be found in references (2), (3), (5), (7) and (9).]

degree 3 and so only the first, or any, four points need to be used when finding f(x).

Find f(2) for the simplest curve satisfying the points

x	0	1	-2	3	-3	6	10
f(x)	3	0	21	6	36	45	153

****

44A

The divided difference table

0	*3*		
		-3	
1	*0*		*2*
		-7	
-2	*21*		*2*
		-3	
3	*6*		*2*
		-5	
-3	*36*		*2*
		1	
6	*45*		*2*
		27	
10	*153*		

indicates that f(x) is of degree 2.   Consequently only three points need be used, say (0, 3), (1, 0) and (3, 6).   Then

$$\frac{f(2)}{2 \times 1 \times (-1)} = \frac{3}{2 \times (-1) \times (-3)} + \frac{0}{1 \times 1 \times (-2)} + \frac{6}{-1 \times 3 \times 2}$$

from which f(2) = 1.

Certain modifications to the Lagrange method have been suggested, for example, that due to Aitken.   This will not be considered here, but, if you are interested, you will find it dealt with in several of the books on numerical work.   One such book is "Computers and Computing" by J.M. Rushforth and J.Ll. Morris (John Wiley & Sons).

A Word of Warning

In all our work on interpolation it has been assumed that, if y is a function of x, then a polynomial f(x) exists which is such that y = f(x) is an adequate representation of the curve on which the points lie.   This may not be so.   For example, there is no polynomial in x which adequately represents f(x) as given in the following table:

x	0	0·5	1	1·5	2	2·5	3·0
f(x)	0	0·8409	1	1·1066	1·1892	1·2574	1·3161

Form a difference table from these values and see what happens when you do so.

****

0	0						
		8409					
0·5	0·8409		-6818				
		1591		6293			
1·0	1·0000		-525		-6008		
		1066		285		5819	
1·5	1·1066		-240		-189		
		826		96		142	
2·0	1·1892		-144		-47		
		682		49			
2·5	1·2574		-95				
		587					
3·0	1·3161						

You will notice that there is no column containing only small differences.

A difference table like this would give us the clue that it is not reasonable to try and proceed further. The values tabulated above were taken from the function $f(x) = x^{\frac{1}{4}}$ and here we have a situation somewhat analogous to that in analytical work if an attempt is made to obtain a Maclaurin series for $x^{\frac{1}{4}}$. If Lagrange's method is used there is no built in early warning system, because, although this method is based <u>on</u> divided differences, these do not <u>have</u> to be found.

Inverse Interpolation

The process we have concentrated on so far is: Given a table of values of x and f(x), what is the value of f(x) at some other value of x? The inverse problem is: Given a table of values of x and f(x), what value of x will correspond to some other value of f(x)? The problem of finding such an x is known as INVERSE INTERPOLATION. As an example of this, one might be asked: From the table in FRAME 3, at what time would the temperature have been 70° C? As you might suspect from previous work, there are certain methods of dealing with a problem of this sort when the original table is given at unequal intervals of the independent variable and others which can only be used if the original table is an equal interval one.

To take an example, suppose a curve passes through the points (0, -4), (0·6, -3·64), (1, -3) and it is necessary to find the value of x when y = -3·5.

You will notice that the intervals between successive values of x are unequal. One way in which we can proceed is to find y using the Lagrangian interpolation formula in FRAME 40, put the result equal to -3·5 and solve the equation so formed for x.

What will you get if you do it this way?
****

$$\frac{y}{x(x - 0\cdot6)(x - 1)} = \frac{-4}{x(-0\cdot6)(-1)} + \frac{-3\cdot64}{(x - 0\cdot6)\,0\cdot6\,(-0\cdot4)} + \frac{-3}{(x - 1)0\cdot4} \,,$$

i.e., $y = x^2 - 4$ (49A.1)

$x^2 - 4 = -3\cdot5$ gives $x \simeq \pm0\cdot71$.   To keep x within the range 0 to 1, i.e., within the range of the given three points, take $x = 0\cdot71$.

If $f(x)$ is of higher degree than two, then the Newton-Raphson process can be used to solve the equation obtained for x.

A second method is to interchange the roles of x and y.   This means that an equation is found for x as a polynomial function of y.   Once this has been done, the given value of y can be inserted into it to give the required value of x.   These two steps can be combined into one as in FRAME 42, but in order that you can see fully what is involved, this will not be done in this example.

Use the three points $(0, -4)$, $(0\cdot6, -3\cdot64)$, $(1, -3)$ to obtain a Lagrangian interpolation formula for x in terms of y.

****

$$\frac{x}{(y + 4)(y + 3\cdot64)(y + 3)} = \frac{0}{(y + 4)(-0\cdot36)(-1)} + \frac{0\cdot6}{(y + 3\cdot64)0\cdot36(-0\cdot64)}$$

$$+ \frac{1}{(y + 3)0\cdot64}$$

$x = -1\cdot041\,67y^2 - 6\cdot291\,67y - 8\cdot5$ (50A.1)

When $y = -3\cdot5$,   this gives $x \simeq 0\cdot76$,   which is not the same as that obtained before.   So the question now is: Why the difference?

Can you spot the reason?   [Don't worry if you can't, it is a bit tricky. As a hint, it is suggested that you consider sketches of the two curves represented by (49A.1) and (50A.1).]

****

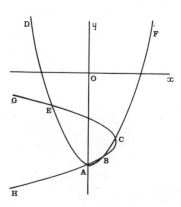

In both 49A and 50A you found a parabola passing through the three given points, A, B and C.   However the one found in 49A had its axis parallel to Oy (the parabola DEABCF) while that found in 50A had its axis parallel to Ox (GECBAH), and these two parabolas are not the same.   For an intermediate value of y, they give different values for x.

As only three points are known, and there is actually no information about what happens in between these points, one cannot be definite and say, with no shadow of doubt, which, if indeed either, should be taken. As we have based our direct interpolation on a polynomial of the form y = f(x), it would seem logical to continue to use one of this form for inverse interpolation and so quote the answer as approximately 0·71.

Sometimes an answer can be immediately discarded. For example, in 41A you found a polynomial to fit the points (−1, −3), (0, −1), (1, 1) and (3, 29). Suppose you were required to find the value of x when y = 20, knowing in addition that, over the range given, y is increasing as x increases. The value you would find by the second method is approximately 6·4 and so you would immediately reject it under the conditions given.

When the original table is given at equal intervals of x, both of the methods just given can be used, care being taken to reject any ridiculous answers. Alternatively, finite differences can now be used, the method, however, being basically similar to that in FRAME 49 and 49A. To see how it works, let us return to the difference table in 20A and seek the value of x for which f(x) = 0·08. Obviously the value required will lie between 2·0 and 2·2. For this calculation, x_0 is now put equal to 2·0 and x_1, 2·2. Any convenient interpolation formula is now used, putting in it f_p = 0·08, f_0 = 0·0739, etc., but leaving p as p.

Suppose we use Bessel's formula. What equation will you get when you make these substitutions?

****

53A

$$0·08 = 0·0821 + (p - \tfrac{1}{2}) \times 0·0164 + \frac{p(p - 1)}{2} \times 0·003\,25$$

$$+ \frac{p(p - 1)(p - \tfrac{1}{2})}{6} \times 0·0005 + \frac{(p + 1)p(p - 1)(p - 2)}{24} \times 0·0002,$$

stopping at the fourth differences as no more are reliable.

This is now a quartic equation in p, but instead of solving it, as is possible, by Newton-Raphson, an alternative iterative technique is often used at this stage. By solving for the p in the term $(p - \tfrac{1}{2}) \times 0·0164$, the quartic can be rewritten as

$$p = 0·371\,95 - 0·099\,09p(p - 1) - 0·005\,08p(p - 1)(p - \tfrac{1}{2})$$

$$- 0·000\,51(p + 1)p(p - 1)(p - 2) \qquad (54.1)$$

A first estimate of p can be taken as p_0 = 0·372, say (i.e. the constant term on the R.H.S.). A second estimate, p_1, is obtained by substituting this value of p into the R.H.S. Thus

$$p_1 = 0·371\,95 - 0·099\,09 \times 0·372 \times (-0·628)$$
$$- 0·005\,08 \times 0·372 \times (-0·628) \times (-0·128)$$
$$- 0·000\,51 \times 1·372 \times 0·372 \times (-0·628) \times (-1·628)$$

$$= 0·371\,95 + 0·023\,15 - 0·000\,15 - 0·000\,27 = 0·394\,68$$

229

FRAME 54   (continued)

A third estimate, p_2, is obtained by using this value of p_1 in the R.H.S. of (54.1).   What do you get when you do this?

****

54A

$0 \cdot 395\ 23$

FRAME 55

Using this value in the R.H.S. of (54.1) produces no further change in the estimate of p and so this gives the value we want.

Then, as $x = x_0 + ph$, $x = 2 \cdot 0 + 0 \cdot 395\ 23 \times 0 \cdot 2 = 2 \cdot 079\ 046$.

There is no reason why it has to be Bessel's formula which is used for this process.   Repeat this example using Everett's formula (FRAME 18). (q must be replaced immediately by $1 - p$).

****

55A

$$0 \cdot 08 = (1 - p) \times 0 \cdot 0739 + \frac{(2 - p)(1 - p)(-p)}{6} \times 0 \cdot 0030$$

$$+ \frac{(3 - p)(2 - p)(1 - p)(-p)(-1 - p)}{24} \times (-0 \cdot 0001) + p \times 0 \cdot 0903$$

$$+ \frac{(p + 1)p(p - 1)}{6} \times 0 \cdot 0035 + \frac{(p + 2)(p + 1)p(p - 1)(p - 2)}{24} \times 0 \cdot 0005$$

Solving for p from the two linear terms,

$$p = 0 \cdot 371\ 95 + p(p - 1)(p - 2)0 \cdot 030\ 49 - p(p^2 - 1)(p - 2)(p - 3)0 \cdot 0003$$
$$- p(p^2 - 1)0 \cdot 035\ 57 - p(p^2 - 1)(p^2 - 4)0 \cdot 0013$$

Successive estimates for p are $0 \cdot 372$, $0 \cdot 394\ 83$, $0 \cdot 395\ 37$, $0 \cdot 395\ 38$. *x is then* $2 \cdot 079\ 076$.

FRAME 56

In the last frame, the two values of p do not agree beyond the third decimal place.   The main cause of this was the division by $0 \cdot 0164$, the first difference coefficient of p.   (In 53A, $0 \cdot 0164$ appeared directly; in 55A it comes from $0 \cdot 0903 - 0 \cdot 0739$.)   As this division is equivalent to multiplying by approximately 61, it can cause quite a difference to the final answer.

FRAME 57

Summary of Interpolation Formulae

For your convenience, the various interpolation formulae mentioned in this programme are listed below:

Newton–Gregory Forward

$$f_p = f_0 + \binom{p}{1} \Delta f_0 + \binom{p}{2} \Delta^2 f_0 + \binom{p}{3} \Delta^3 f_0 + \ldots\ldots\ldots\ldots$$

Newton–Gregory Backward

$$f_p = f_0 + p\nabla f_0 + \binom{p+1}{2} \nabla^2 f_0 + \binom{p+2}{3} \nabla^3 f_0 + \ldots\ldots\ldots\ldots$$

Bessel

$$f_p = \mu f_{\frac{1}{2}} + (p - \tfrac{1}{2})\delta f_{\frac{1}{2}} + \frac{p(p-1)}{2!}\mu\delta^2 f_{\frac{1}{2}} + \frac{p(p-1)(p-\frac{1}{2})}{3!}\delta^3 f_{\frac{1}{2}}$$

$$+ \frac{(p+1)p(p-1)(p-2)}{4!}\mu\delta^4 f_{\frac{1}{2}} + \ldots\ldots\ldots\ldots$$

Stirling

$$f_p = f_0 + p\mu\delta f_0 + \frac{p^2}{2!}\delta^2 f_0 + \frac{p(p^2-1)}{3!}\mu\delta^3 f_0 + \frac{p^2(p^2-1)}{4!}\delta^4 f_0 + \ldots\ldots$$

Everett

$$f_p = \binom{q}{1}f_0 + \binom{q+1}{3}\delta^2 f_0 + \binom{q+2}{5}\delta^4 f_0 + \ldots\ldots\ldots\ldots$$

$$+ \binom{p}{1}f_1 + \binom{p+1}{3}\delta^2 f_1 + \binom{p+2}{5}\delta^4 f_1 + \ldots\ldots\ldots\ldots$$

$$(q = 1 - p)$$

Lagrange

$$\frac{f(x)}{(x-x_0)(x-x_1)\ldots(x-x_n)} = \frac{f(x_0)}{(x-x_0)(x_0-x_1)(x_0-x_2)\ldots(x_0-x_n)}$$

$$+ \frac{f(x_1)}{(x-x_1)(x_1-x_0)(x_1-x_2)\ldots(x_1-x_n)} + \ldots\ldots\ldots\ldots$$

$$+ \frac{f(x_n)}{(x-x_n)(x_n-x_0)(x_n-x_1)\ldots(x_n-x_{n-1})}$$

Miscellaneous Examples

In this frame a collection of miscellaneous examples is given for you to try.   Answers are provided in FRAME 59, together with such working as is considered helpful.

Note: The data from some of these questions will be used in subsequent programmes.   You will find it helpful then to be able to refer quickly to the difference tables that you will now be forming.

1.   The following readings of current, i, against deflection, θ, were obtained for a certain galvanometer.

θ	0·40	0·45	0·50	0·55	0·60	0·65
i	1·268	1·449	1·639	1·839	2·052	2·281

What will be the current when $\theta = 0\cdot536$?

2.   i)   A polynomial f(x) of low degree is tabulated as follows:

x	-3	-2	-1	0	1	2	3	4	5	6
f(x)	-14	-2	2	4	10	25	59	112	194	310

Errors in two values of f(x) are suspected.   Locate and correct them.

ii) Derive the Newton interpolation formulae:

a) $f_p = f_0 + \binom{p}{1} \Delta f_0 + \binom{p}{2} \Delta^2 f_0 + \binom{p}{3} \Delta^3 f_0 + \dots$

b) $f_p = f_0 + \binom{p}{1} \nabla f_0 + \binom{p+1}{2} \nabla^2 f_0 + \binom{p+2}{3} \nabla^3 f_0 + \dots$

iii) Obtain the values of $f(x)$ as given in (i) after correction when $x = -4$, $-2 \cdot 5$, $5 \cdot 5$. (L.U.)

3. A fourth degree polynomial is tabulated as follows:

x	0	0·1	0·2	0·3	0·4
y	1·0000	0·9208	0·6928	0·3448	−0·0752

x	0·5	0·6	0·7	0·8	0·9
y	−0·5000	−0·8452	−0·9992	−0·8432	−0·2312

Show from a difference table that there is an error and use the corrected table with the Stirling interpolation formula

$$f_p = f_0 + \tfrac{1}{2}p(\delta f_{\frac{1}{2}} + \delta f_{-\frac{1}{2}}) + \tfrac{1}{2}p^2 \delta^2 f_0 + \frac{p(p^2 - 1)}{2(3!)} (\delta^3 f_{\frac{1}{2}} + \delta^3 f_{-\frac{1}{2}})$$

$$+ \frac{p^2(p^2 - 1)}{4!} \delta^4 f_0 + \dots$$

to find the value of y when $x = 0 \cdot 45$. (L.U.)

(It was pointed out in FRAME 18 that interpolation formulae are not always quoted in exactly the same form. You should check that the form given in this question agrees with that in FRAME 18.)

4. The following table gives readings of the temperature (θ°) recorded at given times (t):

t	0	1	2	3	4	5	6	7	8	9	10
θ	80·00	70·48	61·87	54·08	47·03	40·65	34·88	29·66	24·93	20·66	16·79

Using (i) Bessel's formula, (ii) Everett's formula, find θ at $t = 4 \cdot 3$. At what time would you expect θ to be 50?

5. The deflection, y, measured at various distances, x, from one end of a cantilever is given by

x	0·0	0·2	0·4	0·6	0·8	1·0
y	0·0000	0·0347	0·1173	0·2160	0·2987	0·3333

For what value of x is $y = 0 \cdot 2$?

6. Write down the Lagrange interpolation polynomial which fits the points $(0, y_0)$, $(1, y_1)$, $(2, y_2)$. Express y_1 and y_2 in terms of y_0 and its differences and hence obtain the Gregory-Newton polynomial.

Write down the Lagrange polynomial which fits the points $(1, 4)$, $(3, 7)$, $(4, 8)$, $(6, 11)$ and use it to interpolate values of y at $x = 2$ and $x = 5$. Check these values by differencing. (L.U.)

7. The following table gives values of the current i in a certain L,C,R circuit at various times t.

t	0·000	0·002	0·004	0·006	0·008	0·010	0·012
i	0·000 00	0·015 63	0·024 63	0·029 31	0·031 23	0·031 41	0·030 54

t	0·014	0·016	0·018	0·020
i	0·029 05	0·027 23	0·025 29	0·023 31

Find i when t = 0·003, 0·0125 and 0·0184 and t when i = 0·030 00.

8. The following table contains one incorrect entry for f(x).   Locate
    the error; suggest a possible reason for its occurrence and a
    suitable correction.   Using this correction, draw up a corrected
    difference table.

x	1·1	1·2	1·3	1·4	1·5	1·6	1·7	1·8	1·9
f(x)	5·743	3·959	2·511	1·486	0·969	1·046	1·800	3·341	5·673

x	2·0	2·1
f(x)	8·965	13·274

Use your corrected table and an interpolation formula to calculate
f(1·45).   Any interpolation formula may be used but it should be
clearly stated before use and all symbols defined.   (L.U.)

Answers to Miscellaneous Examples

1. 1·782

2. i)   25 and 59 should be 26 and 58 respectively.
    ii)   −40, −6·625, 247·375.   These three are obtained by extending
          the difference table back one step and by the use of formulae
          (ii) (a) and (ii) (b) respectively.

3. Corrected value: x = 0·6, y = −0·8432
    When x = 0·45, y = −0·291 95.

4. (i) 45·05   (ii) 45·05

    Using the Newton–Gregory forwards difference formula (any other may be
    chosen) and solving iteratively for p, taking t_0 = 3, gives
    p = 0·57, and so t = 3·57.

5. 0·572

6. $$\frac{y}{x(x-1)(x-2)} = \frac{y_0}{2x} - \frac{y_1}{x-1} + \frac{y_2}{2(x-2)}$$

    As x_0 = 0 and h = 1, x = p.   Using E = 1 + Δ, $y_1 = y_0 + \Delta y_0$,
    $y_2 = y_0 + 2\Delta y_0 + \Delta^2 y_0$ and then $y = y_0 + p\Delta y_0 + \binom{p}{2}\Delta^2 y_0$,

    i.e., the first three terms of the Newton–Gregory formula.

    $$\frac{y}{(x-1)(x-3)(x-4)(x-6)} = -\frac{2}{15(x-1)} + \frac{7}{6(x-3)} - \frac{4}{3(x-4)} + \frac{11}{30(x-6)}$$

    When x = 2, y = 5·8.   When x = 5, y = 9·2.

233

7.   0·020 79,   0·030 22, 0·024 90;   0·006 64,   0·012 80

8.   The difference table suggests that the last two digits of 3·341 are
      in the wrong order and that this figure should be 3·314.   This   ·
      points to a copying error.   A revised difference table corroborates
      this.

      As p = ½, Bessel's formula is a reasonable choice.   Using this
      gives   f(1·45) = 1·159.

Numerical Differentiation

Introduction

It has already been indicated in this book that you may sometimes need to
find the value of a derivative at a point of some function y of x, even
although the only information you are given (or can find experimentally)
is a table showing values of y for specified values of x.   If this
happens, the analytical process of differentiation breaks down as, for
this, it is necessary to know the functional relation between y and x in
the form of an equation.

One way in which you can proceed to estimate a derivative in these
circumstances is to plot the points on a graph, join them by a smooth
curve, draw the tangent at the required point and measure its slope.
This method is not entirely satisfactory as the best smooth curve would be
to a certain extent open to conjecture and also drawing an exact tangent
to a curve is not easy.   It will be better if some numerical process can
be evolved whereby the derivative can be actually calculated from the
known values of y.

Doubtless you have already met many cases where derivatives have been
required in practical problems.   So here we shall just be content with a
few illustrations to indicate the type of problem where numerical
differentiation might arise.

Freudenstein's equation $R_1\cos \theta - R_2\cos \phi + R_3 - \cos(\theta - \phi) = 0$ for a
certain crank mechanism was mentioned in Unit 1.   Now suppose that values
of ϕ, the output lever angle, have been calculated for a series of values
of θ, the input crank angle.   If the crank rotates with constant
angular velocity ω, what will be the angular velocity and acceleration
of the output lever for the given values of θ?   With the information now
available, these can be found by analytical differentiation.   An
alternative method is to use a numerical approach for the actual
differentiation itself.   In this particular example, this has the
advantage that the calculations are simpler although the results are not
quite so accurate.

The set of miscellaneous examples at the end of the programme on
interpolation (see page 231) describe situations where numerical
differentiation would be used to obtain rates of change.   For the
situation described in question 4, it might be necessary to find the rate
of decrease of temperature with respect to time (i.e $-\dfrac{d\theta}{dt}$) when, say,

t = 3·5 or when t = 4·2.   In the question (No. 5) on the cantilever,
the slope at the point where x = 0·8 might be required.   Finally, in
question 7 we may be asked to find the maximum current.   This will
involve finding the time at which $\dfrac{di}{dt} = 0$ and then interpolating for i.

During the course of this programme, the solutions to the three problems
posed in the previous paragraph will be found.

NUMERICAL DIFFERENTIATION

A Basic Process

If you turn back to FRAME 14, page 7, you will see that some estimates were obtained for $\frac{d}{dx}\left(\frac{e^x}{x}\right)$ when $x = 2$ by approximating $f'(x)$ at $x = a$, i.e. $f'(a)$, to $\frac{f(a + h) - f(a)}{h}$. Four values of h, i.e., 0·2, 0·1, 0·05 and 0·01 were taken and it was seen that the estimate of $f'(x)$ at $x = 2$ was becoming more accurate as h was reduced. This is, of course, what one would expect from theory. A table can now be started, showing the various estimates of $f'(2)$ for the different values of h. The following table shows these values and at the same time, extends the range of values of h taken.

h	0·2	0·1	0·05	0·01	0·005	0·002	0·001	0·0001
$f'(2)$	2·038	1·941	1·892	1·85	1·84	1·8	1·8	1

In calculating these values, exponential tables to four places of decimals were used. The only rounding that was done during the calculation was in the division of e^{a+h} by $a + h$, this rounding being to 4 decimal places. Now theoretically, calculus wise, the estimates should continue to improve as h is decreased. As the actual value of the derivative is 1·8473, you will notice that improvement only took place down to $h = 0·01$.

Taking the value of $e^{2·0001}$ as 7·3898, work through the calculation giving $f'(2)$ when $h = 0·0001$. Then suggest two reasons why the result is not as accurate as it should be.

****

$e^{a+h}/(a + h) = 3·6947$, $\quad \frac{e^{a+h}}{a + h} - \frac{e^a}{a} = 0·0001$, $\quad f'(2) \simeq 1$

$e^a/a = 3·6946$ and forming $\frac{e^{a+h}}{a + h} - \frac{e^a}{a}$ thus involved the subtraction of two very nearly equal numbers. This introduced an error which was large in comparison with the true value. This error was then multiplied by 10 000 when the division by $h(= 0·0001)$ was done.

This example thus shows that, when working to a fixed number of decimal places, a stage is reached where accuracy cannot be improved just by decreasing the value of h. It is therefore necessary to look for some other means of increasing the accuracy. In practice, it is desirable to do this for another reason as well - if you have values of a function known only at regular intervals, h is fixed. Any attempt to decrease it would involve doing an interpolation calculation.

A variation of the basic process is shown in the accompanying figure. If the value of $\frac{dy}{dx}$ at P is required, two points Q and R are taken, one on each side of P and equidistant from it x-wise. The slope of the chord

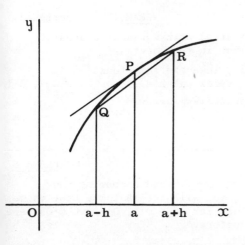

QR is then found, it being assumed that this chord is nearly parallel to the tangent at P.   (It is certainly more so than the chord PR, whose slope the formula

$$\frac{f(a + h) - f(a)}{h}$$ finds.)   This means that

$$\frac{f(a + h) - f(a - h)}{2h}$$ (6.1)

is taken as an estimate of f'(x) at x = a.

Now, if f(a + h) is expanded by means of Taylor's series, the original formula, i.e. $\dfrac{f(a + h) - f(a)}{h}$ gives

$$\frac{\left\{ f(a) + hf'(a) + \dfrac{h^2}{2!} f''(a) + \dfrac{h^3}{3!} f'''(a) + \dots \right\} - f(a)}{h}$$

$$= f'(a) + \frac{h}{2} f''(a) + \frac{h^2}{6} f'''(a) + \dots$$

and, assuming h small, the difference between this and f'(a) is approximately $\dfrac{h}{2} f''(a)$.

Now take the formula (6·1) and, treating this in a similar way, obtain the main term in the difference between it and f'(a).

****

6A

$$\left[\left\{ f(a) + hf'(a) + \frac{h^2}{2!} f''(a) + \frac{h^3}{3!} f'''(a) + \dots \right\} \right.$$

$$\left. - \left\{ f(a) - hf'(a) + \frac{h^2}{2!} f''(a) - \frac{h^3}{3!} f'''(a) + \dots \right\} \right] / 2h$$

$$= f'(a) + \frac{h^2}{6} f'''(a) + \dots$$

$$Difference \simeq \frac{h^2}{6} f'''(a)$$

FRAME 7

This suggests that, for reasonably small h, the result should be more accurate than that obtained previously as the main term in the error now involves h^2 instead of h.   Consequently it should not be necessary to reduce h to the same degree as before to get a similar accuracy.   This will have two advantages from the numerical point of view.   Firstly, unless we are near a maximum or minimum, the difference between f(a + h) and f(a - h) will be larger than that between f(a + h) and f(a).

FRAME 7   (continued)

Secondly the final division is by 2h instead of by h and so does not magnify to quite the same extent any error in the numerator.

To test how well the formula (6·1) works, take the same function as before, i.e. $y = e^x/x$, and calculate dy/dx at x = 2 for h = 0·2, 0·1, 0·05, 0·01.   ***************************************

7A

1·8535, 1·849, 1·848, 1·85

FRAME 8

Remembering that the theoretical value is 1·8473 and comparing your results now with those given in FRAME 4, you will see that the best value is given by h = 0·05 and that the values for h = 0·2, 0·1 and 0·05 are all considerably better than they were before.

If h is further reduced, the table now corresponding completely to that given in FRAME 4 is

h	0·2	0·1	0·05	0·01	0·005	0·002	0·001	0·0001
f'(2)	1·8535	1·849	1·848	1·85	1·85	1·85	1·85	1·5

These figures corroborate very well the remarks made in FRAME 7.

FRAME 9

Even although the results obtained using formula (6·1) are an improvement over the previous ones, there are still some disadvantages inherent in the method.   One of these has been mentioned before - the required values of the function may not be known and hence must be found by interpolation. Secondly, as the theoretical value will not be known in any actual example (otherwise we shouldn't be looking for an approximation to it) it will not be obvious which value of h will give the best answer.   In view of these difficulties, other methods of numerical differentiation are more often used.   These methods are based on the interpolation formulae that were given in the previous programme.

FRAME 10

Differentiation Based on Equal Interval Interpolation Formulae

Effectively, an equal interval interpolation formula assumes that a curve can be adequately represented over a certain range of values by a polynomial in p.   It is thus perhaps natural to enquire whether differentiation of the function represented by an interpolating polynomial formula can be achieved in some way by differentiation of the actual interpolating formula itself w.r.t. p.

Effectively this question means:

If y = f(x) is represented by y = F(p), where $x = x_0 + ph$, can dy/dx be obtained from dy/dp?

What will be the relation between $\frac{dy}{dx}$ and $\frac{dy}{dp}$?
   ****

238

$$\frac{dy}{dx} = \frac{dy}{dp} \cdot \frac{dp}{dx} = \frac{dy}{dp} \cdot \frac{1}{h}$$

So, once $\frac{dy}{dp}$ has been found, $\frac{dy}{dx}$ follows immediately.   You will notice that h still appears as a divisor and so there is good reason for its not being too small a decimal if possible (remember what happened in 4A).

Any interpolation formula can be used for differentiation w.r.t. p.   For example, taking the Newton-Gregory forward difference formula,

$$f_p = f_0 + \binom{p}{1}\Delta f_0 + \binom{p}{2}\Delta^2 f_0 + \binom{p}{3}\Delta^3 f_0 + \binom{p}{4}\Delta^4 f_0 + \dotsb\dotsb\dotsb$$

$$= f_0 + p\Delta f_0 + \frac{p(p-1)}{2}\Delta^2 f_0 + \frac{p(p-1)(p-2)}{6}\Delta^3 f_0$$

$$+ \frac{p(p-1)(p-2)(p-3)}{24}\Delta^4 f_0 + \dotsb\dotsb\dotsb$$

$$= f_0 + p\Delta f_0 + \left(\frac{1}{2}p^2 - \frac{1}{2}p\right)\Delta^2 f_0 + \left(\frac{1}{6}p^3 - \frac{1}{2}p^2 + \frac{1}{3}p\right)\Delta^3 f_0$$

$$+ \left(\frac{1}{24}p^4 - \frac{1}{4}p^3 + \frac{11}{24}p^2 - \frac{1}{4}p\right)\Delta^4 f_0 + \dotsb\dotsb\dotsb$$

$$\therefore \quad f'_p = \Delta f_0 + \left(p - \frac{1}{2}\right)\Delta^2 f_0 + \left(\frac{1}{2}p^2 - p + \frac{1}{3}\right)\Delta^3 f_0$$

$$+ \left(\frac{1}{6}p^3 - \frac{3}{4}p^2 + \frac{11}{12}p - \frac{1}{4}\right)\Delta^4 f_0 + \dotsb\dotsb\dotsb$$

and so $\quad f'(x_p) = \frac{1}{h}\left\{\Delta f_0 + \left(p - \frac{1}{2}\right)\Delta^2 f_0 + \left(\frac{1}{2}p^2 - p + \frac{1}{3}\right)\Delta^3 f_0\right.$

$$\left. + \left(\frac{1}{6}p^3 - \frac{3}{4}p^2 + \frac{11}{12}p - \frac{1}{4}\right)\Delta^4 f_0 + \dotsb\dotsb\dotsb\right\} \qquad (11.1)$$

Taking the data from the example in FRAME 8, page 209, the value of the derivative at $x = 2{\cdot}31$ is

$$\frac{1}{1}\left\{19 + \left(0{\cdot}31 - \frac{1}{2}\right) \times 18 + \left(\frac{1}{2} \times 0{\cdot}31^2 - 0{\cdot}31 + \frac{1}{3}\right) \times 6\right\} \simeq 19 - 3{\cdot}42 + 0{\cdot}428$$
$$= 16{\cdot}008$$

In this example, the table represented the cubic $x^3 + 2$, the derivative of which is $3x^2$.   When $x = 2{\cdot}31$, $3x^2 \simeq 16{\cdot}008$ which agrees with the value just found.

Equation (11.1) takes on a particularly simple form if $p = 0$.   Then

$$f'(x_0) = \frac{1}{h}\left(\Delta f_0 - \frac{1}{2}\Delta^2 f_0 + \frac{1}{3}\Delta^3 f_0 - \frac{1}{4}\Delta^4 f_0 + \dotsb\dotsb\dotsb\right) \qquad (12.1)$$

What will this formula give for the derivative at $x = 3$ of the function defined by the data in the example in FRAME 8, page 209?
****

Here it is necessary to take $x_0 = 3$, *then*

$$f'(3) = \frac{1}{1}\left(37 - \frac{1}{2} \times 24 + \frac{1}{3} \times 6\right) = 27$$

Again this agrees with the value of $3x^2$ *when* $x = 3$.

What will be the differentiation formula corresponding to the Newton-Gregory backward difference formula?

****

$$f_p = f_0 + p\nabla f_0 + \left(\frac{1}{2}p^2 + \frac{1}{2}p\right)\nabla^2 f_0 + \left(\frac{1}{6}p^3 + \frac{1}{2}p^2 + \frac{1}{3}p\right)\nabla^3 f_0$$

$$+ \left(\frac{1}{24}p^4 + \frac{1}{4}p^3 + \frac{11}{24}p^2 + \frac{1}{4}p\right)\nabla^4 f_0 + \dots\dots\dots$$

$$\therefore\; f'(x_p) = \frac{1}{h}\left\{\nabla f_0 + \left(p + \frac{1}{2}\right)\nabla^2 f_0 + \left(\frac{1}{2}p^2 + p + \frac{1}{3}\right)\nabla^3 f_0\right.$$

$$\left.+ \left(\frac{1}{6}p^3 + \frac{3}{4}p^2 + \frac{11}{12}p + \frac{1}{4}\right)\nabla^4 f_0 + \dots\dots\dots\dots\right\} \qquad (13A.1$$

If $p = 0$, this gives the very simple result

$$f'(x_0) = \frac{1}{h}\left(\nabla f_0 + \frac{1}{2}\nabla^2 f_0 + \frac{1}{3}\nabla^3 f_0 + \frac{1}{4}\nabla^4 f_0 + \dots\dots\right)$$

What will be the slope, to 2 decimal places, of the cantilever in question 5, FRAME 58, page 232 when $x = 0 \cdot 8$?

****

$0 \cdot 32$

When interpolating, we used the forward difference formula when near the beginning of a difference table and the backward difference formula when near the end. A similar distinction has been followed when differentiating. But if you are working near the middle of a difference table, it is, as before, better to use a central difference formula. Any one of the central difference interpolation formulae can be used to give a central difference differentiation formula. What such formula will arise if Bessel's interpolation formula is used?

****

$$f_p = \mu f_{\frac{1}{2}} + \left(p - \frac{1}{2}\right)\delta f_{\frac{1}{2}} + \left(\frac{1}{2}p^2 - \frac{1}{2}p\right)\mu\delta^2 f_{\frac{1}{2}} + \left(\frac{1}{6}p^3 - \frac{1}{4}p^2 + \frac{1}{12}p\right)\delta^3 f_{\frac{1}{2}}$$

$$+ \left(\frac{1}{24}p^4 - \frac{1}{12}p^3 - \frac{1}{24}p^2 + \frac{1}{12}p\right)\mu\delta^4 f_{\frac{1}{2}} + \dots\dots\dots$$

$$f'(x_p) = \frac{1}{h}\left\{\delta f_{\frac{1}{2}} + \left(p - \frac{1}{2}\right)\mu\delta^2 f_{\frac{1}{2}} + \left(\frac{1}{2}p^2 - \frac{1}{2}p + \frac{1}{12}\right)\delta^3 f_{\frac{1}{2}}\right.$$

$$\left. + \left(\frac{1}{6}p^3 - \frac{1}{4}p^2 - \frac{1}{12}p + \frac{1}{12}\right)\mu\delta^4 f_{\frac{1}{2}} + \ldots\ldots\ldots\ldots\right\} \qquad (15A.1)$$

FRAME 16

If   p = 0,   this becomes

$$f'(x_0) = \frac{1}{h}\left(\delta f_{\frac{1}{2}} - \frac{1}{2}\mu\delta^2 f_{\frac{1}{2}} + \frac{1}{12}\delta^3 f_{\frac{1}{2}} + \frac{1}{12}\mu\delta^4 f_{\frac{1}{2}} + \ldots\ldots\right) \qquad (16.1)$$

It becomes even more simple if p happens to be $\frac{1}{2}$.   Then

$$f'(x_{\frac{1}{2}}) = \frac{1}{h}\left(\delta f_{\frac{1}{2}} - \frac{1}{24}\delta^3 f_{\frac{1}{2}} \ldots\ldots\ldots\right) \qquad (16.2)$$

Incidentally the next term in the brackets is $\frac{3}{640}\delta^5 f_{\frac{1}{2}}$.

You should now be able to find the rate of decrease of temperature at
t = 3·5 and at t = 4·2 for the cooling curve given in question 4 of
FRAME 58 on page 232.

****

16A

Using (16.2) with θ_0 = 3 *gives* $\frac{d\theta}{dt}$ = −7·047.

Using (15A.1) with θ_0 = 4 *and* p = 0·2 *gives* $\frac{d\theta}{dt}$ = −6·572.

FRAME 17

Let us now have a look at the last problem posed in FRAME 3.   (The other
two have now been dealt with.)   As was mentioned there, it is necessary
to find the time at which $\frac{di}{dt}$ is zero.   This can be done by forming a
table showing the values of $\frac{di}{dt}$ at various times and then using inverse
interpolation to find t when $\frac{di}{dt}$ = 0.   Having found this value of t,
ordinary interpolation will give us the required value of i.

Start by calculating the values of $\frac{di}{dt}$ at t = 0·000, 0·002, 0·004 and
0·006, using differences up to the fourth.   (The required table of values
is given in question 7, FRAME 58, page 232.)

****

17A

9·95,   5·91,   3·26,   1·55

FRAME 18

For the first two of these, the formula (12.1) (but for $\frac{di}{dt}$ instead of $\frac{dy}{dx}$)
was used as we have been using a forward difference formula at the
beginning of a difference table.   For the other two we used (16.1).
This is better than (12.1) as it depends less on the third and fourth
differences.   As h = 1/500, a multiplying factor of 500 is involved in

FRAME 18 (continued)

the 1/h. Any errors in the figures resulting from the terms inside the brackets in (12.1) and (16.1) are therefore multiplied by this amount. Consequently the final figures cannot be relied on to any great degree of accuracy and so have only been quoted to 2 decimal places.

FRAME 19

Continuing the calculation of the values of di/dt gives the following table:

t	0•000	0•002	0•004	0•006	0•008	0•010	0•012	0•014	0•016
di/dt	9•95	5•91	3•26	1•55	0•46	-0•22	-0•64	-0•85	-0•96

t	0•018	0•020
di/dt	-0•98	-1•00

di/dt is obviously zero for a value of t between 0•008 and 0•010. It is also obvious from the original table for i that the maximum value of the current occurs when t is in the region of 0•010.

FRAME 20

In order to use inverse interpolation to find t more accurately a difference table is necessary for di/dt. A section of this is shown below:

t	$\frac{di}{dt}$			
0•000	9•95			
		-404		
0•002	5•91		139	
		-265		-45
0•004	3•26		94	
		-171		-32
0•006	1•55		62	
		-109		-21
0•008	0•46		41	
		-68		-15
0•010	-0•22		26	
		-42		-5
0•012	-0•64		21	
		-21		-11
0•014	-0•85		10	
		-11		-1
0•016	-0•96		9	
		-2		-9
0•018	-0•98		0	
		-2		
0•020	-1•00			

It has only been taken as far as the third differences as these are fluctuating somewhat and consequently any more would not be reliable.

Now taking t_0 = 0•008, write down the appropriate equation for p when $\frac{di}{dt} = 0$, using Bessel's interpolation formula.

****

$$0 = 0 \cdot 12 + (p - \tfrac{1}{2})(-0 \cdot 68) + \frac{p(p - 1)}{2} (0 \cdot 335) + \frac{p(p - 1)(p - \tfrac{1}{2})}{6} (-0 \cdot 15)$$

FRAME 21

What value, to 2 decimal places, does this equation give for p?
****

21A

$p = 0 \cdot 62$

FRAME 22

The maximum value of the current therefore occurs at t = 0•009 24 and
then interpolation in the t - i table yields i = 0•031 50.

FRAME 23

Differentiation Based on Lagrange's Interpolation Polynomial

You will remember that this formula has to be used for interpolation when
a table is given at unequal intervals of x.   It is also necessary, under
similar circumstances, to use it for differentiation.   For this purpose
it is better to use it in the second form given in the previous programme,
i.e., as

$$f(x) = \frac{(x - x_1)(x - x_2) \ldots \ldots (x - x_n)}{(x_0 - x_1)(x_0 - x_2) \ldots (x_0 - x_n)} f(x_0)$$

$$+ \frac{(x - x_0)(x - x_2)(x - x_3) \ldots \ldots (x - x_n)}{(x_1 - x_0)(x_1 - x_2)(x_1 - x_3) \ldots (x_1 - x_n)} f(x_1)$$

$$+ \frac{(x - x_0)(x - x_1)(x - x_3)(x - x_4) \ldots \ldots (x - x_n)}{(x_2 - x_0)(x_2 - x_1)(x_2 - x_3)(x_2 - x_4) \ldots (x_2 - x_n)} f(x_2) + \ldots \ldots$$

$$+ \frac{(x - x_0)(x - x_1) \ldots \ldots (x - x_{n-1})}{(x_n - x_0)(x_n - x_1) \ldots (x_n - x_{n-1})} f(x_n) \qquad (23.1)$$

Differentiation is now straightforward but, as you will realise, the
general expression for f'(x) is rather cumbersome.   Consequently we shall
not obtain it here, but will just do one example for illustration purposes.

FRAME 24

As an example, suppose that the value of f'(x) is required at x = 0•12,
f(x) being given by the table

x	0•05	0•10	0•20	0•26
f(x)	0•0500	0•0999	0•1987	0•2571

Write down the equation resulting from applying (23.1) to this data.
****

$$f(x) \;=\; \frac{0 \cdot 0500(x - 0 \cdot 10)(x - 0 \cdot 20)(x - 0 \cdot 26)}{(0 \cdot 05 - 0 \cdot 10)(0 \cdot 05 - 0 \cdot 20)(0 \cdot 05 - 0 \cdot 26)}$$

$$+ \;\frac{0 \cdot 0999(x - 0 \cdot 05)(x - 0 \cdot 20)(x - 0 \cdot 26)}{(0 \cdot 10 - 0 \cdot 05)(0 \cdot 10 - 0 \cdot 20)(0 \cdot 10 - 0 \cdot 26)}$$

$$+ \;\frac{0 \cdot 1987(x - 0 \cdot 05)(x - 0 \cdot 10)(x - 0 \cdot 26)}{(0 \cdot 20 - 0 \cdot 05)(0 \cdot 20 - 0 \cdot 10)(0 \cdot 20 - 0 \cdot 26)}$$

$$+ \;\frac{0 \cdot 2571(x - 0 \cdot 05)(x - 0 \cdot 10)(x - 0 \cdot 20)}{(0 \cdot 26 - 0 \cdot 05)(0 \cdot 26 - 0 \cdot 10)(0 \cdot 26 - 0 \cdot 20)}$$

Before differentiating, it is a help to simplify this. It reduces to
$f(x) = -0 \cdot 119x^3 - 0 \cdot 025x^2 + 1 \cdot 004x$ from which
$f'(x) = -0 \cdot 357x^2 - 0 \cdot 050x + 1 \cdot 004$.

When $x = 0 \cdot 12$, this becomes $0 \cdot 993$.

Higher Order Derivatives

All the methods used so far in this programme to obtain $\dfrac{dy}{dx}$ can be extended
to find the higher derivatives. Taking the first method (in FRAME 4),

$$f'(a) \simeq \frac{f(a + h) - f(a)}{h} \qquad (26.1)$$

and so
$$f''(a) \simeq \frac{f'(a + h) - f'(a)}{h} \qquad (26.2)$$

$\Bigg($If you can't see this, let $f'(a) = F(a)$, then
$f''(a) = F'(a) = \dfrac{F(a + h) - F(a)}{h} = \dfrac{f'(a + h) - f'(a)}{h}$.$\Bigg)$

Now if, in (26.1), a is replaced by a + h,

$$f'(a + h) \simeq \frac{f(a + 2h) - f(a + h)}{h}$$

Substituting into (26.2)

$$f''(a) \simeq \frac{1}{h}\left\{ \frac{f(a + 2h) - f(a + h)}{h} - \frac{f(a + h) - f(a)}{h} \right\}$$

$$= \frac{1}{h^2}\left\{ f(a + 2h) - 2f(a + h) + f(a) \right\} \qquad (26.3)$$

What will be the formula obtained for $f''(a)$ when (6·1) is used for $f'(a)$
****

$$\frac{1}{2h}\left\{ f'(a + h) - f'(a - h) \right\} \simeq \frac{1}{2h}\left\{ \frac{f(a + 2h) - f(a)}{2h} - \frac{f(a) - f(a - 2h)}{2h} \right\}$$

$$= \frac{1}{4h^2}\left\{ f(a + 2h) - 2f(a) + f(a - 2h) \right\} \qquad (26A.)$$

If h is small, what sources of error can you see in (26.3)?
****

i) $f(a + 2h) + f(a)$ *will be nearly equal to* $2f(a + h)$ *and so the loss of significant figures consequent upon the subtraction of two nearly equal numbers will be involved.*

ii) *The division by* h^2.

Similar sources of error exist in (26A.1) but the effects are likely to be smaller due to the usually greater difference between functional values a distance 2h apart over those only a distance h apart and also due to the presence of the 4 in the denominator.

Returning now to the expression used earlier, i.e. e^x/x, it can quite easily be shown analytically that $\dfrac{d^2}{dx^2}\left(\dfrac{e^x}{x}\right)$ at x = 2 is 1·8473.   Use each of the formulae (26.3) and (26A.1) to estimate this derivative numerically, taking h = 0·1.
****

1·95,   1·85

As you will notice, (26A.1) gives a more accurate result which, you will remember, was forecast in the last frame.

In FRAMES 10-22, differentiation formulae were obtained, based on equal interval interpolation formulae.   There it was found that $\dfrac{dy}{dx} = \dfrac{1}{h}\dfrac{dy}{dp}$.
Extending the ideas developed there enables us to find $\dfrac{d^2y}{dx^2}$ from a knowledge of $\dfrac{d^2y}{dp^2}$.

What is the relation between $\dfrac{d^2y}{dx^2}$ and $\dfrac{d^2y}{dp^2}$?
****

$$\frac{d^2y}{dx^2} \;=\; \frac{d}{dx}\left(\frac{1}{h}\frac{dy}{dp}\right) \;=\; \frac{d}{dp}\left(\frac{1}{h}\frac{dy}{dp}\right)\frac{dp}{dx} \;=\; \frac{1}{h^2}\frac{d^2y}{dp^2}$$

Then, using for example the Newton-Gregory forward difference formula, $\dfrac{d^2y}{dx^2}$ can be obtained from (11.1) by differentiating its R.H.S. w.r.t. p and multiplying by 1/h.

What will be the resulting formula for $\dfrac{d^2y}{dx^2}$ and also the simplified version when p = 0?
****

$$\frac{1}{h^2}\left\{\Delta^2 f_0 + (p-1)\Delta^3 f_0 + \left(\frac{1}{2}p^2 - \frac{3}{2}p + \frac{11}{12}\right)\Delta^4 f_0 \cdots\cdots\cdots\cdots\right\}$$

$$\frac{1}{h^2}\left(\Delta^2 f_0 - \Delta^3 f_0 + \frac{11}{12}\Delta^4 f_0 \cdots\cdots\cdots\cdots\right)$$

Using the data in the example in FRAME 8, page 209, what will be the value of $\frac{d^2y}{dx^2}$ at   (i)   x = 2·31,   (ii)   x = 3?

****

i) $\frac{1}{1^2}\left\{18 + (0 \cdot 31 - 1) \times 6\right\} = 13 \cdot 86$ *ii)* $\frac{1}{1^2}\left(24 - 6\right) = 18$

As the table represented the polynomial $x^3 + 2$, you can immediately see that these values are correct analytically.

What expression will Bessel's interpolation formula give for $\frac{d^2y}{dx^2}$?

****

$$\frac{1}{h^2}\left\{\mu\delta^2 f_{\frac{1}{2}} + \left(p - \frac{1}{2}\right)\delta^3 f_{\frac{1}{2}} + \left(\frac{1}{2}p^2 - \frac{1}{2}p - \frac{1}{12}\right)\mu\delta^4 f_{\frac{1}{2}} \cdots\cdots\cdots\cdots\right\}$$

You will notice that the first term in any of these formulae involves a second difference. When finding dy/dx, it involved a first difference. Similarly that for a third derivative would commence with a third difference and so on. Also the formula for a third difference would involve $\frac{1}{h^3}$. As

i)   only the higher order differences are involved when calculating the higher derivatives,

ii)   these differences are more subject to error when round-off is involved,

iii)   division by a higher power of h takes place,

the values of the higher derivatives as found from these formulae are less accurate than those of the lower ones.

The following table gives the distance d travelled from rest by a car at various times t. What was the acceleration of the car when t = 0, 2·4, 4·6? Work as far as 4th differences.

t(s)	0	0·5	1·0	1·5	2·0	2·5	3·0	3·5	4·0
d(ms⁻¹)	0·00	0·07	0·53	1·60	3·61	6·61	10·62	15·62	21·54

t(s)	4·5	5·0
d(ms⁻¹)	28·09	35·00

****

$1 \cdot 08, \quad 4 \cdot 05, \quad 0 \cdot 04$

For the last of these results, we used the Newton-Gregory backward interpolation formula. *From (13A.1) this gives*

$$f''(x_p) = \frac{1}{h^2}\left\{\nabla^2 f_0 + (p+1)\nabla^3 f_0 + \left(\frac{1}{2}p^2 + \frac{3}{2}p + \frac{11}{12}\right)\nabla^4 f_0 + \cdots\cdots\right\}$$

Lastly with respect to higher derivatives, we will mention the use of Lagrange's interpolation formula. As you saw in FRAME 25, f'(x) can be found directly once the formula has been used to produce f(x). Obviously one can continue in order to find the higher derivatives. Taking the figures quoted in FRAME 24 and using f'(x) from FRAME 25, f"(x) = -0·714x - 0·050 and so, when x = 0·12, f"(x) = -0·136.

Miscellaneous Examples

In this frame a collection of miscellaneous examples is given for you to try. Answers are provided in FRAME 36, together with such working as is considered helpful.

1. Obtain formulae for $\frac{dy}{dx}$ and $\frac{d^2y}{dx^2}$ from the interpolation formula (due to Gauss) $f_p = f_0 + \binom{p}{1}\delta f_{\frac{1}{2}} + \binom{p}{2}\delta^2 f_0 + \binom{p+1}{3}\delta^3 f_{\frac{1}{2}} + \binom{p+1}{4}\delta^4 f_0 + \cdots$

2. The points (x_0, y_0), (x_1, y_1), (x_2, y_2) and (x_3, y_3), the x's being equally spaced, lie on a cubic curve. Show that (12.1) leads to $-\frac{1}{6h}(y_3 - 6y_2 + 3y_1 + 2y_0)$ as the value of dy/dx at (x_1, y_1).

   (This question illustrates that a derivative can be expressed in terms of functional values instead of differences.)

3. Express (16.2) in terms of functional values instead of differences, retaining the terms in (16.2) up to and including that involving $\delta^3 f_{\frac{1}{2}}$.

4. At what time will the rate of decrease of temperature be 8 for the cooling situation, given in question 4, FRAME 58, page 232?

5. Starting with Stirling's formula

$$f_p = f_0 + p\mu\delta f_0 + \frac{1}{2}p^2\delta^2 f_0 + \frac{p(p^2-1)}{3!}\mu\delta^3 f_0 + \frac{p^2(p^2-1)}{4!}\delta^4 f_0$$

$$+ \frac{p(p^2-1)(p^2-4)}{5!}\mu\delta^5 f_0 + \frac{p^2(p^2-1)(p^2-4)}{6!}\delta^6 f_0 + \cdots\cdots$$

   deduce that

$$hf'_0 = \mu\delta f_0 - \frac{1}{6}\mu\delta^3 f_0 + \frac{1}{30}\mu\delta^5 f_0 - \cdots\cdots\cdots$$

$$h^2 f''_0 = \delta^2 f_0 - \frac{1}{12}\delta^4 f_0 + \frac{1}{90}\delta^6 f_0 - \cdots\cdots\cdots$$

where $hf'_p = h \dfrac{d}{dx} f(x_0 + ph) = \dfrac{d}{dp} f_p$.

The following table gives the coordinates (x, y) of points on a certain polynomial curve.

x	0	0·2	0·4	0·6	0·8	1·0	1·2
y	0·710	1·175	1·811	2·666	3·801	5·292	7·232

Calculate the radius of curvature at the point $x = 0·6$ (L.U.)

Note: In certain problems (e.g. deflection of beams, transition curves for railway tracks) the idea of curvature is used.   The radius of curvature of a curve at a point is the radius of the circle which coincides with the curve over a very small length of arc at the point. The smaller the radius of curvature the sharper the bending.   The radius of curvature is given by the formula

$$\left\{ 1 + \left(\frac{dy}{dx}\right)^2 \right\}^{3/2} \Bigg/ \frac{d^2 y}{dx^2}$$

Answers to Miscellaneous Examples

1. Gauss' formula can be written as

$$f_p = f_0 + p\delta f_{\frac{1}{2}} + \frac{1}{2}p(p - 1)\delta^2 f_0 + \frac{1}{6}(p + 1)p(p - 1)\delta^3 f_{\frac{1}{2}}$$

$$+ \frac{1}{24}(p + 1)p(p - 1)(p - 2)\delta^4 f_0 + \dots\dots\dots\dots$$

$$= f_0 + p\delta f_{\frac{1}{2}} + \left(\frac{1}{2}p^2 - \frac{1}{2}p\right)\delta^2 f_0 + \left(\frac{1}{6}p^3 - \frac{1}{6}p\right)\delta^3 f_{\frac{1}{2}}$$

$$+ \left(\frac{1}{24}p^4 - \frac{1}{12}p^3 - \frac{1}{24}p^2 + \frac{1}{12}p\right)\delta^4 f_0 + \dots\dots\dots\dots$$

$$\frac{dy}{dx} = \frac{1}{h}\left\{ \delta f_{\frac{1}{2}} + \left(p - \frac{1}{2}\right)\delta^2 f_0 + \left(\frac{1}{2}p^2 - \frac{1}{6}\right)\delta^3 f_{\frac{1}{2}} \right.$$

$$\left. + \left(\frac{1}{6}p^3 - \frac{1}{4}p^2 - \frac{1}{12}p + \frac{1}{12}\right)\delta^4 f_0 + \dots\dots\dots\dots \right\}$$

$$\frac{d^2 y}{dx^2} = \frac{1}{h^2}\left\{ \delta^2 f_0 + p\delta^3 f_{\frac{1}{2}} + \left(\frac{1}{2}p^2 - \frac{1}{2}p - \frac{1}{12}\right)\delta^4 f_0 + \dots\dots\dots\dots \right\}$$

2. The difference table is

x_0	y_0			
		$y_1 - y_0$		
x_1	y_1		$y_2 - 2y_1 + y_0$	
		$y_2 - y_1$		$y_3 - 3y_2 + 3y_1 - y_0$
x_2	y_2		$y_3 - 2y_2 + y_1$	
		$y_3 - y_2$		*
x_3	y_3			

At (x_1, y_1),   (12.1) gives dy/dx to be

$$\frac{1}{h}\left\{\left(y_2 - y_1\right) - \frac{1}{2}\left(y_3 - 2y_2 + y_1\right) + \frac{1}{3}\left(y_3 - 3y_2 + 3y_1 - y_0\right)\right\}$$

= quoted result.

Note that the formula requires the use of the starred difference. As the curve is a cubic, all third order differences are the same and so that immediately above * has been used in its place.

3. $\delta^3 f_{\frac{1}{2}} = \delta^2 f_1 - \delta^2 f_0$

$\quad\quad = (\delta f_{1\frac{1}{2}} - \delta f_{\frac{1}{2}}) - (\delta f_{\frac{1}{2}} - \delta f_{-\frac{1}{2}})$

$\quad\quad = \delta f_{1\frac{1}{2}} - 2\delta f_{\frac{1}{2}} + \delta f_{-\frac{1}{2}}$

$\quad\quad = (f_2 - f_1) - 2(f_1 - f_0) + (f_0 - f_{-1})$

$\quad\quad = f_2 - 3f_1 + 3f_0 - f_{-1}$

or

$\delta^3 f_{\frac{1}{2}} = (E^{\frac{1}{2}} - E^{-\frac{1}{2}})^3 f_{\frac{1}{2}}$

$\quad\quad = (E^{1\frac{1}{2}} - 3E^{\frac{1}{2}} + 3E^{-\frac{1}{2}} - E^{-1\frac{1}{2}}) f_{\frac{1}{2}}$

$\quad\quad = f_2 - 3f_1 + 3f_0 - f_{-1}$

$\frac{df}{dx} = \frac{1}{h}\left\{f_1 - f_0 - \frac{1}{24}\left(f_2 - 3f_1 + 3f_0 - f_{-1}\right)\right\}$

$\quad\quad = \frac{1}{24h}(-f_2 + 27f_1 - 27f_0 + f_{-1})$

4. Calculating $d\theta/dt$ and differencing gives the table

0	-10·008			
		959		
1	-9·049		-96	
		863		11
2	-8·186		-85		-1
		778		10
3	-7·408		-75		2
		703		12
4	-6·705		-63	
		640		
5	-6·065			

It is not necessary to take this table any further for the result required. Using inverse interpolation leads to $p = 0·23$ where $t_0 = 2$. Hence required $t = 2·23$.

5. At $x = 0·6$, $\frac{dy}{dx} = 4·918$, $\frac{d^2y}{dx^2} = 6·969$, radius of curvature = 18·1.

Note that the leading term in the expression quoted for hf_0' effectively gives (6.1) and that in the expression for $h^2 f_0''$ leads to (26.3).

Numerical Integration

Introduction

You have no doubt already learnt how to integrate a variety of functions
and seen how integration results from a number of different physical
situations.   We shall commence this programme by looking at some problems
which involve integrals but which either cannot be solved by analytical
techniques or, even if they can, are more easily done by other means.

The first problem is this:

A coach accelerates from rest to 100 km/h in 90 s.   Its speed, v km/h,
measured at five second intervals, is given by the table

t	0	5	10	15	20	25	30	35	40	45	50	55	60	65	70	75	80
v	0	5	19	22	32	40	43	52	61	65	66	71	77	85	90	95	98

t	85	90
v	99	100

What distance will it travel in this time?

As $v = \dfrac{ds}{dt}$, $s = \displaystyle\int_0^{90} v\,dt$ and difficulty arises in that the formula for v
in terms of t is not known. Note that, in this formula, v must be in
km/s.

The second problem is:

A solid, made of material of density $0 \cdot 02 \text{ g/mm}^3$ is formed by rotating the
shape shown through 360° about OA.   If O is
taken to be the origin and the coordinates of P
are (x, y), the following pairs of values of x
and y are known.

y (mm)	0	5	10	15	20	25	30	35	40
x (mm)	0	22	48	50	49	45	30	11	7

y	45	50	55	60	65	70	75	80	85
x	5	5	5	5	5	5	5	4	0

What is the moment of inertia of the solid about
OA?   If you are familiar with the ideas involved
in moments of inertia, then, by way of revision,
find the integral formula for this M. of I.,
remembering that k^2 for a circle of radius a
about its axis is $\frac{1}{2}a^2$.   If you are not familiar with moments of inertia,
proceed directly to FRAME 3.

****

Taking a slice, perpendicular to Oy, thickness δy,

Volume of slice $\simeq \pi x^2 \delta y$, Mass of slice $\simeq 0 \cdot 02 \pi x^2 \delta y$
M. of I. of slice $\simeq (0 \cdot 02 \pi x^2 \delta y) \frac{1}{2} x^2 = 0 \cdot 01 \pi x^4 \delta y$

$$\text{Total M. of I. of solid} = \int_0^{85} 0 \cdot 01 \pi x^4 dy = 0 \cdot 01 \pi \int_0^{85} x^4 dy.$$

The units will be $g\,mm^2$.

Again there is difficulty in proceeding further as the formula for x *in terms of* y *is not known.*

FRAME 3

If a body is rotating about an axis, the moment of inertia of the body about that axis is used when finding the kinetic energy due to the rotation.   If you have not met the concept of moment of inertia before, then simply note that

i)   for the body described in FRAME 2, the moment of inertia about OA is

given, in appropriate units, by $0 \cdot 01\pi \displaystyle\int_0^{8\,5} x^4 dy$ and

ii)   as the formula connecting x and y is not known, there is difficulty in proceeding further.

FRAME 4

In Unit 1 it was mentioned that the integral $\displaystyle\int_0^{x_0} \dfrac{x^4 e^x}{(e^x - 1)^2}\, dx$ occurs in

finding the heat capacity of a solid by a method which is based on the vibrational frequencies of the crystal.   Here you would have difficulty in finding the indefinite integral, even although you know the function to be integrated.   But assuming that x_0 is known, it is possible to

calculate the values of $y \left[= \dfrac{x^4 e^x}{(e^x - 1)^2} \right]$ for various values of x in

the range $0 \leqslant x \leqslant x_0$.   The problem can then be regarded as that of

finding $\displaystyle\int_0^{x_0} y\,dx$ from a table of values of x and y .   This is similar

to what has to be done for the other two problems.

FRAME 5

Even where an indefinite integral can be found, it may be such a beast that evaluating it at the limits turns out to be a major operation. Under these circumstances, it may be better to forget completely about the indefinite integral and proceed as described in FRAME 4.

Last, but by no means least, we will mention the situation where it is

desirable to evaluate $\displaystyle\int_a^b f(x)dx,$ $f(x)$ being known analytically, by

means of a computer.   The computer will not try to find the indefinite integral but will proceed by converting it into an entirely numerical problem.

NUMERICAL INTEGRATION

In the first example (in FRAME 1) the distance travelled is given, with a suitable conversion of units, by the area under the distance-time graph. In the second example the final integral is not the area contained by the x - y graph but can be interpreted as the area enclosed by the graph of X and the y-axis where X = x⁴. Similarly any integral can be interpreted as representing an area, even although it may be giving us something else, for example, a first or second moment. It is the purpose of this programme to obtain methods of estimating definite integrals of functions for which either analytical formulae are not known or, even if they are, it is desirable to proceed by non-analytical methods. The methods that will be described are those of NUMERICAL INTEGRATION, or, as it is often known, QUADRATURE.

Counting Squares

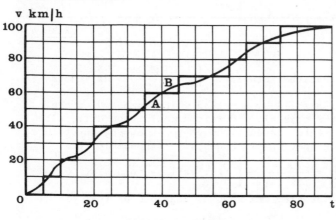

This method you may have learnt at school and the figure shows the original curve and the squares that would be included in the count for the coach problem in FRAME 1, assuming squares of the size shown. What would be the estimate of the distance travelled by this method?

****

Number of squares = 108.

Each square represents 5 s horizontally and 10 km/h = $\frac{1}{360}$ km/s vertically

i.e. a distance $\frac{1}{72}$ km.   Total distance ≃ 1·5 km.

This result is not accurate as some squares, e.g. A, have contributed more to the total than they should, while others, e.g. B, have been totally ignored. It is, of course, hoped that these 'gains' and 'losses' cancel each other out reasonably well so that the net result is not too far out.

Effectively this method replaces the actual curve by a step function, and the area under this step function found instead of that actually required

The Rectangular Rule

An alternative way of replacing the true curve by a step function is the RECTANGULAR RULE which is illustrated in the accompanying diagram.

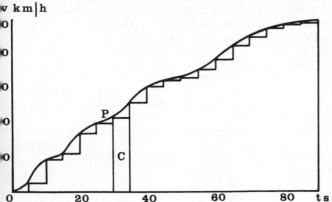

For each tabular point P (except the final one) a rectangle C is constructed as shown. The sum of the areas of these rectangles is then taken as the required area. As the conversion factor from km/h to km/s is 1/3600, the result will now be given by

$$5 \times \frac{0 + 5 + 19 + 22 + 32 + 40 + 43 + 52 + 61 + 65 + 66 + 71 + 77 + 85 + 90 + 95 + 98 + 99}{3600}$$

$$= \frac{5 \times 1020}{3600} = 1 \cdot 417 \text{ km.}$$

The answer obtained previously was 1·50. Which of these two figures do you think is more accurate and why?

****

9A

1·50. A certain amount of compensation was effected in this calculation by including squares like A and excluding those like B in the diagram in FRAME 7. In the second calculation all rectangles C were under-estimates.

Will it always happen, for any curve, that rectangles such as C will be under-estimates?

****

10A

No. They might always be over-estimates as in Fig.(i) or a mixture of over- and under-estimates as in Fig.(ii).

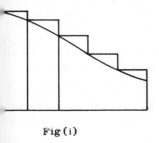

Fig (i)

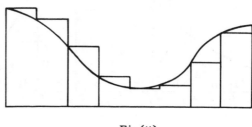

Fig (ii)

It is therefore only in cases similar to that shown in Fig.(ii) that any sort of compensation is produced by this method.   You might then very well ask the question - "If this method only gives any sort of compensation in certain cases, why bother with it at all when counting squares will very often produce some compensation?"   The answer to this question is that it enables us to produce a formula for finding the area under a curve.   Admittedly it may not be a very accurate formula but it will have the advantage that no judgment on the part of the user will be necessary.   When counting squares it is sometimes necessary to judge whether a particular square should or should not be included.   Once we have seen how a formula can be found, it may then be possible to go on to incorporate refinements in order to make it more accurate.

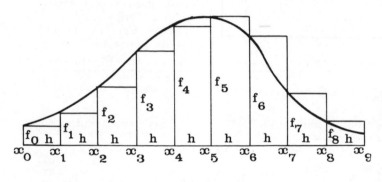

If ordinates are drawn to the curve as shown at x_0, x_1, x_2, etc., the distance between consecutive x's being constant at h, then the area under the stepped function is given by

$$h(f_0 + f_1 + \ldots + f_8) \qquad (12.1)$$

More generally, if there are n + 1 points taken along the x-axis, x_0, x_1, $\ldots$, x_n, all distance h apart, the formula will become $h(f_0 + f_1 + \ldots + f_{n-1})$ i.e.,

$$h \sum_{r=0}^{n-1} f_r \qquad (12.2)$$

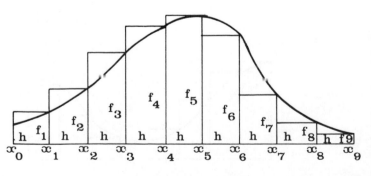

An alternative way in which a rectangular rule could be applied to the area depicted in the last frame is shown here.   What will be the new formulae, corresponding to (12.1) and (12.2), for the new stepped area if this is done?

****

$$h(f_1 + f_2 + \ldots\ldots\ldots + f_9) \qquad (13A.1)$$

$$h \sum_{r=1}^{n} f_r \qquad (13A.2)$$

What will this alternative rectangular rule give as the answer to the coach problem?

$$**************************************$$

$1 \cdot 556$

Now, in FRAME 9, it so happened that all rectangles such as C were under-estimates. If you constructed the diagram corresponding to that in FRAME 13, all the rectangles would be over-estimates. It would, therefore, seem logical to take the average of the two figures $1 \cdot 417$ and $1 \cdot 556$ as being more likely than either of them. This would give the result $1 \cdot 486$. If this is done then we are effectively applying what is known as the TRAPEZIUM RULE (or very frequently, but less accurately, the TRAPEZOIDAL RULE) as will be seen shortly.

The Trapezium Rule

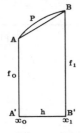

In this rule, the area under the curve APB between the two ordinates at x_0 and x_1 is replaced by that under the chord AB, i.e., by the trapezium A'ABB'. One would naturally expect this to be a better approximation than either of the two rectangles used previously. If f_0 and f_1 are the corresponding ordinates, then

$$\text{area} \simeq \frac{h}{2} (f_0 + f_1) \qquad (16.1)$$

h having the same meaning as before.

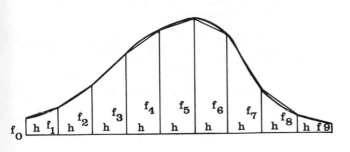

To apply this rule to finding an area such as that shown in FRAMES 12 and 13, the area is first replaced by a number of trapezia as shown here. The rule is then applied to each trapezium in turn and the results added together.

FRAME 17   (continued)

What will be the formula for the complete area shown here when you do
this?   Having obtained this formula verify that it is the average of
formulae (12.1) and (13A.1).
                   ****

17A

$$\frac{h}{2}(f_0 + 2f_1 + 2f_2 + 2f_3 + 2f_4 + 2f_5 + 2f_6 + 2f_7 + 2f_8 + f_9)$$

FRAME 18

Extending the formula to n (instead of nine) intervals and writing it in
a slightly different form gives

$$h(\tfrac{1}{2}f_0 + f_1 + f_2 + f_3 + \ldots\ldots f_{n-2} + f_{n-1} + \tfrac{1}{2}f_n) \qquad (18.1)$$

Now use formulae (12.2), (13A.2) and (18.1) to find estimates of the area
under the curves, from x = 0 to x = 10, given by the following tables:

i)

x	0	2	4	6	8	10
y	0	12	48	108	192	300

ii)

x	0	1	2	3	4	5	6	7	8	9	10
y	0	3	12	27	48	75	108	147	192	243	300

                   ****

18A

i)   720, 1320, 1020
ii) 855, 1155, 1005

Note that, for each table, $f_0 = 0$.

FRAME 19

The figures in the previous frame were taken from the curve $y = 3x^2$.

The actual area $= \int_0^{10} 3x^2 dx = \left[x^3\right]_0^{10}$

$$= 1000$$

Two things are noticeable from the results:

a)   In both cases (i) and (ii), (18.1) gives the best answer.
b)   Each answer in (ii) is better than the corresponding one in (i).

From what has been said previously (a) was to be expected.   (b) is also
to be expected because in (i), h = 2 but in (ii) h = 1.   It has been
remarked in earlier programmes that, unless other considerations are
involved, the smaller the value of h the better the result.

What is the value, to 4 significant figures, of the best estimate you can
expect to obtain so far for the answer to the problem given in FRAME 2?
The values of $x^4 (=X)$ are given in the following table.

y	0	5	10	15	20	25	30
$x^4 (=X)$	0	234 256	5308 416	6250 000	5764 801	4100 625	810 000

y	35	40	45 - 75	80 85	
$x^4 (=X)$	14 641	2401	625	256 0	

                   ****

$0 \cdot 01\pi \times 5 \times 22\,486\,021 \simeq 3532\,000\,g\,mm^2$ *(using the trapezium rule)*

Integration Formulae via Interpolation Formulae

An alternative method will now be used to obtain (16.1).   This method may seem to be very complicated in comparison with that used in FRAME 16 but it will show us a way which can be extended to get other, more accurate, formulae than those we have already.

The Newton-Gregory forward difference interpolation formula was obtained in the programme on Interpolation and is

$$y = f_0 + \binom{p}{1}\Delta f_0 + \binom{p}{2}\Delta^2 f_0 + \binom{p}{3}\Delta^3 f_0 + \cdots\cdots$$

or $$y = f_0 + p\Delta f_0 + \frac{p(p-1)}{2}\Delta^2 f_0 + \frac{p(p-1)(p-2)}{6}\Delta^3 f_0 + \cdots \quad (20.1)$$

where $x = x_0 + ph$.

What will be the connection between $\int y dx$ and $\int y dp$?

****

20A

$$\int y dx = \int y \frac{dx}{dp} dp = h \int y dp$$

Referring to the figure in FRAME 16, what will be the values of p corresponding to A' and B' ?
****

21A

0 and 1.

Thus $$\int_{x_0}^{x_1} y dx = h \int_0^1 y dp$$

$$= h \int_0^1 \left\{ f_0 + p\Delta f_0 + \frac{p^2 - p}{2}\Delta^2 f_0 + \frac{p^3 - 3p^2 + 2p}{6}\Delta^3 f_0 + \cdots \right\} dp$$

$$= h \left[pf_0 + \frac{1}{2}p^2\Delta f_0 + \left(\frac{1}{6}p^3 - \frac{1}{4}p^2\right)\Delta^2 f_0 \right.$$
$$\left. + \left(\frac{1}{24}p^4 - \frac{1}{6}p^3 + \frac{1}{6}p^2\right)\Delta^3 f_0 + \cdots \right]_0^1$$

$$= h \left\{ f_0 + \frac{1}{2}\Delta f_0 - \frac{1}{12}\Delta^2 f_0 + \frac{1}{24}\Delta^3 f_0 + \cdots\cdots \right\}$$

Now, as you know, for well behaved curves, the higher order differences are small compared with the lower ones.   Round-off errors do, of course, upset things, but differences rendered unreliable because of these errors

FRAME 22   (continued)

are ignored.   If second and higher order differences are ignored, the formula for $\displaystyle\int_{x_0}^{x_1} y\,dx$ reduces to $h(f_0 + \frac{1}{2}\Delta f_0)$.   What does this become when it is expressed in terms of functional values only?

****

22A

$$h\left\{f_0 + \frac{1}{2}(f_1 - f_0)\right\} = \frac{h}{2}(f_0 + f_1).$$

(This is the same result as (16.1).   The reason for this is that the same approximation has effectively been made – ignoring second and higher differences replaces a curve between two points by the straight line joining them.)

FRAME 23

The main term ignored in the formula was $-\frac{1}{12}h\Delta^2 f_0$.   This gives us an estimate of the <u>error</u> involved in the formula.

FRAME 24

Simpson's Rule

This is a very frequently used rule in numerical integration and you may already have met it.   It gives a formula for an estimate of the area under the portion of a curve between x_0 and x_2 in terms of the three ordinates f_0, f_1 and f_2. Can you see how to get it by a process similar to that just adopted for the trapezium rule? (Give the step corresponding to the first line in FRAME 22 but don't continue with the actual working yet.)

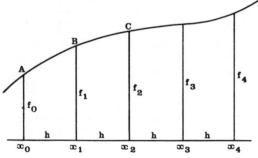

****

24A

$$\int_{x_0}^{x_2} y\,dx = h\int_0^2 y\,dp$$

All that is now necessary is to integrate (20.1) between 0 and 2.

FRAME 25

Using the working done at the beginning of FRAME 22 will give

$$h\left[pf_0 + \frac{1}{2}p^2\Delta f_0 + \left(\frac{1}{6}p^3 - \frac{1}{4}p^2\right)\Delta^2 f_0 + \left(\frac{1}{24}p^4 - \frac{1}{6}p^3 + \frac{1}{6}p^2\right)\Delta^3 f_0 + \cdots\cdots\right]_0^2$$

$$= h\left\{2f_0 + 2\Delta f_0 + \frac{1}{3}\Delta^2 f_0 + 0\Delta^3 f_0 + \cdots\cdots\cdots\right\}$$

Taking the first three terms of this formula, what will the result be when expressed in terms of functional values only?
****

25A

$$h\left\{2f_0 + 2(f_1 - f_0) + \frac{1}{3}(f_2 - 2f_1 + f_0)\right\} = \frac{h}{3}(f_0 + 4f_1 + f_2)$$

(This formula can also be obtained from Taylor's series. This was done in our "Mathematics for Engineers and Scientists" Vol. 1 as an illustration of one of the uses of that series.)

FRAME 26

The first term neglected now is that involving $\Delta^4 f_0$ due to the coefficient of $\Delta^3 f_0$ being zero. It can be shown that the Δ^4 term is $-\frac{h}{90}\Delta^4 f_0$ and so it is to be expected that Simpson's rule gives a much more accurate result than the trapezium rule.

FRAME 27

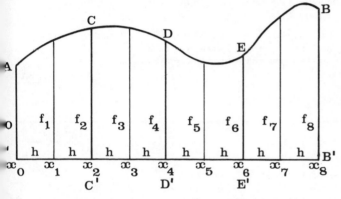

If the range of integration is large, then h will also be large. The figure shows how such an integral can be split up into a number of sub-integrals, for each of which h is smaller. Making h smaller means that you expect the result to be more accurate. In the figure as shown, the area A'ABB' has been divided into four smaller areas A'ACC', C'CDD', D'DEE' and E'EBB'. This has been done in such a way that the width of each area is the same. The formula that you obtained in 25A is now applied to each of these smaller areas in turn. What will you get when you do this for the three areas C'CDD', D'DEE' and E'EBB'?
****

27A

$$\frac{h}{3}(f_2 + 4f_3 + f_4), \quad \frac{h}{3}(f_4 + 4f_5 + f_6), \quad \frac{h}{3}(f_6 + 4f_7 + f_8)$$

FRAME 28

The total area is thus

$$\frac{h}{3}(f_0 + 4f_1 + f_2) + \frac{h}{3}(f_2 + 4f_3 + f_4) + \frac{h}{3}(f_4 + 4f_5 + f_6) + \frac{h}{3}(f_6 + 4f_7 + f_8)$$

$$= \frac{h}{3}(f_0 + 4f_1 + 2f_2 + 4f_3 + 2f_4 + 4f_5 + 2f_6 + 4f_7 + f_8)$$

It is easy to see how this can be extended to an area subdivided into a different number of small sections.   As each sub-area has to be divided into two, it follows that the total number of strips <u>must be even</u> (8 in the example just done).   If this number is n the formula becomes

$$\frac{h}{3}(f_0 + 4f_1 + 2f_2 + 4f_3 + \ldots\ldots + 4f_{n-3} + 2f_{n-2} + 4f_{n-1} + f_n)$$

Alternatively, it can be written as

$$\frac{h}{3}\left\{ f_0 + f_n + 4(f_1 + f_3 + \ldots + f_{n-1}) + 2(f_2 + f_4 + \ldots + f_{n-2})\right\}$$

$$(28.1)$$

To illustrate the rule, if it is applied to the coach problem in FRAME 1, the distance travelled is

$$\frac{5}{3}\left\{ \frac{(0+100) + 4(5+22+40+52+65+71+85+95+99) + 2(19+32+43+61+66+77+90+98)}{3600}\right\}$$

$$= 1 \cdot 485 \text{ km.}$$

In this example h = 5 and you remember that the factor 1/3600 is necessary to change km/h into km/s.

Now apply the formula to set (ii) of data in FRAME 18.   Compare your answers with those in 18A and the exact figure of 1000.

****

28A

$$\frac{1}{3}\left\{ (0+300) + 4(3+27+75+147+243) + 2(12+48+108+192)\right\} = 1000$$

FRAME 29

Two questions now:

a)   Why do you think the answer is exactly right?
b)   Why couldn't you apply the formula to set (i) of data in FRAME 18?

****

29A

a)   The data came from a quadratic, for which all differences higher than the second are zero.   Thus all the terms ignored in arriving at the formula are zero in this case.

b)   The number of strips into which the area is divided is odd (5).

FRAME 30

b)   can be partially overcome by using the formula for the first four strips and applying the trapezium rule to the fifth.   What will be the answer if you do this?

****

30A

$$\frac{2}{3}\left\{ (0+192) + 4(12+108) + 2\times 48\right\} + \frac{2}{2}(192+300) = 512 + 492 = 1004$$

Alternatively the trapezium rule could have been used for the first strip and Simpson's for the other four.   The result would then also have worked

out to 1004.   Usually one would expect different answers in the two
cases.   They are equal here because it happens to be a parabola you are
working with.

A flow diagram for evaluating $\displaystyle\int_a^b f(x)\,dx$ by Simpson's rule using n

strips (n assumed even) is shown on page 262.

The Three-Eighths Rule

The basic trapezium rule $\left\{\dfrac{h}{2}(f_0 + f_1)\right\}$ involves the use of two

functional values, f_0 and f_1.   The basic Simpson's rule involves the
use of three such values, f_0, f_1 and f_2.   Because of this, the
trapezium rule is called a two-point formula and Simpson's rule a
three-point formula.   Other formulae for numerical integration exist
which use more functional values.   One of these is the THREE-EIGHTHS RULE
which uses four functional values in its basic form and so is a four-

point formula.   It gives $\displaystyle\int_{x_0}^{x_3} y\,dx$ which is equivalent to $h\displaystyle\int_0^3 y\,dp.$

In working out the latter integral, it is necessary to retain one more
term in the integrated interpolation formula than for Simpson's rule.
Thus, in finding the trapezium rule from an interpolation formula, it is
necessary to retain the first difference term.   For Simpson's rule it is
necessary to retain the first two difference terms.   For the three-
eighths rule the first three difference terms must be kept.

See if you can obtain this rule, expressing the result in terms of h and
the functional values f_0, f_1, f_2 and f_3.
****

$$h\left[pf_0 + \frac{1}{2}p^2\Delta f_0 + \left(\frac{1}{6}p^3 - \frac{1}{4}p^2\right)\Delta^2 f_0 + \left(\frac{1}{24}p^4 - \frac{1}{6}p^3 + \frac{1}{6}p^2\right)\Delta^3 f_0 + \ldots\ldots\right]_0^3$$

$$= h\left\{3f_0 + \frac{9}{2}\Delta f_0 + \frac{9}{4}\Delta^2 f_0 + \frac{3}{8}\Delta^3 f_0 + \ldots\ldots\ldots\right\}$$

Retaining terms up to and including Δ^3 *gives*

$$h\left\{3f_0 + \frac{9}{2}(f_1 - f_0) + \frac{9}{4}(f_2 - 2f_1 + f_0) + \frac{3}{8}(f_3 - 3f_2 + 3f_1 - f_0)\right\}$$

$$= \frac{3}{8}h(f_0 + 3f_1 + 3f_2 + f_3)$$

It can be shown that the dominant term in the error in this formula is
$-\dfrac{3}{80}h\Delta^4 f_0$.   It is therefore not quite so accurate as Simpson's Rule, but
on the other hand it enables us to find the value of an integral for which

Flow diagram for FRAME 31.

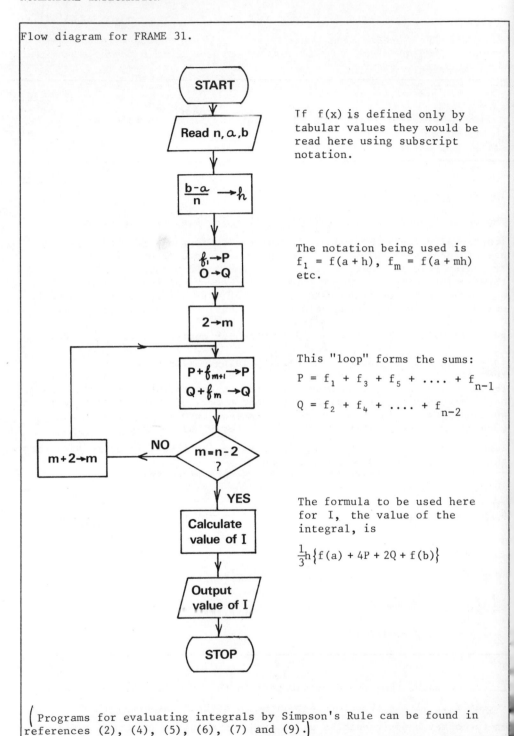

If $f(x)$ is defined only by tabular values they would be read here using subscript notation.

The notation being used is $f_1 = f(a+h)$, $f_m = f(a+mh)$ etc.

This "loop" forms the sums:

$P = f_1 + f_3 + f_5 + \ldots + f_{n-1}$

$Q = f_2 + f_4 + \ldots + f_{n-2}$

The formula to be used here for I, the value of the integral, is

$$\frac{1}{3}h\{f(a) + 4P + 2Q + f(b)\}$$

(Programs for evaluating integrals by Simpson's Rule can be found in references (2), (4), (5), (6), (7) and (9).)

the corresponding area is split into three strips instead of only two. It can be extended to any area which is split into a multiple of three strips.

What will it become for an area split into 9 strips?
****

33A

$$\frac{3}{8}h(f_0 + 3f_1 + 3f_2 + f_3) + \frac{3}{8}h(f_3 + 3f_4 + 3f_5 + f_6) + \frac{3}{8}h(f_6 + 3f_7 + 3f_8 + f_9)$$

$$= \frac{3}{8}h(f_0 + 3f_1 + 3f_2 + 2f_3 + 3f_4 + 3f_5 + 2f_6 + 3f_7 + 3f_8 + f_9)$$

FRAME 34

Now use a combination of Simpson's and the three-eighths rules to obtain an estimate of the moment of inertia of the solid described in FRAME 2. (The required definite integral is in 2A and the table of values of x^4 in FRAME 19.)   It is suggested that you use Simpson's rule for the first fourteen strips and the three-eighths rule for the remaining three.
****

34A

$$\int_0^{85} x^4 dy = \frac{5}{3}\{0 + 625 + 4(234\,256 + 6\,250\,000 + 4\,100\,625 + 14\,641 + 625$$
$$+ 625 + 625) + 2(5\,308\,416 + 5\,764\,801 + 810\,000 + 2401$$
$$+ 625 + 625)\} + \frac{3 \times 5}{8}\{625 + 3(625 + 256) + 0\}$$

$$= 110\,299\,915 + 6128 = 110\,306\,043$$

M. of I. $\simeq 3465\,000\,g\,mm^2$.

FRAME 35

Other Integration Formulae

You will realise that the way in which the Newton-Gregory forward difference formula is used can be varied by integrating between other sets of limits.   We have integrated w.r.t. p between 0 and 1, 0 and 2, 0 and 3 to give, respectively, the trapezium rule, Simpson's rule and the three-eighths rule.

BOOLE'S RULE is obtained by integrating between 0 and 4, retaining differences up to the fourth.   It gives

$$\int_{x_0}^{x_4} y\,dx = h\int_0^4 y\,dp$$

$$= \frac{2h}{45}(7f_0 + 32f_1 + 12f_2 + 32f_3 + 7f_4)$$

This is a five-point formula and as the coefficient of $\Delta^5 f_0$ turns out to be zero, the leading term omitted is that involving $\Delta^6 f_0$.   It is actually $-\frac{8h}{945}\Delta^6 f_0$.

WEDDLE'S RULE is obtained by integrating between 0 and 6, retaining differences up to the sixth.   But in order to give a simple formula, only part of the sixth difference is retained.   The rest is included in the error.   The formula, which is a seven-point one, is

$$\int_{x_0}^{x_6} y\,dx = h\int_0^6 y\,dp$$

$$= \frac{3h}{10}(f_0 + 5f_1 + f_2 + 6f_3 + f_4 + 5f_5 + f_6)$$

The part of the sixth difference term omitted is $-\frac{h}{140}\Delta^6 f_0$.

Each of these formulae can be extended to areas split into more strips. The number of strips will have to be a multiple of four in the case of Boole's rule and a multiple of six for Weddle's.

In the moment of inertia problem there were 17 strips.   These were divided in FRAME 34 into seven twos and a three.   Another way in which they could be split is into two sixes, a three and a two.   Using this split, Weddle's rule can be used for the two sixes, the three-eighths rule for the three and Simpson's rule for the two.

The total area will then be

$$\frac{3\times 5}{10}\Big\{(0 + 5\times 234\,256 + 5308\,416 + 6\times 6250\,000 + 5764\,801 + 5\times 4100\,625$$

$$+ 810\,000) + (810\,000 + 5\times 14\,641 + 2401 + 6\times 625 + 625 + 5\times 625$$

$$+ 625)\Big\} \ + \frac{3\times 5}{8}(625 + 3\times 625 + 3\times 625 + 625) + \frac{5}{3}(625 + 4\times 256 + 0)$$

$$= \ 106\,598\,556$$

$\therefore$ M. of I. $\simeq$ 3349 000 g mm^2

What result do you get for the coach problem using Weddle's rule?
            ****

1·482 km.

From the nature of the formulae, one would expect the last two results to be more accurate than those obtained previously.   In practice this is unlikely to be so as Weddle's rule takes into account higher order differences than, say, Simpson's rule.   From the natures of the two problems these higher differences are probably not very reliable.

Generally speaking, it is reasonable to say that Simpson's rule will probably give you sufficient accuracy in most problems.   If the number of strips is not even, then a combination of Simpson and three-eighths will generally suffice.   You will only need to use Boole and Weddle if you have very accurate data and require a similarly very accurate result.

Errors

So far all that has been done about errors has been to give an estimate of the predominant term omitted from an interpolation formula when finding a basic integration formula. Taking, for example, Simpson's rule this term is $-\dfrac{1}{90}h\Delta^4 f_0$. But when the basic formula is extended as in FRAMES 27 and 28, then a similar error is introduced on each application of this basic formula. Taking the situation described in FRAME 27, the leading term in the error will be

$$-\frac{1}{90}h\Delta^4 f_0 \quad \text{on finding} \quad \text{A'ACC'}$$

$$-\frac{1}{90}h\Delta^4 f_2 \quad \text{on finding} \quad \text{C'CDD'}$$

$$-\frac{1}{90}h\Delta^4 f_4 \quad \text{on finding} \quad \text{D'DEE'}$$

$$\text{and} \quad -\frac{1}{90}h\Delta^4 f_6 \quad \text{on finding} \quad \text{E'EBB'}$$

An estimate of the magnitude of the total error introduced is thus $\dfrac{1}{90}h(|\Delta^4 f_0| + |\Delta^4 f_2| + |\Delta^4 f_4| + |\Delta^4 f_6|)$, the modulus signs being introduced as the differences may be positive or negative.

Now the contents of the brackets will be less than four times the value of the maximum term inside them and so an estimate of the magnitude of the total error is

$$\frac{1}{90}h \times 4M$$

where M is the maximum absolute value of the fourth differences of the function. Now $4h = \frac{1}{2}(x_8 - x_0)$ and so the maximum absolute error is

$$\frac{M(x_8 - x_0)}{180}$$

In a similar way, for the more general formula (28.1), an estimate of the maximum absolute error is

$$\frac{M(x_n - x_0)}{180} \qquad (38.1)$$

Sometimes it may be more convenient, as, for example, if the analytical formula for the function is known, to express the error in terms of a derivative instead of a difference.

Now $f'(x_0) = \lim\limits_{h\to 0} \dfrac{f(x_0 + h) - f(x_0)}{h} = \lim\limits_{h\to 0} \dfrac{\Delta f_0}{h}$

and so, approximately, $\Delta f_0 = hf'(x_0)$.

In a similar way $\Delta^2 f_0 \simeq h^2 f''(x_0)$, $\Delta^3 f_0 \simeq h^3 f'''(x_0)$, $\Delta^4 f_0 \simeq h^4 f^{iv}(x_0)$, and so an approximate value for the maximum fourth difference of a function in a given range is $h^4 \times$ the maximum value of the fourth derivative of the function in the range.   An alternative expression for (38.1) is therefore

$$\frac{mh^4(x_n - x_0)}{180} \qquad (39.1)$$

where m is the maximum absolute value of $f^{iv}(x)$ in the range.

In a similar manner, results corresponding to (38.1) and (39.1) can be obtained for each of the other integration formulae.

FRAME 40

As an example on the use of this, let us look at the following problem:

Determine numerically $\displaystyle\int_0^{\frac{1}{4}\pi} \sin x \, dx$ using Simpson's rule with a step length so chosen that the truncation error is less than 0·001.   Find a bound for the round-off error in your answer.   (C.E.I.)

First of all, it is obvious that Simpson's rule is not really necessary to find $\displaystyle\int_0^{\frac{1}{4}\pi} \sin x \, dx$,   as this can be done directly.   However, an easy example like this will serve to illustrate the ideas involved.

To start with, the maximum value of the fourth derivative of sin x in the range 0   to $\frac{1}{4}\pi$ is required.   What will this be?

****

40A

$\frac{d^4}{dx^4} (\sin x) = \sin x.$ *This increases steadily from 0 to $1/\sqrt{2}$ as x increases from 0 to $\pi/4$.*

Thus m = $1/\sqrt{2}$ = 0·7071.

Note that the greatest value of sin x in this range is not at a turning point and hence cannot be found by putting its first derivative equal to zero.

FRAME 41

For the error to be less than 0·001   requires $\dfrac{0\cdot7071 \; h^4 \left\{\frac{\pi}{4} - 0\right\}}{180} < 0\cdot001,$ using (39.1).   From this $h < 0\cdot76.$

The least number of strips that can be used in Simpson's rule is 2.   If this number is used, then, as the range of integration is $\frac{\pi}{4} = 0\cdot7854$, h = 0·3927   and this satisfies the required condition.

A table of values is then

x	0	0·3927	0·7854
sin x	0	0·3827	0·7071

Using these figures, what does Simpson's rule give for the integral?

****

$$\frac{0 \cdot 3927}{3} \ (0 \ + \ 4 \times 0 \cdot 3827 \ + \ 0 \cdot 7071) \ = \ 0 \cdot 2929$$

Finally, what will be a bound for the round-off error introduced?

****

Maximum error in each decimal = $0 \cdot 000\ 05$
Maximum error in contents of brackets in 41A = $4 \times 0 \cdot 000\ 05\ +\ 0 \cdot 000\ 05$
$= 0 \cdot 000\ 25$
Maximum error in multiplication $\simeq 0 \cdot 4 \times 0 \cdot 000\ 25\ +\ 2 \cdot 2 \times 0 \cdot 000\ 05 = 0 \cdot 000\ 21$
Maximum error on division by 3 $\simeq 0 \cdot 000\ 07$.

Summary of Integration Formulae

Rule	Formula	Leading Error Term
Trapezium	$\displaystyle\int_{x_0}^{x_1} ydx = \frac{h}{2}(f_0 + f_1)$	$-\dfrac{1}{12}h\Delta^2 f_0$
Simpson	$\displaystyle\int_{x_0}^{x_2} ydx = \frac{h}{3}(f_0 + 4f_1 + f_2)$	$-\dfrac{1}{90}h\Delta^4 f_0$
Three-Eighths	$\displaystyle\int_{x_0}^{x_3} ydx = \frac{3h}{8}(f_0 + 3f_1 + 3f_2 + f_3)$	$-\dfrac{3}{80}h\Delta^4 f_0$
Boole	$\displaystyle\int_{x_0}^{x_4} ydx = \frac{2h}{45}(7f_0 + 32f_1 + 12f_2 + 32f_3 + 7f_4)$	$-\dfrac{8}{945}h\Delta^6 f_0$
Weddle	$\displaystyle\int_{x_0}^{x_6} ydx = \frac{3h}{10}(f_0 + 5f_1 + f_2 + 6f_3 + f_4 + 5f_5 + f_6)$	$-\dfrac{1}{140}h\Delta^6 f_0$

These are all examples of what are known as formulae of the NEWTON–COTES type.

Romberg Integration

If the integral of a linear function is required between given limits, the trapezium rule will give it exactly. Similarly, for a quadratic or cubic function, Simpson's rule is exact. The other integral formulae which have been found are exact for certain other polynomial curves. Otherwise the use of a numerical integration formula leads to an error. The magnitude of the error may be reduced by making h smaller but this is not always desirable. For example, if tabular values are given, any reduction in h involves a number of interpolations. An alternative

method, popular for computer work, of increasing the accuracy is known as
ROMBERG INTEGRATION.

This method starts off with the trapezium rule, which, in view of the fact
that this rule is not generally very accurate, may seem surprising.   But
the process then adopted rapidly gives rise to a very accurate answer.

It will be illustrated by applying it to find $\int_0^2 e^x \, dx$, which we know to
be $\left[e^x\right]_0^2$ = 6·389 056 to 6 decimal places.

The following table gives the value of the function e^x for x = 0(0·125)2.

x	0	0·125	0·25	0·375	0·5	0·625
e^x	1	1·133 148	1·284 025	1·454 991	1·648 721	1·868 246

x	0·75	0·875	1	1·125	1·25	1·375
e^x	2·117 000	2·398 875	2·718 282	3·080 217	3·490 343	3·955 077

x	1·5	1·625	1·75	1·875	2	
e^x	4·481 689	5·078 419	5·754 603	6·520 819	7·389 056	

The values of x chosen may appear to be unusual, but they have been
selected so that the interval 0 - 2 can be taken as a whole or divided
into 2, 4, 8 or 16 parts.   This will be equivalent to taking h = 2, 1,
0·5, 0·25 or 0·125, i.e., successively halving its value.   Obviously
one could take this halving further, but it only entails more work and may
not be necessary.   Indeed, we may discover that we have gone further than
necessary already.

The trapezium rule (16.1) and its extended form (18.1) are now used to
give estimates of the integral, taking, in turn, the values of h in the
last frame.

What results will you get when h = 2 and 1?
****

$h = 2$, (16.1) gives $\frac{2}{2}(1 + 7 \cdot 389\,056)$ $= 8 \cdot 389\,056$

$h = 1$, (18.1) gives $1(0 \cdot 5 + 2 \cdot 718\,282 + 3 \cdot 694\,528) = 6 \cdot 912\,810$

Continuing, a table can be built up:

h	Estimate of Integral
2	8·389 056
1	6·912 810
0·5	6·521 610
0·25	6·422 298
0·125	6·397 373

As you would expect, the estimates are increasing in accuracy as one goes
down the table.

Now, in FRAMES 38 and 39, you saw that an estimate of the maximum error for an integral obtained by Simpson's rule is $mh^4(x_n - x_0)/180$. By a similar process, find the corresponding expression for an integral obtained by the trapezium rule.

****

47A

The leading error term for the first strip is $-\frac{1}{12}h\Delta^2 f_0$.

The formula corresponding to (38.1) is $M(x_n - x_0)/12$. *That corresponding to (39.1) is* $mh^2(x_n - x_0)/12$. *This is the required result.*

FRAME 48

For a given function and for fixed values of x_0 and x_n, m will be fixed and so this formula is equivalent to a constant times h^2. Now this is only an estimate of the <u>maximum</u> error. It can be shown, but not very easily, that the <u>actual</u> error itself is of the form $Ah^2 + Bh^4 + Ch^6 + \ldots$ The most important term in this expression, unless h is large, is Ah^2. As   true value − estimated value = error   we now write, approximately,

$$I - I_{est} = Ah^2 \qquad (48.1)$$

If the five successive estimates given in the previous frame are labelled I_1, I_2 etc., then, as h = 2, 1, $\frac{1}{2}$ etc., in turn,

$$I - I_1 = 4A \qquad (48.2)$$
$$I - I_2 = A \qquad (48.3)$$
$$I - I_3 = A/4 \qquad (48.4)$$

What will be the other equations in this list?

****

48A

$I - I_4 = A/16$ *and* $I - I_5 = A/64$

FRAME 49

The next step is to eliminate A between (48.2) and (48.3). Doing this gives

$$\frac{I - I_1}{I - I_2} = 4 \qquad (49.1) \qquad \text{i.e.} \qquad I = (4I_2 - I_1)/3$$

or alternatively $I = I_2 + (I_2 - I_1)/3 \qquad (49.2)$

Note that the denominator here is one less than the number on the R.H.S. of (49.1).

What will you get for I when you eliminate A between (48.3) and (48.4)?

****

49A

$I = (4I_3 - I_2)/3$ *which can also be written as* $I = I_3 + (I_3 - I_2)/3$
$(49A.1)$

Similarly, also,

$$I = I_4 + (I_4 - I_3)/3 \qquad (50.1)$$

$$\text{and} \quad I = I_5 + (I_5 - I_4)/3 \qquad (50.2)$$

Now owing to the fact that equation (48.1) is only an approximation, the I in it will not be exactly the true value of the integral. This means that the I in all equations of the type (49.2) is not quite exact. What we do find, however, is that the I in (49.2) is a better approximation than either I_1 or I_2. Similarly, the I in (49A.1) is a better approximation than either I_2 or I_3 and so on. What will be the values of I, to 6 decimal places, as given by (49.2) and (49A.1).

****

6·420 728   and   6·391 210

The other two values of I, as found from (50.1) and (50.2) are 6·389 194 and 6·389 065. The table in FRAME 47 can now be extended to read

2	8·389 056	
1	6·912 810	6·420 728
0·5	6·521 610	6·391 210
0·25	6·422 298	6·389 194
0·125	6·397 373	6·389 065

Looking to see what was done in (49.2) to give 6·420 728, a relatively small amount was subtracted from 6·912 810. This amount can be regarded as a correction to the value 6·912 810 and so the result has been placed on the same level. Similar remarks apply to the other values obtained at this stage.

In FRAME 47 it was remarked that the estimates there were increasing in accuracy as one went down the table. You will notice that the values in the latest column we have now formed are doing exactly the same.

If we denote the latest estimates by J_1, J_2, J_3, J_4, then

$$J_1 = I_2 + (I_2 - I_1)/3$$
$$J_2 = I_3 + (I_3 - I_2)/3 \quad \text{etc.}$$

These equations can all be embraced by the single equation

$$J_i = I_{i+1} + (I_{i+1} - I_i)/3 \qquad (52.1)$$

where i here takes the values 1, 2, 3, and 4, but can go higher or stop earlier as necessary.

When using a computer, this is a very simple equation to program, incorporating an easy DO loop. But when you are doing examples manually you may prefer to re-write the equation (52.1) as

improved value = more accurate value
+ (more accurate value - less accurate value)/3   (52.2)

The next step in the process is to use the four values in the J column to obtain three, still more improved, estimates of the integral. These are obtained on the basis that all h^2 terms in the errors have now been taken care of and so equation (48.1) is replaced by

$$I - I_{est} = Bh^4$$

and the I_{est} terms are now given by J_1, J_2, etc.

By a process similar to that in FRAMES 48 - 50, still further improved estimates K_1, K_2 and K_3 are found to be given by

$$K_1 = J_2 + (J_2 - J_1)/15$$
$$K_2 = J_3 + (J_3 - J_2)/15$$
$$K_3 = J_4 + (J_4 - J_3)/15$$

The 15's occur (instead of 3's as previously) due to the use of h^4 instead of h^2, the ratio of two successive h^4's being 16 instead of 4. Similarly, if you use the form (52.2) the 3 in the denominator is replaced by 15.

Find the values of the 3 K's and then extend the table in FRAME 51 to include the K column.

****

53A

6·389 242	6·389 060	6·389 056		
2	8·389 056			
1	6·912 810	6·420 728		
0·5	6·521 610	6·391 210	6·389 242	
0·25	6·422 298	6·389 194	6·389 060	
0·125	6·397 373	6·389 065	6·389 056	

If you examine the K values you will see that, once again, the further one goes down the column, the better the estimate of the integral. Not only that, but the column as a whole contains much better values than previous columns.

From the 3 K values, two more estimates can be obtained, L_1 and L_2. Ch^6 is now used instead of Bh^4 and the equations for the L's are

$$L_1 = K_2 + (K_2 - K_1)/63$$
$$L_2 = K_3 + (K_3 - K_2)/63$$

(Note that 64 is the ratio of two successive h^6's.)

From them, we get $L_1 = 6·389 057$ and $L_2 = 6·389 056$.

From the two L's, one further estimate, M_1, can be obtained by using the equation

$$M_1 = L_2 + (L_2 - L_1)/255$$

(256 is the ratio of two successive h^8's.)

This is found to be 6·389 056 and the table is now

271

2	8·389 056			
1	6·912 810	6·420 728		
0·5	6·521 610	6·391 210	6·389 242	
0·25	6·422 298	6·389 194	6·389 060	6·389 057
0·125	6·397 373	6·389 065	6·389 056	6·389 056	6·389 056

Now it has already been pointed out that, for the various columns, as one goes down each column, the estimate of the integral is increasing in accuracy, and it eventually converges to the actual value required. Further, the estimates in the columns on the right are, taking them as a whole, better than those in the columns on the left.   These facts give the clue as to how much or how little of the table is necessary and, also, as to the final value taken for the integral.

In deriving the process, the table was completed as much as possible for the values of h chosen.   In order to obtain more entries, it would be necessary to take the next value of h, i.e. 0·0625.   In practice, one wants to stop, for obvious reasons, as soon as possible.   For this purpose, the entries are calculated in the order

$$I_1; \quad I_2, J_1; \quad I_3, J_2, K_1; \quad I_4, J_3, K_2, L_1; \quad \text{etc.}$$

i.e., in rows as indicated by the arrows.

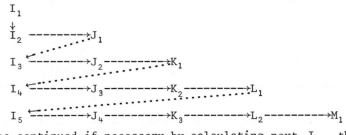

This can be continued if necessary by calculating next I_6, then J_5, K_4 and so on.   One stops when two successive entries in the same column are equal or differ by not more than a pre-specified degree of tolerance, for example, 1 in the last place of decimals.   The lower in position of the two entries concerned, rounded as necessary, gives the value of the integral.   In our case L_1 and L_2 are so close that L_2 can be taken as the value required and the table could stop here.

Working on the same basis, round off the entries at the top of the I column to 3 decimal places and form the table as far as is necessary so that two consecutive entries in a column do not differ by more than 0·001 Form the entries in the order indicated.

****

56A

8·389		
6·913	6·421	
6·522	6·392	6·390
6·422	6·389	6·389

The numbers enclosed indicate the result as 6·389, which, rounding off t 2 decimal places, is 6·39.

Now find, by this method, $\dfrac{1}{\pi}\displaystyle\int_0^{\frac{1}{2}\pi}\sin x\ dx,$ given the following values:

x	0	π/32	π/16	3π/32	π/8
y	0	0·098 017	0·195 090	0·290 284	0·382 683

x	5π/32	3π/16	7π/32	π/4	9π/32
y	0·471 396	0·555 570	0·634 394	0·707 107	0·773 010

x	5π/16	11π/32	3π/8	13π/32	7π/16
y	0·831 470	0·881 922	0·923 880	0·956 940	0·980 785

x	15π/32	π/2
y	0·995 184	1

****

π/2	0·25			
π/4	0·301 777	0·319 036		
π/8	0·314 209	0·318 353	0·318 307	
π/16	0·317 287	0·318 313	0·318 310	0·318 310
π/32	0·318 054	0·318 310	0·318 310	

Integral ≃ *0·318 31.   (To 7 decimal places, the value of the integral obtained analytically, is 0·318 3099.)*

If the table formed in a Romberg integration is very extensive, the notation that has been adopted for the entries becomes rather unwieldy. As the table looks something like a matrix, a double suffix notation such as the following is frequently adopted for labelling the entries:

$$
\begin{array}{ccccc}
T_{11} & & & & \\
T_{21} & T_{22} & & & \\
T_{31} & T_{32} & T_{33} & & \\
T_{41} & T_{42} & T_{43} & T_{44} & \\
T_{51} & T_{52} & T_{53} & T_{54} & T_{55}
\end{array}
$$

etc.   A notation such as this is also easier to use in a flow chart and for computer purposes.

A flow diagram for evaluating $\displaystyle\int_a^b f(x)\,dx$ by Romberg's method is as

shown.   The Romberg table is computed row by row and stops when a value is sufficiently close to the value immediately above it in the same column. If this convergence criterion has not been met when N rows have been formed, the process is halted.

(This flow diagram is shown on page 274.)

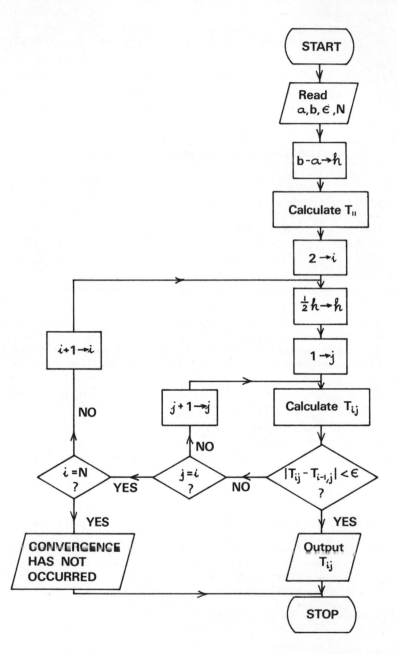

(Programs for Romberg integration can be found in references (2), (3), (8) and (9).)

Unequally Spaced Data - Use of Lagrange's Interpolation Formula

If the points given are not equally spaced in the x-direction, then
Lagrange's interpolation formula can be used to estimate an integral.
You will remember that it was used in the last programme, under similar
circumstances, to estimate the value of a derivative.   Turn to FRAME 24,
page 243 and find $\displaystyle\int_{0\cdot05}^{0\cdot25}$ f(x)dx   for the function f(x) given in the table
in that frame.

****

The Lagrangian Interpolation formula is given in FRAME 25, page 244.
Integrating gives

$$\left[-0\cdot0298x^4 - 0\cdot0083x^3 + 0\cdot502x^2\right]_{0\cdot05}^{0\cdot25} \simeq 0\cdot0299$$

Miscellaneous Examples

In this frame a collection of miscellaneous examples is given for you to
try.   Answers are supplied in FRAME 62, together with such working as is
considered helpful.

1.   A particle moves along a straight line so that at time t its distance
     s from a fixed point of the line is given by $\dfrac{ds}{dt} = t\sqrt{8 - t^3}$.

     Use Simpson's rule with 8 strips to calculate the approximate
     distance travelled by the particle from t = 0 to t = 2.   (L.U.)

2.   The coordinates of points on a curve are given in the following table

x	0	0·2	0·4	0·6	0·8	1·0	1·2
y	1	1·1	1·3	1·5	1·6	1·4	1·3

     Find, using Simpson's rule, the volume of revolution obtained when
     the area under this curve bounded by the lines x = 0,   x = 1·2   and
     the x-axis is rotated through 2π radians about the x-axis.

     Obtain the coordinates of the centroid of this volume.   (L.U.)

3.   Evaluate numerically $\displaystyle\int_0^{\frac{1}{4}\pi}$ cos x dx using the trapezium rule with step

     length so chosen that the truncation error is less than 0·01.   Find
     a bound for the round-off error in your answer.   (C.E.I.)

     (HINT: You will find in 47A the formula you require for the
     truncation error.)

4.   It was mentioned earlier in this programme that $\displaystyle\int_0^{x_0} \dfrac{x^4 e^x}{(e^x - 1)^2}\ dx$

275

occurs in a certain heat capacity problem.   From the following table

of the function $y = \dfrac{x^4 e^x}{(e^x - 1)^2}$,   evaluate this integral when

$x_0 = 0 \cdot 7$ by a combination of Boole's rule for the section from
x = 0 to 0·4   and the three-eighths rule for the rest.

x	0	0·1	0·2	0·3	0·4	0·5	0·6	0·7
y	0	0·0010	0·0399	0·0893	0·1579	0·2449	0·3494	0·4705

5.   The solutions of certain problems in various topics, e.g.,
     gravitational potential, fluid flow, non-linear springs, lead to the
     use of what are known as elliptic functions.   A simple example of

such a function is $\displaystyle\int_0^{\frac{1}{2}\pi} \dfrac{1}{\sqrt{1 - \frac{1}{2}\sin^2\theta}}\, d\theta$.   Find, using Weddle's rule,

the value of this integral from the following table of values of θ

and y where $y = \dfrac{1}{\sqrt{1 - \frac{1}{2}\sin^2\theta}}$.

θ	0	$\pi/24$	$\pi/12$	$\pi/8$	$\pi/6$	$5\pi/24$	$\pi/4$	$7\pi/24$
y	1	1·0043	1·0172	1·0388	1·0690	1·1079	1·1547	1·2080

θ	$\pi/3$	$3\pi/8$	$5\pi/12$	$11\pi/24$	$\pi/2$
y	1·2649	1·3208	1·3692	1·4023	1·4142

6.   Starting from Everett's interpolation formula:

$$y_p = qy_0 + \frac{q(q^2 - 1^2)}{6}\,\delta^2 y_0 + \frac{q(q^2 - 1^2)(q^2 - 2^2)}{120}\,\delta^4 y_0 + \ldots\ldots\ldots$$

$$+ py_1 + \frac{p(p^2 - 1^2)}{6}\,\delta^2 y_1 + \frac{p(p^2 - 1^2)(p^2 - 2^2)}{120}\,\delta^4 y_1 + \ldots\ldots\ldots$$

where q = 1 - p,   derive the formula

$$\frac{1}{h}\int_{x_0}^{x_1} y\,dx = \mu y_{\frac{1}{2}} - \frac{1}{12}\mu\delta^2 y_{\frac{1}{2}} + \frac{11}{720}\mu\delta^4 y_{\frac{1}{2}} - \ldots\ldots\ldots\ldots$$

for integrating over one table interval, and deduce the formula

$$\frac{1}{h}\int_{x_0}^{x_n} y\,dx = \frac{1}{2}y_0 + y_1 + \ldots + y_{n-1} + \frac{1}{2}y_n - \frac{1}{12}(\mu\delta y_n - \mu\delta y_0)$$

$$+ \frac{11}{720}(\mu\delta^3 y_n - \mu\delta^3 y_0) + \ldots\ldots\ldots$$

for integrating over n table intervals.

Use the table provided to evaluate $\displaystyle\int_0^{0\cdot5} e^{-x^2}\, dx$.

x	$y = e^{-x^2}$	Δy	$\Delta^2 y$	$\Delta^3 y$
0	1·000 000		−19 900	
		−9950		589
0·1	0·990 050		−19 311	
		−29 261		1714
0·2	0·960 789		−17 597	
		−46 858		2668
0·3	0·913 931		−14 929	
		−61 787		3373
0·4	0·852 144		−11 556	
		−73 343		3774
0·5	0·778 801		−7782	
		−81 125		3857
0·6	0·697 676		−3925	
		−85 050		3641
0·7	0·612 626		−284	(L.U.)

[There are certain points to notice about this question:

a) In the programme, all integration formulae obtained by
    integrating an interpolation formula used the Newton-Gregory
    interpolation formula. As this question illustrates, other
    interpolation formulae can be used equally as well.

b) When integrating the interpolation formula you will find it
    easier to leave it in terms of p and q than to convert it
    all into terms of p, and, for the q terms, use

$$\int f(q)\,dp = \int f(q)\frac{dp}{dq}\,dq = -\int f(q)\,dq, \qquad \text{as} \quad \frac{dp}{dq} = -1.$$

c) You will find it necessary to extend the table slightly
    backwards in order to get all the necessary differences.   As
    the function e^{-x^2} is symmetrical about x = 0, this will be
    quite easy.]

7. Using the values given in question 6 together with y = 0·527 292 when

    x = 0·8,   find $\displaystyle\int_0^{0·8} e^{-x^2}\,dx$ by Romberg integration.

8.

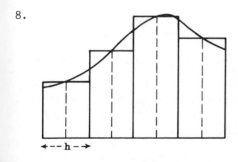

The MID-ORDINATE RULE approximates an
area to the sum of a number of
rectangles, the height of each
rectangle being that of the height of
the ordinate in the centre of each
strip.   Use this rule to find

$$\int_0^{10} 3x^2\,dx, \quad \text{taking} \quad h = 2.$$

(You
will find the values of $3x^2$ given in
FRAME 18.)

Answers to Miscellaneous Examples

1.

t	0	0·25	0·50	0·75	1	1·25	1·50	1·75	2
$\dfrac{ds}{dt}$	0	0·706	1·403	2·065	2·646	3·074	3·225	2·844	0

s = 4·109

2. Volume = $\displaystyle\int_0^{1·2} \pi y^2 dx$

First moment about plane passing through origin and perpendicular to

Ox = $\displaystyle\int_0^{1·2} \pi xy^2 dx$.   Integrals required by numerical means are

$\displaystyle\int_0^{1·2} y^2 dx$ and $\displaystyle\int_0^{1·2} xy^2 dx$.

These two integrals give 2·191 33 and 1·445 60.

Volume ≃ 6·8843 $\bar{x} \simeq 0·6597$ $\bar{y} = 0$ from symmetry.

3. For error to be less than that given, $\dfrac{h^2}{12} \times \dfrac{\pi}{4} < 0·01$; h < 0·3909.

As range of integration = 0·7854, it is therefore necessary to take
3 strips and so h = 0·2618.

Trapezium rule then gives value of integral = 0·7031.

Round-off error bound ≃ 0·0002.

4. 0·1116

5. 1·8541

6. Integrating w.r.t. p gives

$$\int y_p dp = -\frac{q^2}{2} y_0 - \frac{\dfrac{q^4}{4} - \dfrac{q^2}{2}}{6} \delta^2 y_0 - \frac{\dfrac{q^6}{6} - \dfrac{5q^4}{4} + 2q^2}{120} \delta^4 y_0 - \cdots\cdots$$

$$+ \frac{p^2}{2} y_1 + \frac{\dfrac{p^4}{4} - \dfrac{p^2}{2}}{6} \delta^2 y_1 + \frac{\dfrac{p^6}{6} - \dfrac{5p^4}{4} + 2p^2}{120} \delta^4 y_1 + \cdots\cdots$$

Limits of p are 0 and 1.   Those of q are 1 and 0 as $q = 1 - p$.
The first integration formula results.

The corresponding formula for the integral from x_1 to x_2 will be

$$\frac{1}{h} \int_{x_1}^{x_2} y dx = \mu y_{1\frac{1}{2}} - \frac{1}{12} \mu \delta^2 y_{1\frac{1}{2}} + \frac{11}{720} \mu \delta^4 y_{1\frac{1}{2}} - \cdots\cdots$$

and similarly for the other intervals up to that for x_{n-1} to x_n.

Then $\dfrac{1}{h}\displaystyle\int_{x_0}^{x_n} y\,dx = \dfrac{1}{h}\left\{\displaystyle\int_{x_0}^{x_1} y\,dx + \displaystyle\int_{x_1}^{x_2} y\,dx + \ldots + \displaystyle\int_{x_{n-1}}^{x_n} y\,dx\right\}$

Adding the first terms in each integral gives

$\mu y_{\frac{1}{2}} + \mu y_{1\frac{1}{2}} + \mu y_{2\frac{1}{2}} + \ldots + \mu y_{n-\frac{1}{2}}$

$\qquad = \frac{1}{2}(y_0 + y_1) + \frac{1}{2}(y_1 + y_2) + \frac{1}{2}(y_2 + y_3) + \ldots + \frac{1}{2}(y_{n-1} + y_n)$

$\qquad = \frac{1}{2}y_0 + y_1 + y_2 + \ldots + y_{n-1} + \frac{1}{2}y_n$

Adding the second terms gives

$-\dfrac{1}{12}\left(\mu\delta^2 y_{\frac{1}{2}} + \mu\delta^2 y_{1\frac{1}{2}} + \ldots + \mu\delta^2 y_{n-1\frac{1}{2}} + \mu\delta^2 y_{n-\frac{1}{2}}\right)$

$= -\dfrac{1}{12}\left(\frac{1}{2}\delta^2 y_0 + \delta^2 y_1 + \ldots + \delta^2 y_{n-1} + \frac{1}{2}\delta^2 y_n\right)$

$= -\dfrac{1}{12}\left\{\frac{1}{2}\left(\delta y_{\frac{1}{2}} - \delta y_{-\frac{1}{2}}\right) + \left(\delta y_{1\frac{1}{2}} - \delta y_{\frac{1}{2}}\right) + \left(\delta y_{2\frac{1}{2}} - \delta y_{1\frac{1}{2}}\right) + \ldots\right.$

$\qquad\qquad \left. + \left(\delta y_{n-1\frac{1}{2}} - \delta y_{n-2\frac{1}{2}}\right) + \left(\delta y_{n-\frac{1}{2}} - \delta y_{n-1\frac{1}{2}}\right) + \frac{1}{2}\left(\delta y_{n+\frac{1}{2}} - \delta y_{n-\frac{1}{2}}\right)\right\}$

$= -\dfrac{1}{12}\left\{\left(-\frac{1}{2}\delta y_{-\frac{1}{2}} - \frac{1}{2}\delta y_{\frac{1}{2}}\right) + \left(\frac{1}{2}\delta y_{n-\frac{1}{2}} + \frac{1}{2}\delta y_{n+\frac{1}{2}}\right)\right\}$

$= -\dfrac{1}{12}\left(-\mu\delta y_0 + \mu\delta y_n\right)$

The term $\dfrac{11}{720}\left(\mu\delta^3 y_n - \mu\delta^3 y_0\right)$ follows similarly by adding the third terms in each integral.

Extending the quoted difference table backwards slightly,

−0·2	0·960 789			
		29 261		
−0·1	0·990 050		−19 311	
		9950		−589
0	1·000 000		−19 900	
		−9950		589

$\dfrac{1}{0\cdot1}\displaystyle\int_0^{0\cdot5} e^{-x^2}\,dx = \frac{1}{2}\times 1\cdot000\,000 + 0\cdot990\,050 + 0\cdot960\,789 + 0\cdot913\,931$

$\qquad\qquad\qquad + 0\cdot852\,144 + \frac{1}{2}\times 0\cdot778\,801 - \frac{1}{12}(-0\cdot077\,234 - 0)$

$\qquad\qquad\qquad + \dfrac{11}{720}(0\cdot003\,816 - 0)$

$\therefore \quad \displaystyle\int_0^{0\cdot5} e^{-x^2}\,dx = 0\cdot461\,281$

You will recognise the first part of the integration formula i.e.
$h(\frac{1}{2}y_0 + y_1 + \ldots + y_{n-1} + \frac{1}{2}y_n)$ as being the trapezium rule extended

279

to several intervals.   As you know, this is not an accurate formula
for an area.   The complete formula derived in this question
increases the accuracy by including higher order difference terms.
These are omitted when the straightforward simple trapezium rule is
used.

e^{-x^2} is a function whose indefinite integral cannot be found.
Integrals of this kind are needed when finding the area under the
normal curve in probability and also when finding the values of what
is known as the error function.

7.   0·8   0·610 9168

    0·4   0·646 3160   0·658 1157

    0·2   0·654 8510   0·657 6960   0·657 6680

    0·1   0·656 9663   0·657 6714   0·657 6698   0·657 6698

    Integral ≃ 0·657 67

8.   990

1. The acceleration, a, of a rocket at time t, measured from launching, is given by the table

t s	0	10	20	30	40	50	60	70
a ms^{-2}	30•0	31•7	33•6	35•7	38•0	40•7	43•7	47•1

Find the rocket's velocity and height at t = 70.

2. An air cooled engine cylinder is simulated by a cooling fin placed in an air stream, the fin being heated at its centre. The temperature T°C of the fin at various distances r cm from the centre is given by the table.

r	2•5	5•0	7•5	10•0	12•5	15•0	17•5	20•0	22•5
T	58•9	59•3	59•8	60•8	61•9	63•1	64•5	65•7	66•8

r	25•0	27•5
T	67•2	67•5

Find the least squares law of the form $T = a_0 + a_1 r + a_2 r^2 + a_3 r^3$.

3. Prove that $\Delta\sqrt{f_k} = \dfrac{\Delta f_k}{\sqrt{f_k} + \sqrt{f_{k+1}}}$

4. Use Romberg integration to find $\displaystyle\int_0^1 \dfrac{\tan^{-1} x}{x}\,dx$, correct to four decimal places, from the following table of values of $y = \dfrac{\tan^{-1}x}{x}$.

x	0	0•0625	0•125	0•1875	0•25	0•3125
y	1	0•998 70	0•994 84	0•988 52	0•978 36	0•969 23

x	0•375	0•4375	0•5	0•5625	0•625	0•6875
y	0•956 72	0•942 65	0•927 29	0•910 91	0•893 79	0•876 05

x	0•75	0•8125	0•875	0•9375	1
y	0•858 00	0•839 77	0•821 52	0•803 36	0•785 40

5. Find, to 5 decimal places, f'(0•6) for the function given by the following table:

x	0•45	0•50	0•55	0•60	0•65
f(x)	4•069 057	4•053 474	4•035 500	4•014 994	3•991 775

x	0•70	0•75
f(x)	3•965 615	3•936 218

6. Find the value of the normal distribution function, $\phi(x)$, for x = 2•0673 from the following table:

x	2•00	2•05	2•10	2•15	2•20	2•25
$\phi(x)$	0•977 25	0•979 82	0•982 14	0•984 22	0•986 10	0•987 78

x	2•30
$\phi(x)$	0•989 28

7. Find the value of coth 0·102 76 from the table

x	0·100	0·101	0·102	0·103	0·104
coth x	10·033 31	9·934 63	9·837 90	9·743 05	9·650 03

x	0·105
coth x	9·558 78

using (i) Bessel's formula   (ii) Stirling's formula.

8. The points (−1, 8), (0, 5), (1, 4), (3, 56) lie on a certain polynomial curve, y = f(x). Find the value of y when x = 2, assuming that f(x) is of the lowest degree possible.

9. For what value of x, to 6 decimal places, will the function f(x) given below take the value 5?

x	1·595	1·600	1·605	1·610	1·615
f(x)	4·928 329	4·953 032	4·977 860	5·002 811	5·027 888

x	1·620	1·625
f(x)	5·053 090	5·078 419

10. Use (i) Simpson's rule, (ii) Boole's rule to obtain the integral given in question number 4.

ANSWERS

1. 2618 ms^{-1}, 84·75 km. $\left[\text{In order to find the height, it is necessary to complete a table of velocities for } t = 0(10)70.\right]$

2. $T = 59·3 - 0·260r + 0·0519r^2 - 0·001\ 150r^3$

3. $\Delta\sqrt{f_k} = \sqrt{f_{k+1}} - \sqrt{f_k} = \dfrac{f_{k+1} - f_k}{\sqrt{f_{k+1}} + \sqrt{f_k}} = \dfrac{\Delta f_k}{\sqrt{f_{k+1}} + \sqrt{f_k}}$

4. 0·9159

5. −0·436 58

6. 0·980 65

7. 9·765 65 in each case.

8. 17

9. 1·609 438

10. 0·9159 in each case.

UNIT 3
DIFFERENTIAL EQUATIONS

This Unit comprises three programmes:

    (a)   First Order Ordinary Differential Equations

    (b)   Simultaneous and Second Order Differential Equations

    (c)   Partial Differential Equations

Before reading these programmes, it is necessary that you are familiar with the following

Prerequisites

For (a)   Analytical method for solving $\frac{dy}{dx} + ay = bx$.

             Newton-Gregory interpolation formula.

             Taylor's series.

For (b)   The contents of (a).

             Analytical solution of simultaneous first order differential equations by elimination.

             Approximate expressions for first and second derivatives in terms of function values. (See FRAMES 6 and 26, pages 237 and 244.)

             Solution of linear algebraic equations by elimination.

For (c)   Approximate expressions for first and second derivatives in terms of function values. (See FRAMES 6 and 26, pages 237 and 244.)

First Order Ordinary Differential Equations

Introduction

You will already have met differential equations for which a formula can be found giving the relation between the variables, this relation not containing any differential coefficients. For example,

$\frac{dy}{dx} + 2y \tan x = \sin x$ leads to $y = \cos x - A \cos^2 x$ and

$\frac{d^2y}{dx^2} + 4 \frac{dy}{dx} + 4y = 2e^x - 3 \cos x$ leads to

$y = (A + Bx)e^{-2x} + \frac{2}{9} e^x - \frac{3}{25} (3 \cos x + 4 \sin x)$

Many practical situations give rise to differential equations. Fortunately, some of the simpler problems give rise to differential equations for which simple analytical solutions can be found. Thus, if a sinusoidal e.m.f., E sin ωt, is applied to an L, C, R circuit, the

equation $L \frac{di}{dt} + Ri + \frac{1}{C} \int_0^t i \, dt = E \sin \omega t$ (1.1)

results and, if we put $q = \int_0^t i \, dt$, then

$$L \frac{d^2q}{dt^2} + R \frac{dq}{dt} + \frac{1}{C}q = E \sin \omega t,$$ (1.2)

which can easily be solved. i is then given by $\frac{dq}{dt}$.

However, relatively speaking, there are very few differential equations for which an analytical solution can be found. This is really only to be expected because, again relatively speaking, there are very few functions that can be integrated analytically and integration is really the basis of the solution of differential equations. Also, even where an analytical solution can be found, it may be such a beast that it is easier to start with a numerical solution than to insert specific values into the analytical solution itself.

An easy situation leading to an awkward differential equation is the simpl pendulum. Applying the equation Iθ̈ = G gives

$$m\ell^2 \ddot{\theta} = -mg\ell \sin \theta$$

i.e. $\ddot{\theta} + \frac{g}{\ell} \sin \theta = 0$ (2.1)

This equation is not one of the simple types that you know how to solve. If only small oscillations take place, it is common practice to simplify this equation by making the approximation sin θ = θ. If this is done, $\ddot{\theta} + \frac{g}{\ell} \theta = 0$ results which does have a simple solution. If θ is not small enough for this approximation to be sufficiently accurate, it is possible to get part of the way with the

284

solution, i.e., one integration can be performed to give $\dot{\theta}$.   To do this, the equation is multiplied throughout by $2\dot{\theta}$, giving

$2\,\dot{\theta}\,\ddot{\theta} + 2\,\frac{g}{\ell}\,\sin\,\theta\,\dot{\theta} = 0$.   Integration then yields $\dot{\theta}^2 - 2\,\frac{g}{\ell}\,\cos\,\theta = A$.

Assuming that oscillations take place, A can be found by using the condition that $\dot{\theta} = 0$ when the angular displacement is a maximum.   Let this maximum angular displacement be α , then $- \frac{2g}{\ell}\,\cos\,\alpha = A$ and so

$\dot{\theta}^2 = \frac{2g}{\ell}\,(\cos\,\theta - \cos\,\alpha)$.   From this $\frac{d\theta}{dt} = \sqrt{\frac{2g}{\ell}\,(\cos\,\theta - \cos\,\alpha)}$ i.e.,

$dt = \dfrac{1}{\sqrt{\dfrac{2g}{\ell}\,(\cos\,\theta - \cos\,\alpha)}}\,d\theta$ and the R.H.S. cannot be integrated

analytically.

However, you now know how to integrate numerically when limits are given and so, as things stand at the moment, this process would be the one that you would investigate next.   Doing this would enable you to find values of t for specific values of θ.   It would not enable you to find a general formula connecting t and θ.

In FRAME 1, the standard equation was quoted for an L,C,R series circuit. But this equation really represents a somewhat idealised situation, a situation that is only an approximation of what actually does happen in practice.

Now you know that when an electric current is passed through a resistor, heat is generated, the amount being proportional to the square of the current.   This will result in a temperature rise on the part of the resistor, unless there is very efficient cooling.   But this temperature change in turn causes a rise in the resistance offered by the resistor to the current, the change in the resistance being proportional to the rise in temperature.   The net effect is that the actual resistance can be written in the form $A + Bi^2$.   If this, more accurate, value is used instead of R then the equation (1.1) will become

$$L\,\frac{di}{dt} + (A + Bi^2)i + \frac{1}{C}\int_0^t\ idt = E\,\sin\,\omega t \qquad \text{and (1.2) will become}$$

$$L\,\frac{d^2q}{dt} + \left\{A + B\left(\frac{dq}{dt}\right)^2\right\}\frac{dq}{dt} + \frac{1}{C}q = E\,\sin\,\omega t$$

Solving this is an entirely different proposition from solving (1.2).

Time and time again throughout this book we have had recourse to numerical methods when we have been unable to proceed analytically and so it is perhaps natural now to enquire whether anything can be done numerically with a differential equation that cannot be cracked analytically.

As, basically, the solution of a differential equation involves integration, it will perhaps come as no surprise that numerical methods of solution of differential equations are linked with numerical methods of integration.   And as numerical methods of integration only give the values of integrals between specific limits, it will perhaps also not be surprising that numerical methods of solution of differential equations only give us the values of the dependent variable for certain fixed values of the independent variable.   (You will remember that sometimes a function is defined only by a series of tabulated values and it is this situation that will arise here.)

The simplest differential equation you can have is of the form $\frac{dy}{dx} = f(x)$ from which $y = \int f(x)\, dx + c$.   An alternative way of writing this solution is $y - y_0 = \int_{x_0}^{x} f(x)\, dx$ where y_0 is the value of y when $x = x_0$.

If the integration involved on the R.H.S. can be carried out analytically, all well and good.   If this is not the case, the methods of the integration programme in Unit 2 can be used to find y for particular values of x and so nothing new is involved in the solution.

Still keeping to first order equations, $\frac{dy}{dx} = f(y)$ is effectively similar as now $\frac{dx}{dy} = \frac{1}{f(y)} = F(y)$, say, and so $x - x_0 = \int_{y_0}^{y} F(y)\, dy$.

This again just involves straightforward integration – either analytically or numerically.   Once more nothing new is involved.

A slightly more complicated equation, and one that very often cannot be solved by simple direct integration, is $\frac{dy}{dx} = f(x, y)$.

Here, for example, you would have difficulty in solving analytically $\frac{dy}{dx} = x^3 + y^3$ or $\frac{dy}{dx} = e^x + \ln y$.   In this programme we shall have a look at just a few of the numerical methods that are available for equations of this type.

Now, in practice, one would not use numerical methods if $f(x, y)$ is a function such that the equation has a simple analytical solution. However, here it will be a good idea for $f(x, y)$ to be like this so that comparisons can be drawn between the analytical and numerical results.

A very simple equation that we can use is $\frac{dy}{dx} = x - y$ subject to the condition that $y = 1$ when $x = 0$.   By way of revision find analytically the solution of this equation.

****

286

$\dfrac{dy}{dx} + y = x,$ $I.F. = e^x,$ $ye^x = (x - 1)e^x + A,$ $y = 2e^{-x} + x - 1$

FRAME 8

Euler's Method

Turning now to numerical methods of solution, the simplest of these is Euler's. This, in common with the other methods that are available, enables us to estimate the values of y for a set of values of x. It will not enable us to find, as you have just done, a formula such as $y = 2e^{-x} + x - 1$. Let the values of x for which y is required be $0 \cdot 1(0 \cdot 1)0 \cdot 5$.

The first step in Euler's method is to assume that between x = 0 and x = 0·1 the curve is approximately a straight line, the slope of this line being the value of $\dfrac{dy}{dx}$ as given by x - y when x = 0, y = 1, i.e., -1. (You will remember that the equation being solved is $\dfrac{dy}{dx} = x - y$ subject to y = 1 when x = 0.) Then, for the straight line, as x increases by 0·1, y decreases by 0·1 and so goes from 1 down to 0·9. This value is then taken as an estimate of y for the actual curve when x = 0·1.

FRAME 9

The next step is to repeat this process but starting now from the point (0·1, 0·9). For these values of x and y, $\dfrac{dy}{dx} = 0 \cdot 1 - 0 \cdot 9 = -0 \cdot 8$, and between x = 0·1 and 0·2, the curve is assumed to be approximately a straight line with this new slope. Again, in the same way as before, as x increases by 0·1, y now decreases by 0·08 and so goes down from 0·9 to 0·82. 0·82 is taken as an estimate of y for the actual curve when x = 0·2.

For the next stage, $\dfrac{dy}{dx} = 0 \cdot 2 - 0 \cdot 82 = -0 \cdot 62$ when x = 0·2 and y = 0·82. The estimate of y for x = 0·3 is then $0 \cdot 82 - 0 \cdot 1 \times 0 \cdot 62 = 0 \cdot 758$.

What will be the corresponding estimates for y when x = 0·4 and 0·5?
****

9A

$x = 0 \cdot 3,$ $y = 0 \cdot 758$ gives $\dfrac{dy}{dx} = -0 \cdot 458$

When $x = 0 \cdot 4,$ $y = 0 \cdot 758 - 0 \cdot 1 \times 0 \cdot 458 = 0 \cdot 7122$

Then $\dfrac{dy}{dx} = 0 \cdot 4 - 0 \cdot 7122 = -0 \cdot 3122$

When $x = 0 \cdot 5,$ $y = 0 \cdot 7122 - 0 \cdot 1 \times 0 \cdot 3122 = 0 \cdot 680\,98$

FRAME 10

These values can now be compared with the actual values by calculating them from the equation $y = 2e^{-x} + x - 1$ in 7A. A table gives, to not more than 4 decimal places,

FIRST ORDER ORDINARY DIFFERENTIAL EQUATIONS

FRAME 10   (continued)

x	0	0·1	0·2	0·3	0·4	0·5
y (estimated value)	1	0·9	0·82	0·758	0·7122	0·6810
y (true value)	1	0·9097	0·8375	0·7816	0·7406	0·7131
error (i.e. true value – estimated value)	0	0·0097	0·0175	0·0236	0·0284	0·0321

FRAME 11

Two points now arise in connection with this work:

i)   Can the accuracy be improved?
ii)   Can the process be expressed in a formula form so that it can readily be applied to any example?

FRAME 12

The first question has been posed because, as you will see from the table in FRAME 10, the discrepancy between the estimated value and the true value is increasing as x increases.   If this discrepancy continues to get larger, it may not be long before it exceeds that allowable.

Can you suggest any way in which this discrepancy can be reduced?
                  ****

12A

Reduce the width of each step.   Each step length is the equivalent of the h that has been used in previous work.

FRAME 13

Recalculate the value obtained for y when   x = 0·1   if two steps, each 0·05, are taken in order to go from   x = 0   to   x = 0·1.
                  ****

13A

$\frac{dy}{dx}$ *at (0, 1) is −1.   For an increase of 0·05 in x, y decreases by 0·05 and so has value 0·95.*

At   (0·05, 0·95), $\frac{dy}{dx}$ = −0·9.   For a second increase of 0·05 in x, y decreases by   0·05 × 0·9 = 0·045 and hence now y = 0·905.

This is an improvement on the original value of 0·9.

FRAME 14

A second way of improving the accuracy is to make a less crude assumption than that the graph of y against x is a straight line in the interval, the slope of the line being that calculated at the beginning of the interval.   This can be done by means of what is known as the improved Euler method which will be considered shortly.   Before doing so, there is still question (ii) in FRAME 11 to be answered.   In order to do this, we will return to the more general equation $\frac{dy}{dx}$ = f(x, y).

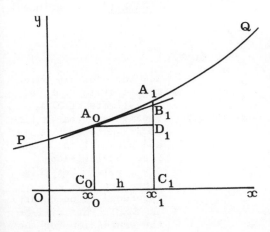

Suppose we are given that $y = y_0$ when $x = x_0$. (In the example worked earlier, $x_0 = 0$, but any value may be given.) Let the step length from x_0 to x_1 be h. Then whereas the true value of y when $x = x_1$ is given by C_1A_1, the estimated value is given by C_1B_1 where, in the diagram, the true curve representing the solution is PA_0A_1Q and A_0B_1 is tangential to this curve at A_0.

What will be the formula giving C_1B_1 in terms of x_0, y_0 and h?

****

15A

$$y_0 + h\left(\frac{dy}{dx}\right)_{A_0} = y_0 + hf(x_0, y_0)$$

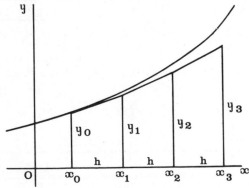

Extending the working beyond the first interval will give the effect shown in the diagram. The various formulae used will be

$$y_1 = y_0 + hf(x_0, y_0)$$
$$y_2 = y_1 + hf(x_1, y_1)$$
$$y_3 = y_2 + hf(x_2, y_2)$$

and, in general,

$$y_n = y_{n-1} + hf(x_{n-1}, y_{n-1})$$

You will see now why the error is likely to increase the further we go. At each stage after the first a spurious value of y is used in two places in the formula. This is in addition to the straight line assumption that is made at each stage.

The Improved Euler Method

The errors introduced by the use of the straightforward Euler method and the build-up of these errors as one proceeds can be reduced by the use of the IMPROVED EULER METHOD. This starts off in exactly the same way as before, i.e., given (x_0, y_0), y_1 is calculated by using $y_1 = y_0 + hf(x_0, y_0)$. $f(x_1, y_1)$ is then found.

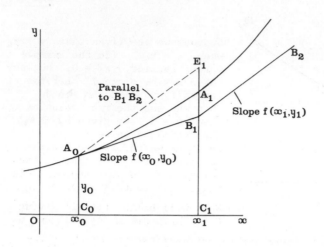

Now the ordinary Euler method gives C_1B_1 as an estimate of C_1A_1. Suppose now that a line is drawn through A_0 and parallel to B_1B_2, i.e., having slope $f(x_1, y_1)$. Let this line meet C_1A_1 (produced if necessary) in E_1. What will be the expression for the height C_1E_1?

****

17A

$y_0 + hf(x_1, y_1)$

FRAME 18

For the figure as shown in FRAME 17, C_1B_1, i.e. $y_0 + hf(x_0, y_0)$, is an under-estimate of C_1A_1 and C_1E_1, i.e. $y_0 + hf(x_1, y_1)$, is an over-estimate. It would therefore seem reasonable to take the mean of these two estimates as being a better approximation to C_1A_1 than either of them separately. This means that

$$y_0 + \frac{h}{2}\left\{ f(x_0, y_0) + f(x_1, y_1) \right\} \qquad (18.1)$$

is taken as a better estimate for C_1A_1.

FRAME 19

Let us now see what effect doing this has on the example worked out in FRAMES 8 - 10. For the first interval, $f(x_0, y_0) = -1$ and $f(x_1, y_1) = -0\cdot8$. Formula (18.1) then gives $1 + \frac{0\cdot1}{2}(-1 - 0\cdot8) = 0\cdot91$ which is much closer to $0\cdot9097$ than the $0\cdot9$ obtained previously.

For the next step in the calculation a start is made from $(0\cdot1, 0\cdot91)$. The value of $f(0\cdot1, 0\cdot91)$ is $-0\cdot81$ and so a first estimate of y when $x = 0\cdot2$ is $0\cdot91 + 0\cdot1(-0\cdot81) = 0\cdot829$.

What will be an improved estimate for this y?
****

19A

$f(0\cdot2, 0\cdot829) = -0\cdot629$

Improved value $= 0\cdot91 + \frac{0\cdot1}{2}(-0\cdot81 - 0\cdot629) = 0\cdot838\,05 = 0\cdot8380$ to 4 *decimal places.   This compares much more favourably with 0·8375 than does 0·82.*

FRAME 20

What will this method give for y when x = 0·3?   Do not take your working beyond 5 decimal places.

****

20A

$f(0·2, 0·838\ 05) = -0·638\ 05$
First estimate $\quad = \quad 0·838\ 05 + 0·1(-0·638\ 05) = 0·774\ 25$

$f(0·3, 0·774\ 25) = -0·474\ 25$
Better estimate $\quad = \quad 0·838\ 05 + \dfrac{0·1}{2}\ (-0·638\ 05 - 0·474\ 25) = 0·782\ 44$

$$= \quad 0·7824 \quad to \ 4 \ decimal \ places.$$

FRAME 21

Continuing gives $y = 0·7416$ when $x = 0·4$ and $y = 0·7142$ when $x = 0·5$.   The errors in y are now, for the values taken in turn from $x = 0$ to $x = 0·5$; 0, −0·0003, −0·0005, −0·0008, −0·0010, −0·0011. These errors are still increasing with x but in magnitude are only about one thirtieth of those previously present.

The Improved Euler method is an example of the technique which is known as a PREDICTOR-CORRECTOR METHOD and in the next frame we shall describe what is meant by this name.

FRAME 22

Predictor-Corrector Methods

Starting from (x_0, y_0), the improved Euler method first requires us to find an initial estimate of y_1 by using the formula $y_1 = y_0 + hf(x_0, y_0)$. This is a relatively inaccurate formula compared with (18.1) and is used to predict an approximate value of y_1.   For obvious reasons it cannot involve y_1 itself as this is unknown.   However the value obtained in this way is then used in a more accurate estimation formula, i.e. (18.1). This now gives a better value for y_1 than that obtained previously. In other words it is used to correct the initially predicted value of y_1. Here $y_1 = y_0 + hf(x_0, y_0)$ is the PREDICTOR FORMULA and

$y_1 = y_0 + \dfrac{h}{2}\left\{f(x_0, y_0) + f(x_1, y_1)\right\}$ the CORRECTOR FORMULA.

FRAME 23

As y_1 occurs with two different meanings in these formulae and also usually has two different values, it is desirable to use some notation that distinguishes between them.   As you would probably suspect, there is no uniform notation that has been adopted by all authors.   One possible notation is y_1^P for the predicted value of y_1 and y_1^C for the corrected value.   With this notation, the formulae of the last frame would become

$$y_1^P = y_0 + hf(x_0, y_0), \qquad y_1^C = y_0 + \dfrac{h}{2}\{f(x_0, y_0) + f(x_1, y_1^P)\} \qquad (23.1)$$

Having found the corrected estimate of y_1, we then go on to calculate y_2, y_3, etc.

What will be the formulae for y_2^p, y_2^c, y_{n+1}^p and y_{n+1}^c?

****

23A

$$y_2^p = y_1^c + hf(x_1, y_1^c) \qquad\qquad y_2^c = y_1^c + \frac{h}{2}\{f(x_1, y_1^c) + f(x_2, y_2^p)\}$$

$$y_{n+1}^p = y_n^c + hf(x_n, y_n^c) \qquad\qquad y_{n+1}^c = y_n^c + \frac{h}{2}\{f(x_n, y_n^c) + f(x_{n+1}, y_{n+1}^p)\}$$

You will realise that when calculating y_2, two values have been found for y_1, i.e. y_1^p and y_1^c. It is obviously preferable to use y_1^c, rather than y_1^p, in the formulae for y_2^p and y_2^c.

FRAME 24

More about the Improved Euler Method - The Modified Euler Method

It is possible to improve quite easily the Improved Euler Method by an iterative process. Having obtained y_1^c, this value is now used in the R.H.S. of (23.1) in place of y_1^p. This will give a second value of y_1^c. This second value is, in its turn, used instead of the first value in place of y_1^p in the R.H.S. of (23.1), giving a third value of y_1^c. This repetition is continued until it produces no change in the value of y_1^c and can be embodied in a formula by changing (23.1) to

$$(y_1^c)_{new} = y_0 + \frac{h}{2}\{f(x_0, y_0) + f[x_1, (y_1^c)_{old}]\}$$

This final value is then taken as the best value and used, where necessary, in the formula for y_2. This repetitive process is sometimes called the MODIFIED EULER METHOD. Unfortunately, however, different writers are not all consistent as to exactly which process is described as 'improved' and which as 'modified'.

To see how well (or otherwise) this works let us return to the example previously used, i.e., $\frac{dy}{dx} = x - y$, given that $y = 1$ when $x = 0$.

We have already calculated $y_1^p = 0 \cdot 9$ and $y_1^c = 0 \cdot 91$. This will be the first value of y_1^c. Now, $f(0 \cdot 1, 0 \cdot 91) = -0 \cdot 81$ and so the second value of y_1^c will be $1 + \frac{0 \cdot 1}{2}(-1 - 0 \cdot 81) = 0 \cdot 9095$. $f(0 \cdot 1, 0 \cdot 9095) = -0 \cdot 8095$ and so the third value of y_1^c will be $1 + \frac{0 \cdot 1}{2}(-1 - 0 \cdot 8095) = 0 \cdot 909\,52$.

The next calculation repeats this value and so, to 4 decimal places, y_1^c is taken as $0 \cdot 9095$.

However, retaining $0 \cdot 909\,52$ for starting the next step, $f(0 \cdot 1, 0 \cdot 909\,52) = -0 \cdot 809\,52$ and so $y_2^p = 0 \cdot 909\,52 + 0 \cdot 1(-0 \cdot 809\,52) = 0 \cdot 828\,57$.

Then $f(0 \cdot 2, 0 \cdot 828\,57) = -0 \cdot 628\,57$ and so $y_2^c = 0 \cdot 909\,52 + \frac{0 \cdot 1}{2}(-0 \cdot 809\,52 - 0 \cdot 628\,57) = 0 \cdot 837\,62$.

Now see whether you can improve on this result.

****

$f(0 \cdot 2, \ 0 \cdot 837 \ 62) \ = \ -0 \cdot 637 \ 62$

$$y_2^c \ = \ 0 \cdot 909 \ 52 \ + \ \frac{0 \cdot 1}{2}(-0 \cdot 809 \ 52 \ - \ 0 \cdot 637 \ 62) \ = \ 0 \cdot 837 \ 16$$

$f(0 \cdot 2, \ 0 \cdot 837 \ 16) \ = \ -0 \cdot 637 \ 16$

$$y_2^c \ = \ 0 \cdot 909 \ 52 \ + \ \frac{0 \cdot 1}{2}(-0 \cdot 809 \ 52 \ - \ 0 \cdot 637 \ 16) \ = \ 0 \cdot 837 \ 19$$

$f(0 \cdot 2, \ 0 \cdot 837 \ 19) \ = \ -0 \cdot 637 \ 19$

$$y_2^c \ = \ 0 \cdot 909 \ 52 \ + \ \frac{0 \cdot 1}{2}(-0 \cdot 809 \ 52 \ - \ 0 \cdot 637 \ 19) \ = \ 0 \cdot 837 \ 18$$

and after this no further change in y_2^c *takes place.*

$\therefore$ To 4 decimal places, $y_2^c = 0 \cdot 8372$.

What is the best estimate you can obtain for y_3^c?

****

$y_3^p \ = \ 0 \cdot 773 \ 46$

Successive values for y_3^c *are* $0 \cdot 781 \ 65, \quad 0 \cdot 781 \ 24, \quad 0 \cdot 781 \ 26.$

To 4 decimal places $y_3^c = 0 \cdot 7813.$

Continuing gives $y_4^c = 0 \cdot 7402$ and $y_5^c = 0 \cdot 7126$.

The errors in the values of y from x = 0 to x = 0·5 are now 0, 0·0002, 0·0003, 0·0003, 0·0004, 0·0005. These errors are approximately, in magnitude, one half of those produced by the straightforward improved Euler method.

Now use all three methods so far discussed to find, to 4 decimal places, y when x = 0·2, given that $\frac{dy}{dx} = 2x + y$ and that y = 2 when x = 0. Take first a single step of length 0·2. Then repeat the calculation for two steps, each of length 0·1. Also solve the differential equation analytically and compare the accuracy you obtain by the various methods.

****

Method	No. of steps	y	Error
Analytical		2·4856	
Euler	1	2·4	0·0856
Improved Euler applied once.	1	2·48	0·0056
Improved Euler applied successively.	1	2·4889	−0·0033
Euler	2	2·44	0·0456
Improved Euler applied once.	2	2·4841	0·0015
Improved Euler applied successively.	2	2·4864	−0·0008

FIRST ORDER ORDINARY DIFFERENTIAL EQUATIONS

In these results, you will notice that the improved Euler formula applied
successively gives the best result, whether one step or two are used and
that the smaller h gives a better result whichever process is used.   For
future reference, the formulae used in the modified Euler method are
collected below:

$$y_{n+1}^p = y_n^c + hf(x_n, y_n^c)$$

$$y_{n+1}^c = y_n^c + \frac{h}{2}\left\{f(x_n, y_n^c) + f(x_{n+1}, y_{n+1}^p)\right\}$$

$$\left(y_{n+1}^c\right)_{new} = y_n^c + \frac{h}{2}\left\{f(x_n, y_n^c) + f\left[x_{n+1}, (y_{n+1}^c)_{old}\right]\right\}$$

the value of y_n^c that is used being the most accurate available.

A flow diagram for the solution of $\frac{dy}{dx} = f(x, y)$, subject to $y = y_0$
when $x = x_0$, for values of y at $x = x_0 + h$, $x_0 + 2h$, ..., $x_0 + nh$,
using the modified Euler method is shown on page 295.

Milne's Method

Another predictor-corrector method for the numerical solution of a
differential equation of the form $\frac{dy}{dx} = f(x, y)$ is that known as MILNE'S
METHOD.   It works on exactly the same principle as Euler's modified
method, but uses different formulae to predict and correct estimates of y.
It also makes use of more values of y and f at each stage than does
Euler.   There is also an inherent difficulty in the method which will
become obvious as we proceed.

You will remember that, in the programme on integration in Unit 2,
integration formulae were obtained by integrating interpolation formulae
between specific limits.   To remind yourself of the process, take the
Newton-Gregory forward difference formula and use it to find a formula for
$\int_{x_0}^{x_4}$ ydx in terms of functional values.   Retain terms in the formula up
to and inculding those in Δ^3.
****

$$\int_{x_0}^{x_4} ydx = h\int_0^4 ydp$$

$$y = \left\{1 + p\Delta + \frac{p(p-1)}{2!}\Delta^2 + \frac{p(p-1)(p-2)}{3!}\Delta^3 + \ldots\right\}f_0$$

$$= \left\{1 + p\Delta + \frac{1}{2}(p^2 - p)\Delta^2 + \frac{1}{6}(p^3 - 3p^2 + 2p)\Delta^3 + \ldots\right\}f_0$$

Flow diagram for FRAME 28.

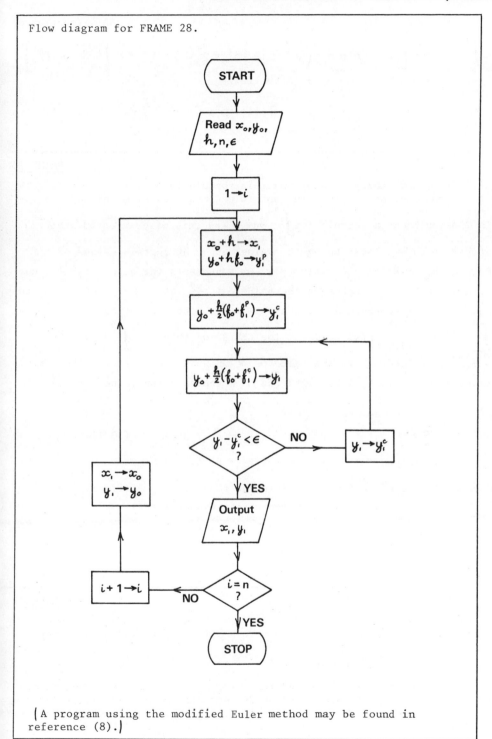

(A program using the modified Euler method may be found in reference (8).)

$$h\int_0^4 y\,dp = h\left[\left\{p + \frac{1}{2}p^2\Delta + \frac{1}{2}\left(\frac{1}{3}p^3 - \frac{1}{2}p^2\right)\Delta^2 + \frac{1}{6}\left(\frac{1}{4}p^4 - p^3 + p^2\right)\Delta^3 + \ldots\right\}f_0\right]_0^4$$

$$= h\left(4 + 8\Delta + \frac{20}{3}\Delta^2 + \frac{8}{3}\Delta^3 + \ldots\right)f_0$$

$$\simeq h\left\{4f_0 + 8(f_1 - f_0) + \frac{20}{3}(f_2 - 2f_1 + f_0) + \frac{8}{3}(f_3 - 3f_2 + 3f_1 - f_0)\right\}$$

$$= \frac{4h}{3}(2f_1 - f_2 + 2f_3) \qquad\qquad (29A.1)$$

Milne's method starts by adapting this formula to predict y_4.   At this stage, you must be very careful not to get confused with the notation.

In 29A you found a formula for $\int_{x_0}^{x_4} y\,dx$, and this was expressed in terms of various f's, these being the values of y at certain values of x. But in the equation $\frac{dy}{dx} = f(x, y)$,   f represents the value of $\frac{dy}{dx}$, not y, and so f_1, f_2 represent the values of $\frac{dy}{dx}$ at x_1, x_2, etc.   If this equation is integrated as it stands w.r.t. x between x_0 and x_4, we get

$$\int_{x_0}^{x_4} \frac{dy}{dx}\,dx = \int_{x_0}^{x_4} f(x, y)\,dx. \quad \text{The L.H.S. of this is} \quad \int_{x_0}^{x_4} dy = \left[y\right]_{x_0}^{x_4}$$

$= y_4 - y_0$ where y_4 is the value of y when $x = x_4$, etc.   The R.H.S. is replaced by (29A.1) but with the f's representing values of $\frac{dy}{dx}$.

Thus
$$y_4 - y_0 = \frac{4h}{3}(2f_1 - f_2 + 2f_3)$$
or
$$y_4 = y_0 + \frac{4h}{3}(2f_1 - f_2 + 2f_3) \qquad (30.1)$$

Then, similarly,
$$y_5 = y_1 + \frac{4h}{3}(2f_2 - f_3 + 2f_4)$$

What will be the corresponding formulae for   (i) y_6 (ii) y_{n+4}?

****

$$y_6 = y_2 + \frac{4h}{3}(2f_3 - f_4 + 2f_5)$$

$$y_{n+4} = y_n + \frac{4h}{3}(2f_{n+1} - f_{n+2} + 2f_{n+3})$$

It was stated at the beginning of the last frame that Milne's method starts by predicting y_4 .   This it does using (30.1) which is better written as

$$y_4^P = y_0 + \frac{4h}{3}(2f_1 - f_2 + 2f_3) \qquad (31.1)$$

to indicate that it is the predicted value of y_4 which is being found. But now the snag mentioned in FRAME 29 rears its ugly head.   The formula for y_4^P involves f_1, f_2 and f_3 which are <u>not known</u> as these are expressed in terms of x_1, x_2, x_3 and y_1, y_2, y_3.   There is no difficulty with

the x's as these are fixed when h is chosen.   Thus $x_1 = x_0 + h$,
$x_2 = x_0 + 2h$, etc.   But the only y known at the start is y_0.   It is
therefore necessary, when making use of this method, to use some other
means for finding y_1, y_2 and y_3.   One way in which this can be done is
by using the Modified Euler method and another way will be mentioned
later.   A method, such as the improved Euler, which does not require any
special treatment to get it going, is described as being SELF-STARTING.

Having predicted the value of y_4, the next step is to improve upon this
estimate.   In other words, having found y_4^P, to calculate y_4^c.   This is
done making use of another integration formula, one with which by this
time you should be very familiar – Simpson's rule.   This is used in the
form

$$y_4^c = y_2 + \frac{h}{3}(f_2 + 4f_3 + f_4^P) \qquad (32.1)$$

and it is the term containing the brackets that you will recognise as
Simpson's rule.   As f_4 is a function of x_4 and y_4, it is only a
predicted value which can be used at this stage as it is only such a
value of y_4 that is known.   Hence the notation f_4^P is used in the
formula.

Having found a corrected value for y_4, the formula (32.1) can be used
with an improved value of f_4, i.c. f_4^c instead of f_4^P Then a better
value of y_4^c is given by

$$y_4^c = y_2 + \frac{h}{3}(f_2 + 4f_3 + f_4^c) \qquad (33.1)$$

The process is now repeated, as was the improved Euler, until two
consecutive values of y_4^c are the same.   This is then taken as the best
value of y_4.

To illustrate the method, let us return to the equation $\frac{dy}{dx} = x - y$ with
y = 1 when x = 0.   This is a convenient equation to take as we have
already worked out values of y_1, y_2 and y_3 and so these can be used in
the Milne formula for y_4.   The best values obtained numerically that we
have so far are $y_0 = 1$, $y_1 = 0.909\,52$, $y_2 = 0.837\,18$ and $y_3 = 0.781\,26$.
These give, in conjunction with $x_0 = 0$, $x_1 = 0.1$, $x_2 = 0.2$ and
$x_3 = 0.3$, $f_0 = -1$, $f_1 = -0.809\,52$, $f_2 = -0.637\,18$ and $f_3 = -0.481\,26$.
Then, (31.1) gives

$$y_4^P = 1 + \frac{4 \times 0.1}{3}\{2 \times (-0.809\,52) - (-0.637\,18) + 2 \times (-0.481\,26)\} = 0.740\,75$$

From this, $f_4^P = 0.4 - 0.740\,75 = -0.340\,75$ and so, using (32.1)

$$y_4^c = 0.837\,18 + \frac{0.1}{3}\{-0.637\,18 + 4 \times (-0.481\,26) - 0.340\,75\} = 0.740\,41.$$

What will be f_4^c and the next estimate of y_4^c?
****

297

$f_4^c = -0 \cdot 340\ 41$

$y_4^c = 0 \cdot 837\ 18 + \dfrac{0 \cdot 1}{3} \left\{ -0 \cdot 637\ 18\ +\ 4 \times (-0 \cdot 481\ 26)\ -\ 0 \cdot 340\ 41 \right\} = 0 \cdot 740\ 43$

The next estimate of y_4^c is again $0 \cdot 740\ 43$ and so, to 4 decimal places, $y_4 = 0 \cdot 7404$.

Now find, using this method, the best value of y_5.

****

$y_5^p = 0 \cdot 909\ 52 + \dfrac{4 \times 0 \cdot 1}{3} \left\{ 2 \times (-0 \cdot 637\ 18)\ -\ (-0 \cdot 481\ 26)\ +\ 2 \times (-0 \cdot 340\ 43) \right\}$

$\qquad = 0 \cdot 712\ 99$

$y_5^c = 0 \cdot 781\ 26 + \dfrac{0 \cdot 1}{3} \left\{ -0 \cdot 481\ 26\ +\ 4 \times (-0 \cdot 340\ 43)\ -\ 0 \cdot 212\ 99 \right\} = 0 \cdot 712\ 73$

$y_5^c = 0 \cdot 781\ 26 + \dfrac{0 \cdot 1}{3} \left\{ -0 \cdot 481\ 26\ +\ 4 \times (-0 \cdot 340\ 43)\ -\ 0 \cdot 212\ 73 \right\} = 0 \cdot 712\ 74$

$y_5^c = 0 \cdot 781\ 26 + \dfrac{0 \cdot 1}{3} \left\{ -0 \cdot 481\ 26\ +\ 4 \times (-0 \cdot 340\ 43)\ -\ 0 \cdot 212\ 74 \right\} = 0 \cdot 712\ 74$

$\therefore$ To 4 decimal places $y_5 = 0 \cdot 7127$.

[*Note that in the successive calculation of* y_5^c, *only the last number in the expression changes, i.e.,* $-0 \cdot 212\ 99$ *to* $-0 \cdot 212\ 73$ *to* $-0 \cdot 212\ 74$. *It follows that if the value of* $0 \cdot 781\ 26 + \dfrac{0 \cdot 1}{3} \left\{ -0 \cdot 481\ 26\ +\ 4 \times (-0 \cdot 340\ 43) \right\}$ *i.e.* $0 \cdot 719\ 83$ *is found, then the last two stages in the calculation can be written* $y_5^c = 0 \cdot 719\ 83 - \dfrac{0 \cdot 1}{3} \times 0 \cdot 212\ 73\ =\ 0 \cdot 712\ 74$

$\qquad\qquad y_5^c = 0 \cdot 719\ 83 - \dfrac{0 \cdot 1}{3} \times 0 \cdot 212\ 74\ =\ 0 \cdot 712\ 74$]

You will notice that the values obtained here are slightly more accurate than any of those obtained previously.

Now although you might have thought that numerical methods would be difficult to apply to the solution of differential equations, you will realise that all the processes that have been discussed are extremely simple, both as regards the formulae that are involved and also the actual working. Also, the most sophisticated methods have been iterative processes. All of these facts mean that it is very easy to program the various methods on to a computer. The program for Milne's method would be the most complicated as it would be necessary to incorporate instructions for a special starting procedure as the Milne method is not self-starting.

Taylor's Series Method for starting the Milne Process

You will recall that it was mentioned in FRAME 31 that another way of starting Milne's method would be discussed later. Taylor's series supplies us with such another method.   You will already have met Taylor's series and one form in which it can be stated is

$$y = y_0 + (x - x_0)y_0' + \frac{(x - x_0)^2}{2!} y_0'' + \frac{(x - x_0)^3}{3!} y_0''' + \ldots \ldots \quad (37.1)$$

Now returning to the equation $\frac{dy}{dx} = x - y$, successive differentiation will give the higher derivatives.   Thus $\frac{d^2y}{dx^2} = 1 - \frac{dy}{dx}$, $\frac{d^3y}{dx^3} = - \frac{d^2y}{dx^2}$, $\frac{d^4y}{dx^4} = - \frac{d^3y}{dx^3}$, etc.

But, as it was given that, when $x = 0$, $y = 1$, we thus have $y_0 = 1$, $y_0' = -1$, $y_0'' = 2$, $y_0''' = -2$, $y_0^{iv} = 2$, $y_0^{v} = -2$, etc.

Thus, when $x = 0 \cdot 1$,

$$y_1 = 1 + (0 \cdot 1)(-1) + \frac{(0 \cdot 1)^2}{2} (2) + \frac{(0 \cdot 1)^3}{6} (-2) + \frac{(0 \cdot 1)^4}{24} (2) + \frac{(0 \cdot 1)^5}{120} (-2) + \ldots$$

$$= 1 - 0 \cdot 1 + 0 \cdot 01 - 0 \cdot 000\,333 + 0 \cdot 000\,008 - \ldots \ldots$$

$$\approx 0 \cdot 909\,67$$

---

y_2 can either be obtained by putting $x = 0 \cdot 2$ in (37.1) or by expanding y in a new Taylor series about $0 \cdot 1$.   The former can be expected to converge more slowly as powers of $0 \cdot 2$ will be involved instead of those of $0 \cdot 1$.   The latter requires the re-calculation of the derivatives as these will be required at $x = 0 \cdot 1$ instead of at $x = 0$.

Using either method, find y_2.
                    ****

0 · 837 46.

In a similar way, $y_3 = 0 \cdot 781\,64$.   Then $f_0 = -1$, $f_1 = -0 \cdot 809\,67$, $f_2 = -0 \cdot 637\,46$ and $f_3 = -0 \cdot 481\,64$ and use can now be made of Milne's method to find y_4.

Thus $y_4^p = 1 + \frac{4 \times 0 \cdot 1}{3} \left\{ 2 \times (-0 \cdot 809\,67) - (-0 \cdot 637\,46) + 2 \times (-0 \cdot 481\,64) \right\}$

$$= 0 \cdot 740\,65$$

$$y_4^c = 0 \cdot 837\,46 + \frac{0 \cdot 1}{3} \left\{ -0 \cdot 637\,46 + 4 \times (-0 \cdot 481\,64) - 0 \cdot 340\,65 \right\}$$

$$= 0 \cdot 740\,64$$

The next estimate of y_4^c is the same as this one and so no further improvement can be made.

What is the best value you can now get for y_5?
                    ****

$$y_5^p = 0 \cdot 909\ 67 + \frac{4 \times 0 \cdot 1}{3} \left\{ 2 \times (-0 \cdot 637\ 46) - (-0 \cdot 481\ 64) + 2 \times (-0 \cdot 340\ 64) \right\}$$

$$= 0 \cdot 713\ 06$$

$$y_5^c = 0 \cdot 781\ 64 + \frac{0 \cdot 1}{3} \left\{ -0 \cdot 481\ 64 + 4 \times (-0 \cdot 340\ 64) - 0 \cdot 213\ 06 \right\}$$

$$= 0 \cdot 713\ 06$$

Rounding off now to 4 decimal places gives

x	0	0·1	0·2	0·3	0·4	0·5
y	1	0·9097	0·8375	0·7816	0·7406	0·7131

On comparison with the table in FRAME 10, it will be seen that all these values are correct to 4 decimal places. The next application of the method gives $y = 0 \cdot 6976$ when $x = 0 \cdot 6$ and again this is correct to 4 decimal places. So although there is the awkwardness in starting the Milne method, the results for this example are the most accurate that have so far been obtained.

Modification of the Milne Method

It would be an improvement to the Milne method if, for each stage, the need for repeated applications of the corrector formula could be overcome or reduced. Fortunately this is possible for the second and higher stages found by this process. Taking the y_5 stage (the second stage at which Milne can be used) the process so far adopted finds y_5^p, f_5^p, y_5^c, f_5^c and then repeats y_5^c, f_5^c as many times as necessary.

The modified method, which is based on the truncation errors in the integration formulae, calculates the values of y_5^p, y_5^m, f_5^m, y_5^c, y_5 where

$$y_5^p = y_1 + \frac{4h}{3}(2f_2 - f_3 + 2f_4^c)$$

$$y_5^m = y_5^p - \frac{28}{29}(y_4^p - y_4^c)$$

f_5^m is found from the differential equation
using x_5 and y_5^m

$$y_5^c = y_3 + \frac{h}{3}(f_3 + 4f_4 + f_5^m)$$

$$y_5 = y_5^c + \frac{1}{29}(y_5^p - y_5^c)$$

and then stops.

The last stage in this process is not universally accepted by all authors. Some give reasons for omitting it and prefer to repeat, as much as necessary, the calculation for y_5^c. The advantage then is that the number of these repetitions is reduced.

What equations would you use to obtain y_6?

****

$$y_6^p = y_2 + \frac{4h}{3}(2f_3 - f_4 + 2f_5^c) \qquad\qquad y_6^m = y_6^p - \frac{28}{29}(y_5^p - y_5^c)$$

f_6^m *from the d.e. using* x_6 *and* y_6^m.

$$y_6^c = y_4 + \frac{h}{3}(f_4 + 4f_5 + f_6^m) \qquad\qquad y_6 = y_6^c + \frac{1}{29}(y_6^p - y_6^c)$$

FRAME 42

Applying the method to the y_6 stage for the equation $\frac{dy}{dx} = x - y$ with $y = 1$ when $x = 0$, it has already been found that $y_2 = 0 \cdot 837\,46$, $y_4 = 0 \cdot 740\,64$, $y_5^p = 0 \cdot 713\,06$, $y_5^c = 0 \cdot 713\,06$, $f_3 = -0 \cdot 481\,64$, $f_4 = -0 \cdot 340\,64$, $f_5 = -0 \cdot 213\,06$.

Then $y_6^p = 0 \cdot 837\,46 + \dfrac{4 \times 0 \cdot 1}{3}\left\{ 2 \times (-0 \cdot 481\,64) - (-0 \cdot 340\,64) + 2 \times (-0 \cdot 213\,06)\right\}$

$\qquad = 0 \cdot 697\,63.$

What will be the values of y_6^m, y_6^c and y_6? (Note that $x_6 = 0 \cdot 6$)
****

42A

$y_6^m = 0 \cdot 697\,63 - \dfrac{28}{29}(0 \cdot 713\,06 - 0 \cdot 713\,06) = 0 \cdot 697\,63; \qquad f_6^m = -0 \cdot 097\,63$

$y_6^c = 0 \cdot 740\,64 + \dfrac{0 \cdot 1}{3}\left\{ -0 \cdot 340\,64 + 4 \times (-0 \cdot 213\,06) - 0 \cdot 097\,63\right\} = 0 \cdot 697\,62$

$y_6 = 0 \cdot 697\,62 + \dfrac{1}{29}(0 \cdot 697\,63 - 0 \cdot 697\,62) = 0 \cdot 697\,62$

FRAME 43

Due to the small difference between the predicted and corrected values, the effect of using the modification is not well shown in the example just taken. It is more apparent if steps of $0 \cdot 2$ are taken instead of $0 \cdot 1$. The initial values would then be

x	0	0·2	0·4	0·6
y	1	0·837 46	0·740 64	0·697 63
f	-1	-0·637 46	-0·340 64	-0·097 63

These values have been extracted from previous results. As the step length is now $0 \cdot 2$, the values of x shown here will now be labelled x_0, x_1, x_2, x_3. Then the values of y shown will be labelled y_0, y_1, y_2, y_3 and those of f; f_0, f_1, f_2, f_3.

The ordinary Milne method for $x = 0 \cdot 8$ then gives $y_4^p = 0 \cdot 698\,79$, $y_4^c = 0 \cdot 698\,65.$

Now carry through the calculation for y_5 using the modified Milne method.
****

43A

$y_5^p = 0 \cdot 735\,87, \quad y_5^m = 0 \cdot 735\,73, \quad f_5^m = 0 \cdot 264\,27, \quad y_5^c = 0 \cdot 735\,77, \quad y_5 = 0 \cdot 735\,77.$

Note: The ordinary Milne method obtains the values $y_5^p = 0 \cdot 735\,87,$

$y_5^c = 0 \cdot 735\,76, \quad 0 \cdot 735\,77, \quad 0 \cdot 735\,77$ *in turn.*

The Accuracy of the Various Methods

The work we have done so far has suggested that the most accurate method is Milne's and that the least accurate is Euler's. The improved Euler method has come somewhere in the middle. If $h < 1$, theory shows that the error introduced at each step is of the order of magnitude

h^2 for Euler, h^3 for improved Euler, h^5 for Milne, and modified Milne.

As $h^5 < h^3 < h^2$ when $h < 1$, it will be seen that what was found in practice in our example agrees with theory. Also, Milne started by Taylor is more accurate than Milne started by improved Euler as Taylor can be made as accurate as desired whereas improved Euler cannot. Unfortunately Taylor is not so good computerwise, as all the derivatives must be found analytically first.

One further point about errors. You will realise that, as each step in the application of any method depends on the results previously obtained, the errors introduced are cumulative. The truth of this statement will be obvious if you examine the errors listed in FRAMES 10, 21 and 26.

Stability of Milne's Method - Hamming's Method

In the example taken to illustrate Milne's method, only two forward steps were taken - the calculation of y_4 and y_5. (y_1, y_2 and y_3 were forward steps, but had to be found by some method other than Milne's.) If it is necessary to compute a large number of steps, Milne's method can lead to trouble in certain circumstances. The trouble, known as INSTABILITY, is that the accuracy of the solution becomes very poor if $\frac{\partial f}{\partial y}$ is negative - and simply reducing the step length doesn't remedy this.

Another modification of Milne's method is that due to Hamming, whose formulae will be quoted but not obtained. This modification is stable when $\frac{\partial f}{\partial y} < 0$, provided h is taken to be less than $0 \cdot 75 \left/ \left| \frac{\partial f}{\partial y} \right| \right.$. If $\frac{\partial f}{\partial y} > 0$, Hamming's method works provided h is taken less than $0 \cdot 4 \left/ \frac{\partial f}{\partial y} \right.$.

HAMMING'S METHOD uses the same predictor formula as Milne's, but replaces the corrector formula $y_4^c = y_2 + \frac{h}{3}(f_2 + 4f_3 + f_4^P)$ by

$$y_4^a = \frac{1}{8}\left\{ 9y_3 - y_1 + 3h(f_4^P + 2f_3 - f_2) \right\} \quad (46.1)$$

Similar expressions will give y_5^c, y_6^c, etc.

What will be the formulae for y_5^c and y_n^c ?

****

$$y_5^c = \frac{1}{8}\left\{ 9y_4 - y_2 + 3h(f_5^p + 2f_4 - f_3) \right\}$$

$$y_n^c = \frac{1}{8}\left\{ 9y_{n-1} - y_{n-3} + 3h(f_n^p + 2f_{n-1} - f_{n-2}) \right\}$$

Applying the formula (46.1) to the example used earlier, i.e., $\frac{dy}{dx} = x - y$ with $y = 1$ when $x = 0$, and inserting into (46.1) the figures in FRAMES 37 - 39, we have

$$y_4^c = \frac{1}{8}\left\{ 9 \times 0 \cdot 781\,64 - 0 \cdot 909\,67 + 3 \times 0 \cdot 1(-0 \cdot 340\,64 - 2 \times 0 \cdot 481\,64 + 0 \cdot 637\,46) \right\}$$

$$= 0 \cdot 740\,64.$$

If necessary the use of the corrector formula can be repeated with f_4^c replacing f_4^p, just as Milne's was, until no further change in the value of y_4^c occurs.

In examples where the question of instability does not arise, the Hamming formula is marginally less accurate than Milne. Hamming has modified it to overcome this and at the same time obviate the need for repeated use of the corrector formula.

Modified Hamming's Method

As with the modification to Milne's method, the Hamming modification can only be used from y_5 onwards. It adopts the following sequence of calculations for y_5:

$$y_5^p = y_1 + \frac{4h}{3}(2f_2 - f_3 + 2f_4)$$

$$y_5^m = y_5^p - \frac{112}{121}(y_4^p - y_4^c)$$

$$y_5^c = \frac{1}{8}\left\{ 9y_4 - y_2 + 3h(f_5^m + 2f_4 - f_3) \right\}$$

$$y_5 = y_5^c + \frac{9}{121}(y_5^p - y_5^c)$$

Again, as with modified Milne, some authors prefer to replace the last step by repeated use, as much as is necessary, of the formula for y_5^c.

Similar formulae will give y_6, y_7, etc.

It is, however, found that the condition on h, when $\frac{\partial f}{\partial y} < 0$, has to be modified to $h < 0 \cdot 65 \bigg/ \left| \frac{\partial f}{\partial y} \right|$.

Applying this method to the example in FRAME 43,

$y_5^p = 0 \cdot 735\,87$ (the same as for Milne), $\qquad y_5^m = 0 \cdot 735\,74$,

$f_5^m = 0 \cdot 264\,26$, $\qquad y_5^c = 0 \cdot 735\,75$, $\qquad y_5 = 0 \cdot 735\,76$

Runge-Kutta Method

Due to its high accuracy, Milne's method is an obvious candidate for the popularity charts. Another popular method is that known as the RUNGE-KUTTA METHOD. This has a similar degree of accuracy to Milne's method and furthermore it is self-starting. As a result, it is sometimes used for starting Milne's method.

As its name suggests, the method was derived by two persons, Runge and Kutta. The derivation is somewhat complicated and so here we shall be content with giving the formulae involved and seeing how it works. However, we would remark that the method effectively involves finding a series of estimates for ordinates and then combining them in a particular way, the combination being based on Taylor's series.

The method again assumes the differential equation is in the form $\frac{dy}{dx} = f(x, y)$ and that y is given as y_0 when $x = x_0$.

To find y_1, the following quantities are calculated:

$$k_1 = hf(x_0, y_0)$$
$$k_2 = hf(x_0 + \tfrac{1}{2}h, y_0 + \tfrac{1}{2}k_1)$$
$$k_3 = hf(x_0 + \tfrac{1}{2}h, y_0 + \tfrac{1}{2}k_2)$$
$$k_4 = hf(x_0 + h, y_0 + k_3)$$

Then y_1 is given by

$$y_1 = y_0 + (k_1 + 2k_2 + 2k_3 + k_4)/6$$

(You will notice that k_1 is what the Euler formula adds to y_0 to give an estimate of y_1.)

Once y_1 has been found, the formulae are used again but with x_0 replaced by x_1 and y_0 by y_1. This second use of the formulae gives an estimate of y_2. y_3, y_4, etc., then follow similarly.

To see how the formula works in practice and also to get an idea as to how well it works, the equation $\frac{dy}{dx} = x - y$ with $y = 1$ when $x = 0$ will again be used. As before, it is desired to find y when $x = 0 \cdot 1$, $0 \cdot 2$, $0 \cdot 3$, etc. As $x_0 = 0$ and $y_0 = 1$, $f(x_0, y_0) = -1$ and so $k_1 = -0 \cdot 1$.

Then $x_0 + \tfrac{1}{2}h = 0 \cdot 05$ and $y_0 + \tfrac{1}{2}k_1 = 0 \cdot 95$ so that
$$k_2 = 0 \cdot 1(0 \cdot 05 - 0 \cdot 95) = -0 \cdot 09$$

This gives $y_0 + \tfrac{1}{2}k_2 = 0 \cdot 955$ and so
$$k_3 = 0 \cdot 1(0 \cdot 05 - 0 \cdot 955) = -0 \cdot 0905$$

Now $x_0 + h = 0 \cdot 1$ and $y_0 + k_3 = 0 \cdot 9095$ and so
$$k_4 = 0 \cdot 1(0 \cdot 1 - 0 \cdot 9095) = -0 \cdot 08095$$

Thus $y_1 = 1 + \left\{-0 \cdot 1 + 2(-0 \cdot 09) + 2(-0 \cdot 0905) - 0 \cdot 080\,95\right\}/6 \simeq 0 \cdot 909\,68$.

To find y_2, $f(x_1, y_1) = -0 \cdot 809\ 68$, and so, for the next round in the calculation, $k_1 = -0 \cdot 080\ 97$, working to 5 decimal places.

Then $x_1 + \frac{1}{2}h = 0 \cdot 15$ and $y_1 + \frac{1}{2}k_1 = 0 \cdot 869\ 20$ and so
$$k_2 = 0 \cdot 1(0 \cdot 15 - 0 \cdot 869\ 20) = -0 \cdot 071\ 92$$
Next $\quad k_3 = 0 \cdot 1(0 \cdot 15 - 0 \cdot 873\ 72) = -0 \cdot 072\ 37$

What will be the values of k_4 and y_2?
****

52A

$k_4 = 0 \cdot 1(0 \cdot 2 - 0 \cdot 837\ 31) = -0 \cdot 063\ 73$
$y_2 = 0 \cdot 909\ 68 + \left\{ -0 \cdot 080\ 97 + 2(-0 \cdot 071\ 92) + 2(-0 \cdot 072\ 37) - 0 \cdot 063\ 73 \right\}/6$
$\quad = 0 \cdot 837\ 47.$

For y_3, $k_1 = -0 \cdot 063\ 75$, $k_2 = -0 \cdot 055\ 56$, $k_3 = -0 \cdot 055\ 97$, $k_4 = -0 \cdot 048\ 15$, $y_3 = 0 \cdot 781\ 64$.

Now use the method to find y_4 and y_5.
****

53A

$y_4 = 0 \cdot 740\ 64$, $\quad y_5 = 0 \cdot 713\ 07$

Similarly, $y_6 = 0 \cdot 697\ 63$ and so, quoting the results to 4 decimal places, $y_0 - 1$, $y_1 = 0 \cdot 9097$, $y_2 = 0 \cdot 8375$, $y_3 = 0 \cdot 7816$, $y_4 = 0 \cdot 7406$, $y_5 = 0 \cdot 7131$ and $y_6 = 0 \cdot 6976$.

These agree with the values found by Taylor–Milne and also with the true values. Actually the Runge–Kutta process, as described, was devised to give a step error of the order of h^5, i.e., of the same order as Milne's step error. Being correct to h^4, it is known as a fourth order process.

A flow diagram showing the solution of $\dfrac{dy}{dx} = f(x, y)$, subject to $y = y_0$ when $x = x_0$, for values of y at $x = x_0 + h$, $x_0 + 2h$, ..., $x_0 + nh$, using the Runge–Kutta fourth-order process is shown on page 306.

Comparison of Methods

Obviously, if great accuracy is required, Milne, Hamming or Runge–Kutta would be used. The advantages and disadvantages of these three methods can be summarised as shown on page 307.

Flow diagram for FRAME 55.

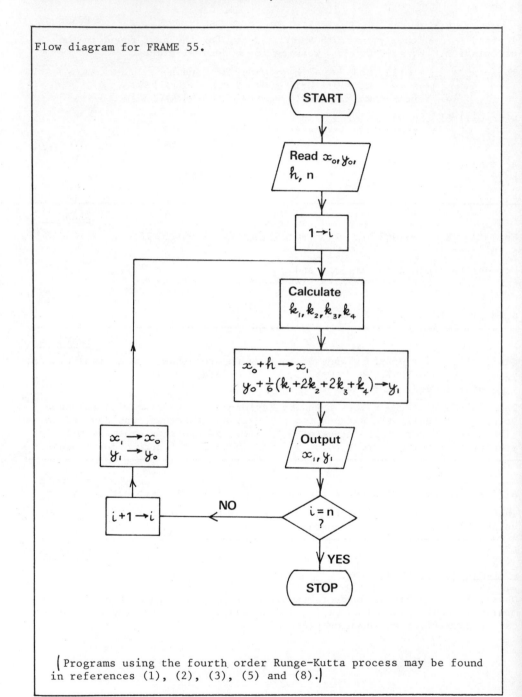

(Programs using the fourth order Runge-Kutta process may be found in references (1), (2), (3), (5) and (8).)

Method	Advantages	Disadvantages
Milne	The method can be modified to eliminate or reduce the repeated use of the corrector formula. Only two evaluations of the f-function may be needed per forward step if the above modification is incorporated.	The method is not self-starting. It becomes unstable in some cases if it has to be used for many steps.
Hamming	As for Milne but, in addition, does not become unstable.	The method is not self-starting. It is not quite so accurate as Milne when this is behaving.
Runge-Kutta	No special procedure necessary for starting. Inherently stable. Easy to change the step size. Particularly advantageous when memory requirements are to be minimised.	Four evaluations of the f-function needed per forward step.

Summary of Formulae for solving $\dfrac{dy}{dx} = f(x, y)$

All of these formulae are for y_n.   Some were only given for specific values of n in the programme or in a slightly different form.   Where a y or an f is quoted without a qualifying p or c, the best value available is to be taken.

Euler

$$y_n = y_{n-1} + hf_{n-1}$$

Improved Euler

$$y_n^p = y_{n-1} + hf_{n-1}$$

$$y_n^c = y_{n-1} + \frac{h}{2}(f_{n-1} + f_n^p)$$

$$\left(y_n^c\right)_{new} = y_{n-1} + \frac{h}{2}\left\{f_{n-1} + \left(f_n^c\right)_{old}\right\}$$

FIRST ORDER ORDINARY DIFFERENTIAL EQUATIONS

Milne
$$y_n^p = y_{n-4} + \frac{4h}{3}(2f_{n-3} - f_{n-2} + 2f_{n-1}) \qquad n > 3$$

$$y_n^c = y_{n-2} + \frac{h}{3}(f_{n-2} + 4f_{n-1} + f_n^p)$$

$$\left(y_n^c\right)_{new} = y_{n-2} + \frac{h}{3}\left\{f_{n-2} + 4f_{n-1} + \left(f_n^c\right)_{old}\right\} \qquad (57.1)$$

Modified Milne
$$y_n^p = y_{n-4} + \frac{4h}{3}(2f_{n-3} - f_{n-2} + 2f_{n-1}) \qquad n > 4$$

$$y_n^m = y_n^p - \frac{28}{29}(y_{n-1}^p - y_{n-1}^c)$$

$$y_n^c = y_{n-2} + \frac{h}{3}(f_{n-2} + 4f_{n-1} + f_n^m)$$

Either
$$y_n = y_n^c + \frac{1}{29}(y_n^p - y_n^c) \qquad or \qquad (57.1)$$

Hamming
$$y_n^p = y_{n-4} + \frac{4h}{3}(2f_{n-3} - f_{n-2} + 2f_{n-1}) \qquad n > 3$$

$$y_n^c = \frac{1}{8}\left\{9y_{n-1} - y_{n-3} + 3h(f_n^p + 2f_{n-1} - f_{n-2})\right\}$$

$$\left(y_n^c\right)_{new} = \frac{1}{8}\left\{9y_{n-1} - y_{n-3} + 3h\left[\left(f_n^c\right)_{old} + 2f_{n-1} - f_{n-2}\right]\right\}$$
$$(57.2)$$

Modified Hamming
$$y_n^p = y_{n-4} + \frac{4h}{3}(2f_{n-3} - f_{n-2} + 2f_{n-1}) \qquad n > 4$$

$$y_n^m = y_n^p - \frac{112}{121}(y_{n-1}^p - y_{n-1}^c)$$

$$y_n^c = \frac{1}{8}\left\{9y_{n-1} - y_{n-3} + 3h(f_n^m + 2f_{n-1} - f_{n-2})\right\}$$

Either
$$y_n = y_n^c + \frac{9}{121}(y_n^p - y_n^c) \qquad or \qquad (57.2)$$

Runge-Kutta
$$k_1 = hf(x_{n-1}, y_{n-1})$$

$$k_2 = hf(x_{n-1} + \tfrac{1}{2}h, y_{n-1} + \tfrac{1}{2}k_1)$$

$$k_3 = hf(x_{n-1} + \tfrac{1}{2}h, y_{n-1} + \tfrac{1}{2}k_2)$$

$$k_4 = hf(x_{n-1} + h, y_{n-1} + k_3)$$

$$y_n = y_{n-1} + (k_1 + 2k_2 + 2k_3 + k_4)/6.$$

Miscellaneous Examples

In this frame a collection of miscellaneous examples is given for you to try.   Answers are supplied in FRAME 59, together with such working as is considered helpful.

Some of these questions indicate that formulae other than those in the text can be adapted to the numerical solution of differential equations. Different mathematicians have done this to produce a variety of processes.

1. Use the modified Euler method to find y when x = 0·1 and 0·2 given that $\frac{dy}{dx} = \sqrt{x + y}$ and y = 0·3600 when x = 0.

2. Solve numerically the differential equation $\frac{dy}{dx} = x - y^2$, y = 1 at x = 0, for the values of x = 0·2 and x = 0·4, specifying your results to 3 decimal places.   (C.E.I.)

    (Note: The Runge-Kutta formulae were quoted.)

3. Derive the following integration formulae:

    a) $\int_0^2 f_\mu \, dp = \frac{1}{3}(f_0 + 4f_1 + f_2)$,   b) $\int_0^4 f_p \, dp = \frac{4}{3}(2f_1 - f_2 + 2f_3)$

    Show how these may be used as predictor-corrector formulae suitable for solving the differential equation $\frac{dy}{dx} = f(x, y)$.

    Show that the solution of $\frac{dy}{dx} - 3x^2 y = 1 - 3x^3$, given y = 1 when x = 0, can be written in the form $y = 1 + x + x^3 + O(x^5)$.

    Obtain starting values for this solution at x = ±0·1, ±0·2 and hence use the Predictor-Corrector formulae to evaluate y at x = 0·3 to four decimal places.   (L.U.)

    (Note: This was the question as originally set. There is no need for you to obtain formulae (a) and (b) unless you wish.   These results have been obtained in either the present programme or that on integration in Unit 2.   The expression $1 + x + x^3 + O(x^5)$ means that, up as far as x^4, the terms are $1 + x + x^3$, the coefficients of x^2 and x^4 being zero.   After that there are terms in x^5 and higher powers.   For the purposes of this question it is unnecessary to know their coefficients.)

4.

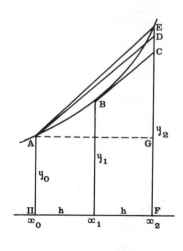

A variation in the straightforward Euler method for predicting the value of a y is shown in the diagram. Starting from y_1, the Euler formula $y_2^P = y_1 + hf(x_1, y_1)$ gives FC as a first estimate of y_2. Let AD be parallel to BC, the tangent to the curve at B.   Then

$$FD = HA + AG \tan D\hat{A}G$$
$$= y_0 + 2hf(x_1, y_1)$$

and this is taken instead of FC as a first estimate of y_2, i.e. the formula $y_2^P = y_0 + 2hf(x_1, y_1)$ is used. Its use means that two initial values are required, y_0 and y_1.   The latter can be obtained from a Taylor series or Runge-Kutta.

309

The same corrector formula is then used as with improved Euler, i.e.

$$y_2^c = y_1 + \frac{h}{2}\{f(x_1, y_1) + f(x_2, y_2^p)\} \; .$$

Iteration can then be carried out as necessary to improve the result still further.

Taking the values $x_0 = 0$, $y_0 = 1$, $h = 0 \cdot 1$, $y_1 = 0 \cdot 909\,68$, obtain the best values, to 4 decimal places, of y_2 and y_3 by this method for the equation $\dfrac{dy}{dx} = x - y$.

5. A function $f(x)$ is given at points $x_n = x_0 + nh$.   If $x - x_0 = ph$, show that the parabola $y = \frac{1}{2}(p+1)(p+2)f_0 - p(p+2)f_{-1} + \frac{1}{2}p(p+1)f_{-2}$ passes through three successive table points.

Hence derive the approximate integration formulae

$$\int_{x_0}^{x_1} f(x)\,dx = \frac{1}{12}h(23f_0 - 16f_{-1} + 5f_{-2})$$

$$\int_{x_{-1}}^{x_0} f(x)\,dx = \frac{1}{12}h(5f_0 + 8f_{-1} - f_{-2})$$

Use these formulae to carry the solution of the differential equation $\dfrac{dy}{dx} = \dfrac{2x-1}{x^2}\,y + 1$ one step forward if the values given below have been calculated already

x	1	1·1	1·2	
y	2	2·3148	2·6589	
$\frac{dy}{dx}$	3	3·2957	3·5851	(L.U.)

Note: In this question $x_n (= x_0 + nh)$ is used to indicate a point at which the value of the function is tabulated. $x(= x_0 + ph)$ is any value of x, which will sometimes be at a tabular point, but more often, not.

6. A first order differential equation may be solved numerically by using one of the pairs (a) or (b) of predictor and corrector formulae.

(a)
$$\begin{cases} y_1 = y_0 + h\left(1 + \frac{1}{2}\nabla + \frac{5}{12}\nabla^2 + \frac{3}{8}\nabla^3 + \frac{251}{720}\nabla^4 + \ldots\ldots\ldots\right)y_0' \\[2mm] y_1 = y_0 + h\left(1 - \frac{1}{2}\nabla - \frac{1}{12}\nabla^2 - \frac{1}{24}\nabla^3 - \frac{19}{720}\nabla^4 + \ldots\ldots\ldots\right)y_1' \end{cases}$$

(b)
$$\begin{cases} y_4 = y_0 + \frac{4}{3}h(2y_1' - y_2' + 2y_3') \\[2mm] y_4 = y_2 + \frac{1}{3}h(y_2' + 4y_3' + y_4') \end{cases}$$

Prove one only of the predictor formulae.

The differential equation $y' = x - \frac{1}{10}y^2$ is to be solved with the initial value $y = 1$ when $x = 0$.   Assuming that the following

310

starting values have been obtained, find the value of y at x = 0·3.

x	−0·2	−0·1	0·1	0·2	
y	1·040 68	1·015 13	0·995 07	1·000 13	(L.U.)

(Note: This was the question as set. Either (a) or (b) could be used for the last part of the question. As predictor formula (b) (it is Milne's) has been obtained in the text, you can now prove the predictor in (a). Then find y when x = 0·3 by the use of both (a) and (b), checking that your answers agree with each other. Formulae (a) are known as the Adams-Bashforth formulae.)

7. State the Runge-Kutta equations that would be used to give x at

t = t_0 + h for the equation $\frac{dx}{dt}$ = f(t, x) given that x = x_0 when t = t_0.

8. The response x of a given hydraulic valve subject to sinusoidal input

variation is given by $\frac{dx}{dt} = \sqrt{2\left(1 - \frac{x^2}{\sin^2 t}\right)}$, with x = 0 at t = 0.

Show that $\left(\frac{dx}{dt}\right)_{x=0} = \sqrt{\frac{2}{3}}$ and hence use the Runge-Kutta fourth order method to obtain a solution at t = 0·2.   (C.E.I.)

(Note: The Runge-Kutta formulae were quoted. These particular formulae have been evolved to give a solution which is equivalent to a Taylor expansion as far as the term involving x^4 - hence the description 'fourth-order'. Other Runge-Kutta formulae of different orders have also been evolved.)

HINT: In order to obtain the value of $\frac{dx}{dt}$ when x = 0, it will be necessary to rearrange the quoted formula without surds or fractions and differentiate twice w.r.t. t.

9. Use Hamming's method to find the value of y when x = 0·8 for the

equation $\frac{dy}{dx}$ = x − 2y, given the following starting values:

x	0	0·2	0·4	0·6
y	0·750 00	0·520 32	0·399 33	0·351 19

Then use the modified Hamming method to find y when x = 1·0.

10. An alternative method to Hamming's that is designed to overcome the instability of Milne's, and is frequently used, is known as the ADAMS-MOULTON method. It uses the formulae

$$y_n^P = y_{n-1} + \frac{h}{24}(55f_{n-1} - 59f_{n-2} + 37f_{n-3} - 9f_{n-4})$$

$$y_n^c = y_{n-1} + \frac{h}{24}(9f_n^P + 19f_{n-1} - 5f_{n-2} + f_{n-3})$$

Use this method to find y_5 for the differential equation and starting values given in question No. 9.

311

Answers to Miscellaneous Examples

1. 0·4263, 0·5045

2. 0·851, 0·780

3. The way of extracting the predictor formula from (b) was given in
    FRAME 30. In a similar way (a) leads to $y_2 = y_0 + \frac{h}{3}(f_0 + 4f_1 + f_2)$
    and adding 2 to each suffix gives $y_4 = y_2 + \frac{h}{3}(f_2 + 4f_3 + f_4)$ i.e.
    the Milne corrector formula.

    By differentiating successively $y' = 3x^2y + 1 - 3x^2$ and then
    substituting $x = 0$, $y = 1$, it is found that $y' = 1$, $y'' = 0$,
    $y''' = 6$, $y^{iv} = 0$.

    Then, from Maclaurin, $y = 1 + x + x^3 + O(x^5)$.

    Using this formula and also that for y' gives the table

x	−0·2	−0·1	0	0·1	0·2
y	0·792	0·899	1	1·101	1·208
f		1·029 97	1	1·030 03	1·120 96

    When $x = 0·3$, (b) gives $y^P = 1·327\ 25$ and hence $f^P = 1·277\ 36$.
    (a) then gives $y^C = 1·327\ 37 = 1·3274$ to 4 decimal
    places.

    You will notice that the value at $x = -0·2$ is not actually needed.

4. $y_2^P = 0·838\ 06$, $y_2^C = 0·837\ 29$ and then $0·837\ 33$

    $y_3^P = 0·782\ 21$ $y_3^C = 0·781\ 35$ and then $0·781\ 39$

    Then, to 4 decimal places $y_2 = 0·8373$ and $y_3 = 0·7814$.

5. Putting $p = 0, -1, -2$, in turn gives $y = f_0, f_{-1}, f_{-2}$, i.e. three
    successive table points.

    $$\int_{x_0}^{x_1} f(x)dx = h\int_0^1 y\,dp \quad \text{and} \quad \int_{x_{-1}}^{x_0} f(x)dx = h\int_{-1}^0 y\,dp. \quad \text{Results}$$
    follow.

    These integration formulae then lead, by a process similar to that in
    FRAME 30, to the formulae

    $$y_1 = y_0 + \frac{h}{12}(23f_0 - 16f_{-1} + 5f_{-2})$$

    $$y_0 = y_{-1} + \frac{h}{12}(5f_0 + 8f_{-1} - f_{-2})$$

    Increasing the suffixes by 2 in the first of these equations and by 3
    in the second, and using p and c in the usual way to denote predicted
    and corrected values, gives

    $$y_3^P = y_2 + \frac{h}{12}(23f_2 - 16f_1 + 5f_0), \qquad y_3^C = y_2 + \frac{h}{12}(5f_3^P + 8f_2 - f_1)$$

Taking $x_0 = 1$, $x_1 = 1 \cdot 1$, etc., the first formula gives

$$y_3^P = 2 \cdot 6589 + \frac{1}{12} \times 0 \cdot 1(23 \times 3 \cdot 5851 - 16 \times 3 \cdot 2957 + 5 \times 3) = 3 \cdot 031\ 62$$

Then $f_3^P = \dfrac{2 \times 1 \cdot 3 - 1}{1 \cdot 3\ 2} \times 3 \cdot 031\ 62 + 1 \simeq 3 \cdot 8702$ and

$$y_3^C = 2 \cdot 6589 + \frac{1}{12} \times 0 \cdot 1(5 \times 3 \cdot 8702 + 8 \times 3 \cdot 5851 - 3 \cdot 2957) = 3 \cdot 0317$$

6. (a) The first formula in (a) is obtained by a process similar to that used in FRAMES 28A and 30 but starting with the Newton-Gregory <u>backward</u> difference formula and integrating only between x_0 and x_1. This means that the limits of p are 0 and 1.   Note that y' has been used in this question instead of f.

The table can be extended to show the values of y' and its differences above the dashed line:

x	y'			
$-0 \cdot 2$	$-0 \cdot 308\ 30$			
		10 525		
$-0 \cdot 1$	$-0 \cdot 203\ 05$		-220	
		10 305		13
0	$-0 \cdot 1$		-207		-5
		10 098		8
$0 \cdot 1$	$0 \cdot 000\ 98$		-199		-7
		9899		1
$0 \cdot 2$	$0 \cdot 099\ 97$		-198	
		9701		
$0 \cdot 3$	$0 \cdot 196\ 98$			

Taking $x_0 = 0 \cdot 2$,

$$y_1^P = 1 \cdot 000\ 13 + 0 \cdot 1\left\{0 \cdot 099\ 97 + \frac{1}{2} \times 0 \cdot 098\ 99 + \frac{5}{12}(-0 \cdot 001\ 99) + \frac{3}{8}(0 \cdot 000\ 08)\right.$$
$$\left. + \frac{251}{720}(-0 \cdot 000\ 05)\right\} = 1 \cdot 014\ 99$$

y' when $x = 0 \cdot 3$ and $y = 1 \cdot 014\ 99$ is $0 \cdot 3 - \frac{1}{10} \times 1 \cdot 014\ 99^2 = 0 \cdot 196\ 98$

The figures in the table below the dashed line can now be added.

Then $y_1^C = 1 \cdot 000\ 13 + 0 \cdot 1\left\{0 \cdot 196\ 98 - \frac{1}{2} \times 0 \cdot 097\ 01 - \frac{1}{12}(-0 \cdot 001\ 98)\right.$
$$\left. - \frac{1}{24}(0 \cdot 000\ 01) - \frac{19}{720}(-0 \cdot 000\ 07)\right\} = 1 \cdot 014\ 99.$$

Although they have been shown, you will realise from the difference table that the fourth differences are not reliable.

(b) Taking $x_0 = -0 \cdot 1$

$$y_4^P = 1 \cdot 015\ 13 + \frac{4 \times 0 \cdot 1}{3}\left\{2(-0 \cdot 1) - 0 \cdot 000\ 98 + 2 \times 0 \cdot 099\ 97\right\} = 1 \cdot 014\ 99$$

Then $f_4^P = 0 \cdot 196\ 98$

$$y_4^c = 0 \cdot 995\,07 + \frac{0 \cdot 1}{3}\{0 \cdot 000\,98 + 4 \times 0 \cdot 099\,97 + 0 \cdot 196\,98\} = 1 \cdot 015\,00$$

7. $k_1 = hf(t_0, x_0)$

$k_2 = hf(t_0 + \tfrac{1}{2}h, x_0 + \tfrac{1}{2}k_1)$

$k_3 = hf(t_0 + \tfrac{1}{2}h, x_0 + \tfrac{1}{2}k_2)$

$k_4 = hf(t_0 + h, x_0 + k_3)$

$x_1 = x_0 + (k_1 + 2k_2 + 2k_3 + k_4)/6$

8. $\dot{x}^2 \sin^2 t = 2\sin^2 t - 2x^2$

Differentiating $2\dot{x}\,\ddot{x}\,\sin^2 t + \dot{x}^2 \sin 2t = 2\sin 2t - 4x\dot{x}$

Differentiating again,

$2(\ddot{x}^2 + \dot{x}\,\dddot{x})\sin^2 t + 4\dot{x}\,\ddot{x}\sin 2t + 2\dot{x}^2 \cos 2t = 4\cos 2t - 4(\dot{x}^2 + x\ddot{x})$

Substituting in this $t = 0$, $x = 0$ gives $\dot{x} = \sqrt{\dfrac{2}{3}}$.

For one step from $t = 0$ to $0 \cdot 2$,

$k_1 = 0 \cdot 163\,30$, $k_2 = 0 \cdot 162\,74$, $k_3 = 0 \cdot 163\,86$, $k_4 = 0 \cdot 159\,93$,
$x = 0 \cdot 1627$.

9. $0 \cdot 351\,82$ $0 \cdot 385\,38$.

10. $0 \cdot 351\,81$.

Simultaneous and Second Order
Differential Equations

Simultaneous Differential Equations - Introduction

Simultaneous differential equations occur in many situations in practice. Two simple examples are given by oscillating masses and coupled electric circuits.

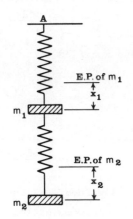

If two masses are attached to springs as shown and x_1 and x_2 are their distances at time t below their equilibrium positions, then the equations giving x_1 and x_2 are

$$m_1 \frac{d^2x_1}{dt^2} + (k_1 + k_2)x_1 - k_2x_2 = 0$$

$$m_2 \frac{d^2x_2}{dt^2} + k_2x_2 - k_2x_1 = 0$$

where k_1 and k_2 are the stiffnesses of the upper and lower springs respectively.

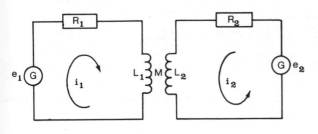

In the coupled circuit shown, the equations giving the currents i_1 and i_2 are

$$L_1 \frac{di_1}{dt} + R_1i_1 + M \frac{di_2}{dt} = e_1$$

$$M \frac{di_1}{dt} + L_2 \frac{di_2}{dt} + R_2i_2 = e_2$$

The equations quoted in the last frame can be solved analytically without any trouble provided that, in the second case, e_1 and e_2 are simple functions of t that we know how to deal with. (Special cases of these equations are actually solved on pages 5:166 - 5:168 and 5:177 - 5:180 in our first volume of Mathematics for Engineers and Scientists.) Our purpose in the first part of this programme is to see how to find numerical solutions for certain types of simultaneous differential equations that we do not know how to solve analytically, or whose analytical solution is too complicated to be of practical use.

SIMULTANEOUS AND SECOND ORDER DIFFERENTIAL EQUATIONS

In the last programme, various methods of tackling the equation $\frac{dy}{dx} = f(x, y)$ were described.   These methods can very easily be extended to simultaneous equations of the form $\frac{dx}{dt} = f(t, x, y)$, $\frac{dy}{dt} = g(t, x, y)$.

As nothing fresh is involved in the way of theory, it will suffice simply to take examples.   Once again, in order to be able to see how our solutions compare with those obtained analytically, our main example will be one that can be solved analytically.

The methods used in the last programme were:-

   Euler,   Improved and Modified Euler,   Milne – and modified Milne, Hamming - and modified Hamming,   Runge-Kutta

and now each of these will be applied to the equations $\frac{dx}{dt} = -t + x + y$, $\frac{dy}{dt} = 2t + 3x - y$, given that when $t = 0$, $x = 1$ and $y = 2$.   The values of x and y at $t = 0 \cdot 1$, $0 \cdot 2$, $0 \cdot 3$, etc., will be sought.

Euler's Method

For a single equation of the form $\frac{dy}{dx} = f(x, y)$,   Euler's method used the equations $y_1 = y_0 + hf(x_0, y_0)$, $y_2 = y_1 + hf(x_1, y_1)$,   etc., to estimate the values of y when $x = x_1$, x_2, etc., y_0 being the given value of y when $x = x_0$.

In a similar way, for the equations $\dot{x} = f(t, x, y)$, $\dot{y} = g(t, x, y)$, successive estimates of x and y can be obtained from

$x_1 = x_0 + hf(t_0, x_0, y_0)$ (4.1), $y_1 = y_0 + hg(t_0, x_0, y_0)$ (4.2),
$x_2 = x_1 + hf(t_1, x_1, y_1)$ $y_2 = y_1 + hg(t_1, x_1, y_1)$ etc.,

where x_0 and y_0 are the values of x and y when $t = t_0$ and h is the interval between consecutive values of t.

What will be the formulae for x_{n+1} and y_{n+1}?

$$**************************************$$

$$x_{n+1} = x_n + hf(t_n, x_n, y_n) \qquad y_{n+1} = y_n + hg(t_n, x_n, y_n)$$

Taking the equations quoted in FRAME 3, we have $f(t, x, y) = -t + x + y$, $g(t, x, y) = 2t + 3x - y$, and as, when $t = 0$, $x = 1$ and $y = 2$, these values of t, x and y are taken as our t_0, x_0 and y_0.

Thus $f(t_0, x_0, y_0) = 0 + 1 + 2 = 3$, $g(t_0, x_0, y_0) = 2 \times 0 + 3 \times 1 - 2 = 1$.

Using the values found above with $h = 0 \cdot 1$, the values of x and y when $t = 0 \cdot 1$, i.e., x_1 and y_1, are, using (4.1) and (4.2),

316

$$x_1 = 1 + 0 \cdot 1 \times 3 = 1 \cdot 3, \qquad\qquad y_1 = 2 + 0 \cdot 1 \times 1 = 2 \cdot 1$$

What will be the values of x_2 and y_2, i.e., those of x and y when t = 0·2?   **

5A

$x_2 = 1 \cdot 3 + 0 \cdot 1(-0 \cdot 1 + 1 \cdot 3 + 2 \cdot 1) = 1 \cdot 63$
$y_2 = 2 \cdot 1 + 0 \cdot 1(2 \times 0 \cdot 1 + 3 \times 1 \cdot 3 - 2 \cdot 1) = 2 \cdot 3$

FRAME 6

Similarly, $x_3 = 1 \cdot 63 + 0 \cdot 1(-0 \cdot 2 + 1 \cdot 63 + 2 \cdot 3) = 2 \cdot 003$
$y_3 = 2 \cdot 3 + 0 \cdot 1(0 \cdot 4 + 4 \cdot 89 - 2 \cdot 3) = 2 \cdot 599$

and so on.   As you would expect from our work in the previous programme, these values are not very accurate, as will be seen later.   They can be made more accurate by using the Improved Euler Method.

FRAME 7

The Improved Euler Method

When solving $\dfrac{dy}{dx} = f(x, y)$ by the improved Euler method, the first step was to predict y_1 using the straightforward ordinary Euler method.   This gave $y_1^P = y_0 + hf(x_0, y_0)$.

The next step was to correct the value predicted for y_1 by means of the formula $y_1^c = y_0 + \dfrac{h}{2}\{f(x_0, y_0) + f(x_1, y_1^P)\}$ (7.1)

A similar process is used for simultaneous equations.   First x_1 and y_1 are predicted by means of the formulae

$$x_1^P = x_0 + hf(t_0, x_0, y_0), \qquad y_1^P = y_0 + hg(t_0, x_0, y_0)$$

and these values are then corrected by formulae similar to (7.1).   Can you suggest what the formulae for x_1^c and y_1^c are?
****

7A

$$x_1^c = x_0 + \frac{h}{2}\{f(t_0, x_0, y_0) + f(t_1, x_1^P, y_1^P)\} \qquad (7A.1)$$

$$y_1^c = y_0 + \frac{h}{2}\{g(t_0, x_0, y_0) + g(t_1, x_1^P, y_1^P)\} \qquad (7A.2)$$

FRAME 8

Now, returning to the equations in FRAME 3, x_1^P and y_1^P have already been found to be 1·3 and 2·1 (the x_1 and y_1 of FRAME 5).   Applying the formulae in 7A then gives

$$x_1^c = 1 + \frac{0 \cdot 1}{2}(3 + 3 \cdot 3) = 1 \cdot 315, \qquad y_1^c = 2 + \frac{0 \cdot 1}{2}(1 + 2) = 2 \cdot 15$$

What will be the values of x_2^c and y_2^c by this method?
****

$$x_2^p = 1 \cdot 315 + 0 \cdot 1 \times 3 \cdot 365 = 1 \cdot 6515 \qquad\qquad y_2^p = 2 \cdot 15 + 0 \cdot 1 \times 1 \cdot 995 = 2 \cdot 3495$$

$$x_2^c = 1 \cdot 315 + \frac{0 \cdot 1}{2}(3 \cdot 365 + 3 \cdot 801) \qquad\qquad y_2^c = 2 \cdot 15 + \frac{0 \cdot 1}{2}(1 \cdot 995 + 3 \cdot 005)$$

$$= 1 \cdot 6733 \qquad\qquad\qquad\qquad\qquad\qquad = 2 \cdot 40$$

The Modified Euler Method

There are two ways in which modifications can be made to the method just described. The first of these follows exactly what was done in the last programme. It takes the values found for x_1^c and y_1^c and replaces x_1^p and y_1^p in (7A.1) and (7A.2) by these new values. Second estimates are thus obtained for x_1^c and y_1^c. Third estimates are then obtained by using these latest values in (7A.1) and (7A.2) and so on until no further changes occur in the estimates of x_1 and y_1.

Using $1 \cdot 315$ and $2 \cdot 15$ in place of x_1^p and y_1^p in (7A.1) and (7A.2) gives second estimates of x_1^c and y_1^c to be

$$x_1^c = 1 + \frac{0 \cdot 1}{2}(3 + 3 \cdot 365) = 1 \cdot 3182, \qquad y_1^c = 2 + \frac{0 \cdot 1}{2}(1 + 1 \cdot 995) = 2 \cdot 1498$$

What will be the third estimates of x_1^c and y_1^c?

****

$$x_1^c = 1 + \frac{0 \cdot 1}{2}(3 + 3 \cdot 3680) = 1 \cdot 3184 \qquad\qquad y_1^c = 2 + \frac{0 \cdot 1}{2}(1 + 2 \cdot 0048) = 2 \cdot 1502$$

The next estimates are $1 \cdot 318\,43$ and $2 \cdot 150\,25$ and after that no further changes take place.

The calculation then proceeds in a similar way to find x_2 and y_2. What are the best values you can get for these?

****

$$x_2^p = 1 \cdot 655\,30, \qquad y_2^p = 2 \cdot 350\,75$$

Successive estimates of x_2^c and y_2^c are then

$1 \cdot 677\,17,$ $2 \cdot 401\,26;$ $1 \cdot 680\,79,$ $2 \cdot 402\,01;$ $1 \cdot 681\,00,$ $2 \cdot 402\,52;$
$1 \cdot 681\,04,$ $2 \cdot 402\,53$

and after this no further changes take place.

The second slight modification that can be made is to formula (7A.2). When using this formula, x_1^c has already been found, and on the basis that this is a better estimate of x_1 than is x_1^p, (7A.2) is changed to

$$y_1^c = y_0 + \frac{h}{2}\{g(t_0, x_0, y_0) + g(t_1, x_1^c, y_1^p)\}.$$

The successive estimates of x_1^c are calculated as before, but when finding successive estimates of y_1^c, the latest value found for x_1^c is always used in the g function.

Thus, when $t = 0 \cdot 1$, $x_1^P = 1 \cdot 3$ and $y_1^P = 2 \cdot 1$ as before.

First estimate of $x_1^C = 1 + \dfrac{0 \cdot 1}{2}(3 + 3 \cdot 3) = 1 \cdot 315$, again as before.

First estimate of $y_1^C = 2 + \dfrac{0 \cdot 1}{2}(1 + 2 \cdot 045) = 2 \cdot 152$.

Continue the process until no further changes take place in x_1 and y_1, working to 5 decimal places.

****

11A

Second estimates of x_1^C and y_1^C are $1 \cdot 318\,35$ and $2 \cdot 150\,15$.
The third estimates are $1 \cdot 318\,42$ and $2 \cdot 150\,26$.   The next are $1 \cdot 318\,43$
and $2 \cdot 150\,25$ and after this no further change takes place.

FRAME 12

For future reference, the formulae used in this process can be collected together.   Thus we have

$$x_1^P = x_0 + hf(t_0, x_0, y_0)$$
$$y_1^P = y_0 + hg(t_0, x_0, y_0)$$

$$x_1^C = x_0 + \frac{h}{2}\{f(t_0, x_0, y_0) + f(t_1, x_1^P, y_1^P)\}$$

$$y_1^C = y_0 + \frac{h}{2}\{g(t_0, x_0, y_0) + g(t_1, x_1^C, y_1^P)\}$$

$$(x_1^C)_{new} = x_0 + \frac{h}{2}\{f(t_0, x_0, y_0) + f[t_1, (x_1^C)_{old}, (y_1^C)_{old}]\}$$

$$(y_1^C)_{new} = y_0 + \frac{h}{2}\{g(t_0, x_0, y_0) + g[t_1, (x_1^C)_{new}, (y_1^C)_{old}]\}$$

Now carry through the calculation for x_2 and y_2.

****

12A

$x_2^P = 1 \cdot 655\,30,$ $y_2^P = 2 \cdot 350\,75$
Successive estimates of x_2^C and y_2^C are

$1 \cdot 677\,17,\quad 2 \cdot 404\,54;\quad\quad 1 \cdot 680\,95,\quad 2 \cdot 402\,42;\quad\quad 1 \cdot 681\,03,\quad 2 \cdot 402\,54;$
$1 \cdot 681\,04,\quad 2 \cdot 402\,53$

with no further changes.

FRAME 13

Runge-Kutta Method

The method described after modified Euler in the previous programme was Milne's.   However, as this was found to be not self-starting, we will here postpone it until after Runge-Kutta as this method is one from which starting values for Milne can be obtained.

You will remember that, for the equation $\dfrac{dy}{dx} = f(x, y)$,

$y_1 = y_0 + (k_1 + 2k_2 + 2k_3 + k_4)/6$ where $k_1 = hf(x_0, y_0)$,

$k_2 = hf(x_0 + \frac{1}{2}h, y_0 + \frac{1}{2}k_1)$, $k_3 = hf(x_0 + \frac{1}{2}h, y_0 + \frac{1}{2}k_2)$,
$k_4 = hf(x_0 + h, y_0 + k_3)$.

When applying Runge–Kutta to the differential equations, $\dot{x} = f(t, x, y)$, $\dot{y} = g(t, x, y)$ the following sets of algebraic equations, found by making use of Taylor's series in two variables, are used:

$k_1 = hf(t_0, x_0, y_0)$ $m_1 = hg(t_0, x_0, y_0)$

$k_2 = hf(t_0 + \frac{1}{2}h, x_0 + \frac{1}{2}k_1, y_0 + \frac{1}{2}m_1)$ $m_2 = hg(t_0 + \frac{1}{2}h, x_0 + \frac{1}{2}k_1, y_0 + \frac{1}{2}m_1)$

$k_3 = hf(t_0 + \frac{1}{2}h, x_0 + \frac{1}{2}k_2, y_0 + \frac{1}{2}m_2)$ $m_3 = hg(t_0 + \frac{1}{2}h, x_0 + \frac{1}{2}k_2, y_0 + \frac{1}{2}m_2)$

$k_4 = hf(t_0 + h, x_0 + k_3, y_0 + m_3)$ $m_4 = hg(t_0 + h, x_0 + k_3, y_0 + m_3)$

$x_1 = x_0 + (k_1 + 2k_2 + 2k_3 + k_4)/6$ $y_1 = y_0 + (m_1 + 2m_2 + 2m_3 + m_4)/6$

The calculations are done in the order k_1, m_1, k_2, m_2, k_3, m_3, k_4, m_4; x_1, y_1.

FRAME 14

Again using the equations $\dot{x} = -t + x + y$, $\dot{y} = 2t + 3x - y$, with $x = 1$, $y = 2$ when $t = 0$, we have $f(t, x, y) = -t + x + y$, $g(t, x, y) = 2t + 3x - y$, and so, to find x and y when $t = 0 \cdot 1$,

$k_1 = 0 \cdot 1\, f(0, 1, 2) = 0 \cdot 1 \times 3 = 0 \cdot 3$
$m_1 = 0 \cdot 1\, g(0, 1, 2) = 0 \cdot 1 \times 1 = 0 \cdot 1$
$k_2 = 0 \cdot 1\, f(0 + \frac{1}{2} \times 0 \cdot 1, 1 + \frac{1}{2} \times 0 \cdot 3, 2 + \frac{1}{2} \times 0 \cdot 1)$
    $= 0 \cdot 1\, f(0 \cdot 05, 1 \cdot 15, 2 \cdot 05) = 0 \cdot 1 \times 3 \cdot 15 = 0 \cdot 315$
$m_2 = 0 \cdot 1\, g(0 \cdot 05, 1 \cdot 15, 2 \cdot 05) = 0 \cdot 1 \times 1 \cdot 50 = 0 \cdot 150$
$k_3 = 0 \cdot 1\, f(0 + \frac{1}{2} \times 0 \cdot 1, 1 + \frac{1}{2} \times 0 \cdot 315, 2 + \frac{1}{2} \times 0 \cdot 150)$
    $= 0 \cdot 1\, f(0 \cdot 05, 1 \cdot 1575, 2 \cdot 075) = 0 \cdot 1 \times 3 \cdot 1825 = 0 \cdot 318\,25$

Now find m_3, k_4 and m_4 to 5 decimal places.

****

14A

$m_3 = 0 \cdot 1\, g(0 \cdot 05, 1 \cdot 1575, 2 \cdot 075) = 0 \cdot 1 \times 1 \cdot 4975 = 0 \cdot 149\,75$

$k_4 = 0 \cdot 1\, f(0 + 0 \cdot 1, 1 + 0 \cdot 318\,25, 2 + 0 \cdot 149\,75)$
    $= 0 \cdot 1\, f(0 \cdot 1, 1 \cdot 318\,25, 2 \cdot 149\,75) = 0 \cdot 1 \times 3 \cdot 3680 = 0 \cdot 336\,80$

$m_4 = 0 \cdot 1\, g(0 \cdot 1, 1 \cdot 318\,25, 2 \cdot 149\,75) = 0 \cdot 1 \times 2 \cdot 0050 = 0 \cdot 200\,50$

FRAME 15

Then $x_1 = 1 + (0 \cdot 3 + 2 \times 0 \cdot 315 + 2 \times 0 \cdot 318\,25 + 0 \cdot 336\,80)/6 = 1 \cdot 317\,22$
and $y_1 = 2 + (0 \cdot 1 + 2 \times 0 \cdot 150 + 2 \times 0 \cdot 149\,75 + 0 \cdot 200\,50)/6 = 2 \cdot 150\,00$.

Now find x_2 and y_2 by this method, working to 5 decimal places.

****

15A

$\begin{cases} k_1 = 0 \cdot 336\,72 \\ m_1 = 0 \cdot 200\,17 \end{cases}$ $\begin{cases} k_2 = 0 \cdot 358\,57 \\ m_2 = 0 \cdot 250\,67 \end{cases}$ $\begin{cases} k_3 = 0 \cdot 362 \cdot 18 \\ m_3 = 0 \cdot 251\,42 \end{cases}$ $\begin{cases} k_4 = 0 \cdot 388\,08 \\ m_4 = 0 \cdot 303\,68 \end{cases}$ $\begin{cases} x_2 = 1 \cdot 678\,27 \\ y_2 = 2 \cdot 401\,34 \end{cases}$

The next stage in the calculation then gives $x_3 = 2 \cdot 098\,65$, $y_3 = 2 \cdot 759\,08$.

Milne's Method

For the single differential equation $y' = f(x, y)$, the algebraic equations used in Milne's method were $y_4^P = y_0 + \frac{4h}{3}(2f_1 - f_2 + 2f_3)$, $y_4^C = y_2 + \frac{h}{3}(f_2 + 4f_3 + f_4^P)$ being, respectively, the predictor and corrector formulae.

For the pair of differential equations $\dot{x} = f(t, x, y)$, $\dot{y} = g(t, x, y)$, the equations of the Milne method are adapted as follows:

$$x_4^P = x_0 + \frac{4h}{3}(2f_1 - f_2 + 2f_3), \qquad y_4^P = y_0 + \frac{4h}{3}(2g_1 - g_2 + 2g_3),$$

$$x_4^C = x_2 + \frac{h}{3}(f_2 + 4f_3 + f_4^P) \quad (17.1), \qquad y_4^C = y_2 + \frac{h}{3}(g_2 + 4g_3 + g_4^P) \quad (17.2)$$

You will realise from these equations that it is necessary to know all of $x_0, x_1, x_2, x_3, y_0, y_1, y_2$ and y_3 before this method can be applied. As before with Milne's method, some other method of starting the solution is necessary. Also, as before, Runge-Kutta or Taylor series are possible ways of doing this.

Collecting up the results from Runge-Kutta for our example $\dot{x} = -t + x + y$, $\dot{y} = 2t + 3x - y$ with $t_0 = 0$, $x_0 = 1$, $y_0 = 2$, we have

t	0	0·1	0·2	0·3
x	1	1·317 22	1·678 27	2·098 65
y	2	2·150 00	2·401 34	2·759 08
f	3	3·367 22	3·879 61	4·557 73
g	1	2·001 66	3·033 47	4·136 87

Then, using the Milne predictor formulae,

$$x_4^P = 1 + \frac{4 \times 0 \cdot 1}{3}(2 \times 3 \cdot 367\,22 - 3 \cdot 879\,61 + 2 \times 4 \cdot 557\,73) = 2 \cdot 596\,04$$

$$y_4^P = 2 + \frac{4 \times 0 \cdot 1}{3}(2 \times 2 \cdot 001\,66 - 3 \cdot 033\,47 + 2 \times 4 \cdot 136\,87) = 3 \cdot 232\,48$$

These values give $f_4^P = 5 \cdot 428\,52$ and $g_4^P = 5 \cdot 355\,64$. Then from the x corrector formula

$$x_4^C = 1 \cdot 678\,27 + \frac{0 \cdot 1}{3}(3 \cdot 879\,61 + 4 \times 4 \cdot 557\,73 + 5 \cdot 428\,52) = 2 \cdot 596\,24.$$

What will be the value of y_4^C?

****

$$2 \cdot 401\,34 + \frac{0 \cdot 1}{3}(3 \cdot 033\,47 + 4 \times 4 \cdot 136\,87 + 5 \cdot 355\,64) = 3 \cdot 232\,56$$

As there is a fair amount of change, particularly in x, from the predicted values, these corrected values should now be used in the calculation of f_4 and g_4 and a second estimate of x_4^c and y_4^c made with (17.1) and (17.2) replaced by

$$x_4^c = x_2 + \frac{h}{3}(f_2 + 4f_3 + f_4^c), \quad y_4^c = y_2 + \frac{h}{3}(g_2 + 4g_3 + g_4^c)$$

Doing this gives $f_4^c = 5 \cdot 428\,80$ and $g_4^c = 5 \cdot 356\,16$, and then second estimates of x_4^c and y_4^c are $x_4^c = 2 \cdot 596\,25$, $y_4^c = 3 \cdot 232\,58$ and after this no further changes take place.

What are the best values you can get for x_5 and y_5 by this method?

****

Best values for f_4 and g_4 are $5 \cdot 428\,83$ and $5 \cdot 356\,17$.

$x_5^p = 3 \cdot 191\,77,$ $y_5^p = 3 \cdot 835\,65,$ $f_5^p = 6 \cdot 527\,42,$ $g_5^p = 6 \cdot 739\,66$

$x_5^c = 3 \cdot 192\,00,$ $y_5^c = 3 \cdot 835\,79,$ $f_5^c = 6 \cdot 527\,79,$ $g_5^c = 6 \cdot 740\,21$

$x_5^c = 3 \cdot 192\,01,$ $y_5^c = 3 \cdot 835\,80.$

Modified Milne Method

Just as in the previous programme, Milne's method can be modified to eliminate or reduce the need to use repeatedly the corrector equation.

Can you write down the sets of equations that would be used in the modified Milne method to give x_n and y_n for $n > 4$?

****

$$x_n^p = x_{n-4} + \frac{4h}{3}(2f_{n-3} - f_{n-2} + 2f_{n-1}), \quad y_n^p = y_{n-4} + \frac{4h}{3}(2g_{n-3} - g_{n-2} + 2g_{n-1})$$

$$x_n^m = x_n^p - \frac{28}{29}(x_{n-1}^p - x_{n-1}^c), \qquad y_n^m = y_n^p - \frac{28}{29}(y_{n-1}^p - y_{n-1}^c)$$

$$x_n^c = x_{n-2} + \frac{h}{3}(f_{n-2} + 4f_{n-1} + f_n^m), \qquad y_n^c = y_{n-2} + \frac{h}{3}(g_{n-2} + 4g_{n-1} + g_n^m)$$

$$x_n = x_n^c + \frac{1}{29}(x_n^p - x_n^c), \qquad y_n = y_n^c + \frac{1}{29}(y_n^p - y_n^c)$$

Again, as was noticed in the previous programme, some authors omit the two equations for x_n and y_n, preferring instead to repeat the use of the equations for x_n^c and y_n^c as much as is necessary.

Now apply the equations to find x_5 and y_5 for the example in FRAME 18.

****

$$\begin{cases} x_5^p = 3 \cdot 191\ 77 \\ y_5^p = 3 \cdot 835\ 65 \end{cases} \quad \begin{cases} x_5^m = 3 \cdot 191\ 97 \\ y_5^m = 3 \cdot 835\ 75 \end{cases} \quad \begin{cases} x_5^c = 3 \cdot 192\ 00 \\ y_5^c = 3 \cdot 835\ 80 \end{cases} \quad \begin{cases} x_5 = 3 \cdot 192\ 01 \\ y_5 = 3 \cdot 835\ 81 \end{cases}$$

Hamming's Method

Again, if many values of x and y are to be calculated, instability may be present in the corrector equations of Milne, and Hamming replaces these by

$$x_4^c = \frac{1}{8}\left\{ 9x_3 - x_1 + 3h(f_4^p + 2f_3 - f_2) \right\}$$

$$y_4^c = \frac{1}{8}\left\{ 9y_3 - y_1 + 3h(g_4^p + 2g_3 - g_2) \right\}$$

Applying these formulae to the figures in FRAME 18 gives

$$x_4^c = \frac{1}{8}\left\{ 9 \times 2 \cdot 098\ 65 - 1 \cdot 317\ 22 + 3 \times 0 \cdot 1(5 \cdot 428\ 52 + 2 \times 4 \cdot 557\ 73 - 3 \cdot 879\ 61) \right\}$$
$$= 2 \cdot 596\ 24$$

$$y_4^c = \frac{1}{8}\left\{ 9 \times 2 \cdot 759\ 08 - 2 \cdot 150\ 00 + 3 \times 0 \cdot 1(5 \cdot 355\ 64 + 2 \times 4 \cdot 136\ 87 - 3 \cdot 033\ 47) \right\}$$
$$= 3 \cdot 232\ 56$$

Now apply these corrector formulae again, but replacing f_4^p and g_4^p by f_4^c and g_4^c, and see if any further difference results.
****

$f_4^c = 5 \cdot 428\ 80,$ $\qquad g_4^c = 5 \cdot 356\ 16$

New $x_4^c = 2 \cdot 596\ 25,$ $\qquad y_4^c = 3 \cdot 232\ 58$

Modified Hamming's Method

What will be the equations for x_n and y_n when Hamming's method is modified as in the last programme?
****

$$x_n^p = x_{n-4} + \frac{4h}{3}(2f_{n-3} - f_{n-2} + 2f_{n-1})$$
$$y_n^p = y_{n-4} + \frac{4h}{3}(2g_{n-3} - g_{n-2} + 2g_{n-1})$$

$$x_n^m = x_n^p - \frac{112}{121}(x_{n-1}^p - x_{n-1}^c)$$
$$y_n^m = y_n^p - \frac{112}{121}(y_{n-1}^p - y_{n-1}^c)$$

$$x_n^c = \frac{1}{8}\left\{ 9x_{n-1} - x_{n-3} + 3h(f_n^m + 2f_{n-1} - f_{n-2}) \right\}$$
$$y_n^c = \frac{1}{8}\left\{ 9y_{n-1} - y_{n-3} + 3h(g_n^m + 2g_{n-1} - g_{n-2}) \right\}$$

$$x_n = x_n^c + \frac{9}{121}(x_n^p - x_n^c)$$

$$y_n = y_n^c + \frac{9}{121}(y_n^p - y_n^c)$$

Applying the formulae to the example under consideration gives, for calculating x_5 and y_5,

$$\begin{cases} x_5^p = 3 \cdot 191\ 77 \\ y_5^p = 3 \cdot 835\ 65 \end{cases} \begin{cases} x_5^m = 3 \cdot 192\ 03 \\ y_5^m = 3 \cdot 835\ 74 \end{cases} \begin{cases} x_5^c = 3 \cdot 192\ 03 \\ y_5^c = 3 \cdot 835\ 82 \end{cases} \begin{cases} x_5 = 3 \cdot 192\ 01 \\ y_5 = 3 \cdot 835\ 81 \end{cases}$$

Comparison of the Results

The example chosen in this programme to illustrate the various methods can be solved analytically.   What do you get when you do this?   (In view of what is coming later, don't use Laplace transforms here.)
****

Eliminating, say, y leads to $\ddot{x} - 4x = t - 1$.
Then $x = Ae^{2t} + Be^{-2t} + \frac{1}{4}(1 - t)$, $y = Ae^{2t} - 3Be^{-2t} - \frac{1}{4}(2 - 5t)$
Using the conditions given
$x = \frac{19}{16} e^{2t} - \frac{7}{16} e^{-2t} + \frac{1}{4}(1 - t)$, $y = \frac{19}{16} e^{2t} + \frac{21}{16} e^{-2t} - \frac{1}{4}(2 - 5t)$

Inserting $t = 0 \cdot 1(0 \cdot 1)0 \cdot 5$ into these equations and collecting up the various values previously obtained, we have, as far as we have taken the working

t		Euler	Modified Euler	Runge-Kutta	Milne or Modified Milne	Hamming or Modified Hamming	Analytical
0·1	x	1·3	1·318 43	1·317 22			1·317 22
	y	2·1	2·150 25 (6)	2·150 00			2·150 00
0·2	x	1·63	1·681 04	1·678 27			1·678 27
	y	2·3	2·402 53 (4)	2·401 34			2·401 33
0·3	x	2·003		2·098 65			2·098 66
	y	2·599		2·759 08			2·759 08
0·4	x				2·596 25	2·596 25	2·596 25
	y				3·232 58	3·232 58	3·232 57
0·5	x				3·192 01	3·192 01	3·192 01
	y				3·835 80*	3·835 81	3·835 80

The figure 2•150 25 (6) means that the result was 2•150 25 or 2•150 26, according to which form of the corrector was used.

* The modified Milne method gave 3•835 81.

More Complicated Systems of Equations

The various methods have been illustrated using a pair of equations of the form $\dot{x} = f(t, x, y)$, $\dot{y} = g(t, x, y)$.

It is a comparatively easy matter to see how the methods can be extended to systems involving more equations, for example $\dot{x} = f_1(t, x, y, z)$, $\dot{y} = f_2(t, x, y, z)$, $\dot{z} = f_3(t, x, y, z)$.

As no new theory is involved, examples of these more complicated systems will not be taken here.   You will, however, come across the idea in the miscellaneous examples at the end of the programme.

Examples

Now try the following examples before continuing with the next part of this programme.   Answers are given in FRAME 29.

1. Working to 5 decimal places, find, as accurately as possible, the values of x and y when t = 0•1 given that $\dot{x} = x^2 + y$, $\dot{y} = 2x - 3y + t$ and that x = 0•2 and y = 0•1 when t = 0.

2. Use Runge-Kutta to solve, at t = 1•1, $\dot{x} = -0•3t + 0•1x + 0•2y$, $\dot{y} = 0•2t + 0•2x - 0•1y$, given that x = 1 and y = 2 when t = 1.

    Then, assuming the values

t	1•2	1•3
x	1•035 23	1•049 29
y	2•044 31	2•069 59

    use Milne to find x and y when t = 1•4 and Hamming to find them when t = 1•5.

3. Given the equations

    $$L_1 \frac{di_1}{dt} + R_1 i_1 + M \frac{di_2}{dt} = e_1, \qquad M \frac{di_1}{dt} + L_2 \frac{di_2}{dt} + R_2 i_2 = e_2$$

    (quoted in FRAME 1 for a coupled electric circuit) how would you rearrange them in order to make them suitable for treatment such as described in this programme?

Answers to Examples

1. Runge-Kutta gives x = 0•214 95, y = 0•114 50

2.   When t = 1·1,   x = 1·018 80,   y = 2·021 09
     When t = 1·4,   x = 1·061 00,   y = 2·096 87
     When t = 1·5,   x = 1·070 38,   y = 2·126 07

3.   It is necessary to solve the two equations simultaneously for $\dfrac{di_1}{dt}$

and $\dfrac{di_2}{dt}$.   Thus $\dfrac{di_1}{dt} = \dfrac{L_2(e_1 - R_1 i_1) - M(e_2 - R_2 i_2)}{L_1 L_2 - M^2}$

$\dfrac{di_2}{dt} = \dfrac{L_1(e_2 - R_2 i_2) - M(e_1 - R_1 i_1)}{L_1 L_2 - M^2}$

Second Order Differential Equations - Initial Value Problems

As this is a programme on numerical methods, you may have wondered why you were asked to find the analytical solution in FRAME 25, instead of just having it quoted.   The reason was to remind you of the fact that when one solves analytically two first order simultaneous differential equations, the method of solution entails finding, and then solving, a second order d.e.   This suggests that if we start off with a second order d.e. then it should be possible to decompose it into two first order simultaneous d.e.'s.

Now, given two simultaneous d.e.'s, say for x and y in terms of t, then the second order equation that results when y, say, is eliminated, is unique.   For example, $\dot{x} = 2x - y$, $\dot{y} = x + 3y$ lead to $\ddot{x} - 5\dot{x} + 7x = 0$.

But $\ddot{x} - 5\dot{x} + 7x = 0$ can be decomposed into an endless variety of simultaneous first order d.e.'s.   Some possibilities are

$$\left.\begin{array}{l} \dot{x} - 2x + y = 0 \\ -x + \dot{y} - 3y = 0 \end{array}\right\}, \quad \left.\begin{array}{l} \dot{x} - 7y = 0 \\ x + \dot{y} - 5y = 0 \end{array}\right\}, \quad \left.\begin{array}{l} 2\dot{x} - x + 38y = 0 \\ -x + 4\dot{y} - 18y = 0 \end{array}\right\}, \quad \left.\begin{array}{l} \dot{x} + x - 13y = 0 \\ x + \dot{y} - 6y = 0 \end{array}\right\}$$

Now, given initial conditions, you know how to solve, numerically, any of these sets of equations.   In other words, you already know how to solve a second order equation. All you have to do is to decompose it into two simultaneous first order differential equations and use the methods of the first part of this programme.

The only remaining question is - which of the many pairs of first order d.e.'s should be used?   Theoretically it doesn't matter, any pair of the forms in the last frame will do.   Practically, one just chooses as simple a combination as possible and the normal procedure is to let one equation be $\dot{x} = y$.

Then $\ddot{x} = \dot{y}$ and so $\ddot{x} - 5\dot{x} + 7x = 0$ becomes $\dot{y} - 5y + 7x = 0$, i.e., $\dot{y} = -7x + 5y$ (other letters, instead of y, may of course be used.)

Working on the same lines, decompose

(i) $\ddot{x} + 3\dot{x} - 4x = 0$,   (ii) $2\ddot{x} - 3\dot{x} + 5x = 0$,   (iii) $\ddot{x} - 3\dot{x} + x = \sin t$,

(iv) $3\ddot{y} - 2\dot{y} - 5y = e^{-t}$,   (v) $\dfrac{d^2u}{dr^2} + 2\dfrac{du}{dr} + 4u = r^2 + 3$

into two simultaneous first order d.e.'s.

****

32A

(i) $\begin{cases} \dot{x} = y \\ \dot{y} = 4x - 3y \end{cases}$ (ii) $\begin{cases} \dot{x} = y \\ \dot{y} = -(5x - 3y)/2 \end{cases}$ (iii) $\begin{cases} \dot{x} = y \\ \dot{y} = -x + 3y + \sin t \end{cases}$

(iv) $\begin{cases} \dot{y} = x \\ \dot{x} = (2x + 5y + e^{-t})/3 \end{cases}$ (v) $\begin{cases} \dfrac{du}{dr} = x \\ \dfrac{dx}{dr} = -4u - 2x + r^2 + 3 \end{cases}$

FRAME 33

As an example, let us return to the equation

$$L\dfrac{d^2q}{dt^2} + \left\{ A + B\left(\dfrac{dq}{dt}\right)^2 \right\}\dfrac{dq}{dt} + \dfrac{1}{C}q \;=\; E \sin \omega t$$

mentioned in the last programme and find q when t = 0·001 and 0·002
for L = 1, A = 10, B = 0·01, C = 2 × 10⁻⁶, E = 100, ω = 100π,
assuming that i = q = 0 when t = 0.

Remembering that i = dq/dt, and inserting the values of the constants,
what will be the linear simultaneous first order equations that will be
used here?   **

33A

$\dfrac{dq}{dt} = i$ $\dfrac{di}{dt} = 100 \sin 100\pi t - (10 + 0\cdot01i^2)i - 5 \times 10^5 q$

FRAME 34

Using the notation $\dfrac{di}{dt} = f(t, i, q)$, $\dfrac{dq}{dt} = g(t, i, q)$,   the equations are
now in the standard form that we had earlier.   However, you will notice
that here g is actually a function of i only.

What method of solution do you suggest to find the required values?
****

34A

*Runge-Kutta.   There is only one pair of starting values, so Milne and
Hamming are no use.   Runge-Kutta is the most accurate of the other
methods.*

FRAME 35

For the first step:

k_1 = 0·001 f(0, 0, 0) = 0
      m_1 = 0·001 g(0, 0, 0) = 0

k_2 = 0·001 f(0·0005, 0, 0)
     = 0·001 × 100 sin $\dfrac{\pi}{20}$ = 0·015 643 5
          m_2 = 0·001 g(0·0005, 0, 0) = 0

$k_3 = 0 \cdot 001 \ f(0 \cdot 0005, \ 0 \cdot 007 \ 821 \ 8, \ 0)$

$\qquad = 0 \cdot 001 \left[100 \ \sin \dfrac{\pi}{10} - (10 + 0 \cdot 01 \times 0 \cdot 007 \ 821 \ 8^2)0 \cdot 007 \ 821 \ 8 \right] = 0 \cdot 015 \ 565 \ 4$

$\qquad\qquad m_3 = 0 \cdot 001 \ g(0 \cdot 0005, \ 0 \cdot 007 \ 821 \ 8, \ 0) = 0 \cdot 000 \ 007 \ 821 \ 8$

$k_4 = 0 \cdot 001 \ f(0 \cdot 001, \ 0 \cdot 015 \ 565 \ 4, \ 0 \cdot 000 \ 007 \ 821 \ 8)$

$\qquad = 0 \cdot 001 \left[100 \ \sin \dfrac{\pi}{10} - (10 + 0 \cdot 01 \times 0 \cdot 015 \ 565 \ 4^2)0 \cdot 015 \ 565 \ 4 \right.$

$\qquad\qquad\qquad\qquad\qquad\qquad \left. - \ 5 \times 10^5 \times 0 \cdot 000 \ 007 \ 821 \ 8 \right]$

$\qquad = 0 \cdot 026 \ 835 \ 2$

$\qquad\qquad m_4 = 0 \cdot 001 \ g(0 \cdot 001, \ 0 \cdot 015 \ 565 \ 4, \ 0 \cdot 000 \ 007 \ 821 \ 8)$

$\qquad\qquad\quad = 0 \cdot 000 \ 015 \ 565 \ 4$

$\qquad\qquad i = 0 \cdot 014 \ 875 \ 5 \qquad\qquad\qquad q = 0 \cdot 000 \ 005 \ 201 \ 5$

Now carry the calculation through the next step, noting that

$$\sin \frac{\pi}{10} = 0 \cdot 309 \ 018, \qquad \sin \frac{3\pi}{20} = 0 \cdot 453 \ 990, \qquad \sin \frac{\pi}{5} = 0 \cdot 587 \ 785$$

****

$k_1 = 0 \cdot 001 \ f(0 \cdot 001, \ 0 \cdot 014 \ 875 \ 5, \ 0 \cdot 000 \ 005 \ 201 \ 5) = 0 \cdot 028 \ 152 \ 3$
$\quad m_1 = 0 \cdot 001 \ g(0 \cdot 01, \ 0 \cdot 014 \ 875 \ 5, \ 0 \cdot 000 \ 005 \ 201 \ 5) = 0 \cdot 000 \ 014 \ 875 \ 5$

$k_2 = 0 \cdot 001 \ f(0 \cdot 0015, \ 0 \cdot 028 \ 951 \ 6, \ 0 \cdot 000 \ 012 \ 639 \ 2) = 0 \cdot 387 \ 899$
$\quad m_2 = 0 \cdot 001 \ g(0 \cdot 0015, \ 0 \cdot 028 \ 951 \ 6, \ 0 \cdot 000 \ 012 \ 639 \ 2) = 0 \cdot 000 \ 028 \ 951 \ 6$

$k_3 = 0 \cdot 001 \ f(0 \cdot 0015, \ 0 \cdot 034 \ 270 \ 4, \ 0 \cdot 000 \ 019 \ 677 \ 3) = 0 \cdot 035 \ 217 \ 6$
$\quad m_3 = 0 \cdot 001 \ g(0 \cdot 0015, \ 0 \cdot 034 \ 270 \ 4, \ 0 \cdot 000 \ 019 \ 677 \ 3) = 0 \cdot 000 \ 034 \ 270 \ 4$

$k_4 = 0 \cdot 001 \ f(0 \cdot 002, \ 0 \cdot 050 \ 093 \ 1, \ 0 \cdot 000 \ 039 \ 471 \ 9) = 0 \cdot 038 \ 541 \ 6$
$\quad m_4 = 0 \cdot 001 \ g(0 \cdot 002, \ 0 \cdot 050 \ 093 \ 1, \ 0 \cdot 000 \ 039 \ 471 \ 9) = 0 \cdot 000 \ 050 \ 093 \ 1$

$\qquad i = 0 \cdot 050 \ 660 \ 3 \qquad\qquad\qquad q = 0 \cdot 000 \ 037 \ 419 \ 3$

The question did not ask us to continue with the solution of this equation. If continuation has to be done, Runge-Kutta would be necessary for one more step, after which a change could be made to Milne or Hamming if desired.

The Special Case when the First Derivative is missing

Another second order equation mentioned at the beginning of the previous programme was the equation for a simple pendulum, i.e. $\ddot{\theta} + \dfrac{g}{\ell} \sin \theta = 0$. This can be solved, given initial conditions and the values of g and ℓ, by letting $\dot{\theta} = p$, say, then the simultaneous equations are: $\dot{p} = - \dfrac{g}{\ell} \sin \theta$, $\dot{\theta} = p$.

Each step will involve finding the values of both p and θ, even although p may not be required. In the example just done the value of the first derivative was required as it occurred in the function f. But

328

here it doesn't and so, if it is possible to find values of θ without the corresponding values of p, it may be possible to save some work.

FRAME 38

Taking a more general second order equation in which the first derivative is missing, let us look for a moment at $y'' = f(x, y)$ (38.1)
A start towards a numerical solution of this is made by taking the formula
$\frac{1}{4h^2}\{f(a + 2h) - 2f(a) + f(a - 2h)\}$ $\left(\text{see (26A.1), page 244}\right)$ for a
second derivative and rewriting it as

$$y_n'' \simeq \frac{1}{h^2}(y_{n+1} - 2y_n + y_{n-1})$$ (38.2)

This is accomplished by replacing h by $\frac{1}{2}$h and using a slightly different notation.   (38.1) now becomes, approximately,

$y_{n+1} - 2y_n + y_{n-1} = h^2 f(x_n, y_n)$ or $y_{n+1} = 2y_n - y_{n-1} + h^2 f(x_n, y_n)$

This equation is used to predict y_{n+1} and so it will be written as

$$y_{n+1}^p = 2y_n - y_{n-1} + h^2 f_n$$

When n = 1, it becomes $y_2^p = 2y_1 - y_0 + h^2 f_1$ and so requires two starting values of y.   The first of these will be given and the second must be obtained by some other method, e.g. a Taylor or Maclaurin series.

FRAME 39

To find a more accurate value of y_{n+1}, a corrector formula has to be used. This is found by taking a more accurate expression for y_n'' than
$\frac{1}{h^2}(y_{n+1} - 2y_n + y_{n-1})$.   To find this more accurate expression, y_{n+1}
and y_{n-1} are expanded in Taylor series about y_n, i.e., in powers of h.
What will $y_{n+1} - 2y_n + y_{n-1}$ become when this is done, retaining powers
of h up to h^5?

****

39A

$$y_{n+1} = y_n + hy_n' + \frac{h^2}{2!} y_n'' + \frac{h^3}{3!} y_n''' + \frac{h^4}{4!} y_n^{iv} + \frac{h^5}{5!} y_n^{v} + \dots\dots$$

$$y_{n-1} = y_n - hy_n' + \frac{h^2}{2!} y_n'' - \frac{h^3}{3!} y_n''' + \frac{h^4}{4!} y_n^{iv} - \frac{h^5}{5!} y_n^{v} + \dots\dots$$

$$y_{n+1} - 2y_n + y_{n-1} = h^2 y_n'' + \frac{h^4}{12} y_n^{iv} + \dots\dots\dots$$

FRAME 40

Extracting y_n'' from this and substituting it into (38.1) gives, for the
point (x_n, y_n), $\dfrac{y_{n+1} - 2y_n + y_{n-1}}{h^2}$ $- \left(\dfrac{h^2}{12} y_n^{iv} + \dots\right) = f_n$ or,

approximately, $y_{n+1} - 2y_n + y_{n-1} = h^2 f_n + \dfrac{h^4}{12} y_n^{iv}$.

Now as $y'' = f(x, y)$, $y^{iv} = f''(x, y)$ and so

$$y_{n+1} - 2y_n + y_{n-1} = h^2 f_n + \frac{h^4}{12} f_n''$$

Just as y_n'' can be expressed approximately in terms of y_{n+1}, y_n, y_{n-1} and h, so also can f_n'' be expressed approximately in terms of f_{n+1}, f_n, f_{n-1} and h.   What will this expression be?

****

40A

$\dfrac{1}{h^2}(f_{n+1} - 2f_n + f_{n-1})$

FRAME 41

Therefore

$y_{n+1} - 2y_n + y_{n-1} = h^2 f_n + \dfrac{h^2}{12}(f_{n+1} - 2f_n + f_{n-1}) = \dfrac{h^2}{12}(f_{n+1} + 10f_n + f_{n-1})$

or $y_{n+1} = 2y_n - y_{n-1} + \dfrac{h^2}{12}(f_{n+1} + 10f_n + f_{n-1})$ and this is used as the corrector formula for y_{n+1}.   As such, it is written as

$$y_{n+1}^c = 2y_n - y_{n-1} + \frac{h^2}{12}(f_{n+1}^p + 10f_n + f_{n-1})$$

and this is used in conjunction with the predictor formula obtained in FRAME 38, i.e.,

$$y_{n+1}^p = 2y_n - y_{n-1} + h^2 f_n$$

FRAME 42

Let us now apply this process to the equation $\ddot{\theta} + \dfrac{g}{\ell} \sin \theta = 0$ with initial conditions $\theta = \dfrac{\pi}{2}$, $\dot{\theta} = 0$ when $t = 0$, taking $\dfrac{g}{\ell} = 5 \text{ s}^{-2}$,   to find θ when $t = 0 \cdot 05, 0 \cdot 10, 0 \cdot 15 \text{ s}$.

The equation then becomes $\ddot{\theta} = -5 \sin \theta$.

To start the process it is necessary to find, by other means, θ when $t = 0 \cdot 05$.   Suppose this is done by a Maclaurin series.   Find the relevant series as far as the term in t^6.   Also, write down the formulae for θ_2^p and θ_2^c .

****

42A

Differentiating $\ddot{\theta} = -5 \sin \theta$ successively w.r.t. t gives
$\dddot{\theta} = -5 \cos \theta \, \dot{\theta}$, etc.

Substitution then gives $(\ddot{\theta})_{t=0} = -5$, $(\dddot{\theta})_{t=0} = 0$, etc.

$\theta = \dfrac{\pi}{2} - \dfrac{5}{2}t^2 + \dfrac{25}{48}t^6$,

$$\theta_2^p = 2\theta_1 - \theta_0 + h^2 f_1, \qquad \theta_2^c = 2\theta_1 - \theta_0 + \frac{h^2}{12}(f_2^p + 10f_1 + f_0)$$

FRAME 43

Thus, when t = 0•05, θ = 1•564 546.

Also, when t = 0, f(θ) = −5 and when t = 0•05, f(θ) = −4•999 902.

So, θ_2^p = 2 × 1•564 546 − 1•570 796 + 0•05² (−4•999 902) = 1•545 796.

Then f_2^p = −4•998 44.

∴. θ_2^c = 2 × 1•564 546 − 1•570 796 + $\dfrac{0•05^2}{12}$ (−4•998 44 − 10 × 4•999 902 − 5)
    = 1•545 797

As before, when using predictor-corrector methods, this new value of θ_2
can be inserted into the corrector formula instead of θ_2^p.   If you do
this, no change takes place in the value of θ_2^c.

FRAME 44

Taking this value for θ_2, now find, by this method, θ_3, i.e., θ when
t = 0•15.   (sin 1•545 797 = 0•999 688,   sin 1•514 552 = 0•998 419.)
        ****

44A

θ_3^p = 1•514 552 θ_3^c = 1•514 553,   *with no further change.*

FRAME 45

Boundary Value Problems

Sometimes, instead of being given initial conditions for a second order
equation, one is given boundary values.   This means that instead of being
given the values of y and y' when x = 0, you know instead the values
of y when x = a (which can be, and often is, zero) and when x = b,
say.   For example, a beam of length ℓ might be freely supported at its
ends, these being at the same horizontal level.   Then, if the origin is
at one end of the beam, the boundary conditions can be y = 0 when
x = 0 and when x = ℓ.   A second illustration is that of a missile
launched from one point and required to hit a target whose position is
known.   The unknown here would be the angle of projection necessary for
this to happen.

The numerical technique for solving such a problem is not as easy as when
initial values are given.   Two methods for such problems will be
considered.   One of these is a trial and error method involving
interpolation and the other replaces the differential equation by a series
of algebraic equations. The use of the latter method is usually
restricted to linear differential equations, as these produce linear
algebraic equations.

FRAME 46

The Trial and Error or Shooting Method

In order to be able to apply the standard method of dealing with a second
order differential equation, it is necessary to have the value of y' as

331

well as that of y at the starting point, i.e. y_0' as well as y_0.   This is now unknown and so a guess is made as to its value.   If one guesses correctly, the solution will pass through the other boundary position. This, however, is extremely unlikely and almost certainly the solution will miss the other boundary point.

A second guess of y_0' is then made.   A new solution is now built up which will also almost certainly miss the second boundary point.   But the two solutions found for the differential equation are then used to give a more likely value of the unknown y_0'.   A third solution, this one using the latest value of y_0', can then be built up and it should land much nearer to the prescribed second boundary condition than either of the others.   One then continues until a solution is found that is satisfactory.

To illustrate the method, let us find the solution of $\dfrac{d^2y}{dx^2} - \left(\dfrac{dy}{dx}\right)^2 + x = 0$ subject to the boundary conditions that y = 1 when x = 0 and y = 1·8 when x = 0·4.

What simultaneous equations would you use to solve the given d.e.?   Also, can you make a guess as to a suitable value to try for y' when x = 0?
****

Let y' = u, *say, then the simultaneous equations are* u' = u² - x, y' = u.

Any guess at this stage is likely to be somewhat out.   The straight line joining the two points (0, 1) and (0·4, 1·8) has slope 2 and so, as a first try, this can quite well be taken.

The two values of x chosen here are not too far apart.   In actual practice the difference between them may be much greater.   These values have been taken so that the range can be divided into reasonably small intervals without having to use very many steps.

In the initial stages, very great accuracy is not necessary.   Taking h = 0·2,   say, there will be two steps between 0 and 0·4.   Write down the relevant Runge–Kutta equations and, working to 3 decimal places, find the value of y when x = 0·4,   taking the value 2 as suggested above for y' at x = 0.
****

Taking f(x, u, y) = u² - x   *and*   g(x, u, y) = u

$k_1 = hf(x_0, u_0, y_0),$ $m_1 = hg(x_0, u_0, y_0)$

$k_2 = hf(x_0 + \tfrac{1}{2}h, u_0 + \tfrac{1}{2}k_1, y_0 + \tfrac{1}{2}m_1),$ $m_2 = hg(x_0 + \tfrac{1}{2}h, u_0 + \tfrac{1}{2}k_1, y_0 + \tfrac{1}{2}m_1)$

$k_3 = hf(x_0 + \tfrac{1}{2}h, u_0 + \tfrac{1}{2}k_2, y_0 + \tfrac{1}{2}m_2),$ $m_3 = hg(x_0 + \tfrac{1}{2}h, u_0 + \tfrac{1}{2}k_2, y_0 + \tfrac{1}{2}m_2)$

$k_4 = hf(x_0 + h, u_0 + k_3, y_0 + m_3),$ $m_4 = hg(x_0 + h, u_0 + k_3, y_0 + m_3)$

$$u_1 = u_0 + (k_1 + 2k_2 + 2k_3 + k_4)/6, \qquad y_1 = y_0 + (m_1 + 2m_2 + 2m_3 + m_4)/6$$

As f is effectively a function of x and u only and g is one of u only, these equations can be abbreviated so that only the relevant items are included, i.e., if we write

$$u^2 - x = f(x, u), \qquad\qquad u = g(u)$$

then $k_1 = hf(x_0, u_0),$ $\qquad\qquad m_1 = hg(u_0)$

$\qquad\quad k_2 = hf(x_0 + \tfrac{1}{2}h, y_0 + \tfrac{1}{2}k_1),$ $m_2 = hg(u_0 + \tfrac{1}{2}k_1)$

$$etc.$$

2·540

An initial slope of 2 at $x = 0$ has produced a value of y which is too high. Although we can't be absolutely certain, this suggests that 2 is too large, so a second attempt is now made taking the initial slope to be 1, say, instead of 2. This would seem reasonable as 2 has taken us up nearly twice the amount required. If this is done, it is found that y at $x = 0·4$ is 1·497. This is too low but it would now seem reasonable that the value required for y' at $x = 0$ is between 1 and 2 as 1·8 is between 1·497 and 2·540.

The next value taken for the unknown y_0' is obtained from simple proportion, this effectively being linear interpolation. The new value of y_0' is thus taken as $1 + \dfrac{1·8 - 1·497}{2·540 - 1·497}(2 - 1) \simeq 1·29$

As this is expected to be a better value than either 1 or 2, the accuracy of the working can now be increased. Taking $h = 0·1$ and consequently four steps leads to the value 1·7094 for y when $x = 0·4$.

The two nearest values to 1·8 are now 1·497 (for $y' = 1$) and 1·7094 (for $y' = 1·29$) and a linear relation can be used again to give a next estimate of the initial value of y'. This is really equivalent to extrapolation which we normally avoid. In this case it may give a better estimate than using 2·540 (for $y' = 2$) in conjunction with 1·7094 as 2·540 is somewhat far away from 1·8 compared with the other values.

What do you get for the next estimate for y_0'?

$$*************************************$$

$$1 + \frac{1·8 - 1·497}{1·7094 - 1·497} \times (1·29 - 1) \simeq 1·414$$

This value of y_0' leads to $y = 1·819\,78$ at $x = 0·4$. We now use 1·29 giving 1·7094 and 1·414 giving 1·819 78 to form a next estimate of y_0'. This comes to be 1·380. Continuing the process along the same lines gives 1·788 79 for y instead of 1·8 and 1·3960 for the next trial value of y_0' .   This now gives 1·799 95 instead of 1·8, which may be sufficiently accurate, but if not, it is easy, by continuing the process, to increase the accuracy still more.

The Simultaneous Algebraic Equations Method

This method replaces

$$y_n' \quad \text{by} \quad (y_{n+1} - y_{n-1})/2h \qquad \text{(From (6.1), page 237)}$$

and $\quad y_n'' \quad$ by $\quad (y_{n+1} - 2y_n + y_{n-1})/h^2, \qquad$ (38.2)

applying these formulae for various values of x.   (Doing this, of course, introduces an error and you saw how to estimate this in the programme on differentiation.)   To see how it works suppose a non-uniform circular shaft, freely supported at its two ends, has masses of 1 t and 1·5 t placed on it at C and D, all dimensions being in mm. The diameters of the thin and thick sections are 50 mm and 100 mm.   The problem is to find the deflections at various points on the shaft.   Taking axes as shown, let the points at which the deflection is required be at

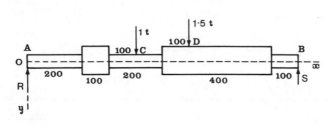

x = 100(100)900, denoted by $x_1, x_2, \ldots\ldots, x_9$ and let the corresponding deflections be $y_1, y_2, \ldots\ldots, y_9$. The deflections at the two ends, i.e., y_0 and y_{10}, will both be zero.   The modulus of elasticity for the material of the shaft will be required and a typical value of this, which will be taken here, is 200 kN mm^{-2} .

The differential equation that has to be solved is $\quad EI \dfrac{d^2y}{dx^2} = M \quad$ where E is the modulus of elasticity, I the second moment of area about a diameter of the cross-section of the shaft and M the anti-clockwise moment of the forces acting to the left of the section, about an axis through the centre of the section and perpendicular to the plane of the page.

It is first necessary to find the reaction at R.   The forces acting at C and D are respectively 1g kN and 1·5g kN, i.e. 9·81 kN and 14·715 kN taking g as 9·81 m s^{-2}.   A simple application of moments then gives R = 11·772 kN.   (Incidentally S = 12·753 kN, but this value will not be required here.)

It is next necessary to calculate the values of I and M at $x_1, x_2, \ldots, x_9$. Taking I first, the formula for this is $I = \dfrac{1}{4} Aa^2,$ A being the area of the cross-section and a the radius.   At points x_1 and x_4 this is

$\frac{1}{4} \pi \times 25^2 \times 25^2 \approx 306\ 796\ mm^4$ and at the points x_6, x_7 and x_8 it is

$\frac{1}{4} \pi \times 50^2 \times 50^2 \approx 4\ 908\ 734\ mm^4$.   At the points x_2, x_3, x_5 and x_9,   where the thickness of the shaft changes, it is taken as the average of these two values, i.e., 2 607 765 mm^4.

The values of M at x_1, x_2, ..., x_9 are

$-11 \cdot 772 \times 100$, $-11 \cdot 772 \times 200$, $-11 \cdot 772 \times 300$,
$-11 \cdot 772 \times 400$, $-11 \cdot 772 \times 500 + 9 \cdot 81 \times 100$,
$-11 \cdot 772 \times 600 + 9 \cdot 81 \times 200$,
$-11 \cdot 772 \times 700 + 9 \cdot 81 \times 300 + 14 \cdot 715 \times 100$,
$-11 \cdot 772 \times 800 + 9 \cdot 81 \times 400 + 14 \cdot 715 \times 200$,
$-11 \cdot 772 \times 900 + 9 \cdot 81 \times 500 + 14 \cdot 715 \times 300$.

i.e. $-1177 \cdot 2$, $-2354 \cdot 4$, $-3531 \cdot 6$, $-4708 \cdot 8$, $-4905 \cdot 0$, $-5101 \cdot 2$, $-3825 \cdot 9$, $-2550 \cdot 6$, $-1275 \cdot 3$ kN mm.   (The diagram shows the forces contributing to M at x_5.)

Applying EI $\dfrac{d^2y}{dx^2} = M$ at x_1 and replacing $\dfrac{d^2y}{dx^2}$ by $(y_0 - 2y_1 + y_2)/100^2$ $\left(\text{using (38.2) with } n = 1 \text{ and } h = 100\right)$ gives
$200 \times 306\ 796\ (y_0 - 2y_1 + y_2)/100^2 = -1177 \cdot 2$ i.e. $-2y_1 + y_2 = -0 \cdot 1919$
as $y_0 = 0$.

Similarly, at x_2, $200 \times 2\ 607\ 740 (y_1 - 2y_2 + y_3)/100^2 = -2354 \cdot 4$ i.e.
$y_1 - 2y_2 + y_3 = -0 \cdot 0451$.

What will be the corresponding equations at x_3 and at x_4?

****

55A

At x_3: $200 \times 2\ 607\ 740 (y_2 - 2y_3 + y_4)/100^2 = -3531 \cdot 6$
          i.e. $y_2 - 2y_3 + y_4 = -0 \cdot 0677$

At x_4: $200 \times 306\ 780\ (y_3 - 2y_4 + y_5)/100^2 = -4708 \cdot 8$
          i.e. $y_3 - 2y_4 + y_5 = -0 \cdot 7674$

Similar equations can be written down at x_5, x_6, x_7, x_8 and x_9.   Doing this and collecting up the four equations already obtained, we get

$$
\begin{aligned}
-2y_1 + y_2 &= -0 \cdot 1919 &&(56.1) \\
y_1 - 2y_2 + y_3 &= -0 \cdot 0451 &&(56.2) \\
y_2 - 2y_3 + y_4 &= -0 \cdot 0677 &&(56.3) \\
y_3 - 2y_4 + y_5 &= -0 \cdot 7674 &&(56.4) \\
y_4 - 2y_5 + y_6 &= -0 \cdot 0940 & \\
y_5 - 2y_6 + y_7 &= -0 \cdot 0520 & \\
y_6 - 2y_7 + y_8 &= -0 \cdot 0390 & \\
y_7 - 2y_8 + y_9 &= -0 \cdot 0260 & \\
y_8 - 2y_9 &= -0 \cdot 0245 &
\end{aligned}
$$

The left hand side of the last equation is actually $y_8 - 2y_9 + y_{10}$, but $y_{10} = 0$.

There are now nine equations for nine unknowns.   The coefficient matrix is sparse and is also a band matrix.   It is very easy to adopt a systematic elimination process and this is the best method here.   Gauss-Seidel is also a possibility but is not recommended as it is found to be very slowly convergent in this case.   (Except for the first and last equations the magnitude of each element on the leading diagonal is only equal to, not greater than, the sum of the others in the same row.)

Starting the elimination process,

$\frac{1}{2} \times$ (56.1) + (56.2)   gives $-1 \cdot 5 y_2 + y_3 = -0 \cdot 1411$ $\qquad$ (57.1)

Next $\frac{1}{1 \cdot 5} \times$ (57.1) + (56.3)   gives $-1 \cdot 3333 y_3 + y_4 = -0 \cdot 1618$ $\qquad$ (57.2)

$\frac{1}{1 \cdot 3333} \times$ (57.2) + (56.4)   gives $-1 \cdot 25 y_4 + y_5 = -0 \cdot 8888$ $\qquad$ (57.3)

Continuing this pattern, finish the elimination process.

****

$-1 \cdot 2 y_5 \quad + y_6 = -0 \cdot 8050$ $\qquad$ (57A.1)
$-1 \cdot 1667 y_6 + y_7 = -0 \cdot 7228$ $\qquad$ (57A.2)
$-1 \cdot 1429 y_7 + y_8 = -0 \cdot 6585$ $\qquad$ (57A.3)
$-1 \cdot 125 y_8 \quad + y_9 = -0 \cdot 6022$ $\qquad$ (57A.4)
$-1 \cdot 1111 y_9 \qquad = -0 \cdot 5598$ $\qquad$ (57A.5)

Back substitution now gives the result.

From (57A.5) $\qquad\qquad$ $y_9 = 0 \cdot 5038$
Using this in (57A.4), $\qquad$ $y_8 = 0 \cdot 9831$
Then, from (57A.3), $\qquad$ $y_7 = 1 \cdot 4363$

Now find the rest of the y's.

****

$y_6 = 1 \cdot 8506,$ $\qquad$ $y_5 = 2 \cdot 2130,$ $\qquad$ $y_4 = 2 \cdot 4814,$ $\qquad$ $y_3 = 1 \cdot 9824,$
$y_2 = 1 \cdot 4157,$ $\qquad$ $y_1 = 0 \cdot 8038.$

Thus, to 3 decimal places,

$y_1 = 0 \cdot 804,$ $\qquad$ $y_2 = 1 \cdot 416,$ $\qquad$ $y_3 = 1 \cdot 982,$ $\qquad$ $y_4 = 2 \cdot 481,$ $\qquad$ $y_5 = 2 \cdot 213,$
$y_6 = 1 \cdot 851,$ $\qquad$ $y_7 = 1 \cdot 436,$ $\qquad$ $y_8 = 0 \cdot 983,$ $\qquad$ $y_9 = 0 \cdot 504.$

As you may have noticed, the coefficient matrix for the equations in FRAME 56 is tridiagonal. (See FRAME 59, page 146, if you have forgotten what a tridiagonal matrix is.) Even if there had been a term in the differential equation that involved $\frac{dy}{dx}$, this would still have been the case. Had any of the terms in y, $\frac{dy}{dx}$ or $\frac{d^2y}{dx^2}$ had variable coefficients, a tridiagonal matrix would still have resulted but would not have exhibited the same simplicity as in our example. In general, the equations in such a case are of the forms

$$
\begin{aligned}
b_1y_1 + c_1y_2 &= d_1 && (60.1)\\
a_2y_1 + b_2y_2 + c_2y_3 &= d_2 && (60.2)\\
a_3y_2 + b_3y_3 + c_3y_4 &= d_3 && (60.3)\\
a_4y_3 + b_4y_4 + c_4y_5 &= d_4 && (60.4)\\
a_5y_4 + b_5y_5 + c_5y_6 &= d_5 && (60.5)
\end{aligned}
$$

$$\cdots\cdots\cdots\cdots\cdots$$

$$a_m y_{m-1} + b_m y_m = d_m \qquad (60.6)$$

m being the number of equations.

As with the set of equations in FRAME 56, a very easy elimination process can be used to solve these equations and gives rise to a simple computer program due to the repetitive nature of the operations involved. A simple flow diagram is given later on but first it is necessary to have a look at the actual method of solution.

(60.1) can be written as $\quad \beta_1y_1 + c_1y_2 = \gamma_1 \qquad (61.1)$
simply by putting $\beta_1 = b_1$ and $\gamma_1 = d_1$. This may seem trivial but is done simply to fit this equation into the pattern that follows.

(60.2) $- \frac{a_2}{\beta_1}$ (61.1) then gives $\left(b_2 - \frac{a_2}{\beta_1}c_1\right)y_2 + c_2y_3 = d_2 - \frac{a_2}{\beta_1}\gamma_1$

which, if we put $\beta_2 = b_2 - \frac{a_2}{\beta_1}c_1$ and $\gamma_2 = d_2 - \frac{a_2}{\beta_1}\gamma_1$ is

$$\beta_2y_2 + c_2y_3 = \gamma_2 \qquad (61.2)$$

Next, (60.3) $- \frac{a_3}{\beta_2}$(61.2) leads to $\left(b_3 - \frac{a_3}{\beta_2}c_2\right)y_3 + c_3y_4 = d_3 - \frac{a_3}{\beta_2}\gamma_2$,

or, if we put $\beta_3 = b_3 - \frac{a_3}{\beta_2}c_2$ and $\gamma_3 = d_3 - \frac{a_3}{\beta_2}\gamma_2$, $\beta_3y_3 + c_3y_4 = \gamma_3$.
The pattern should now be obvious.

What will be the next equation in this set, and the expressions for β_4 and γ_4? You should be able to write these down without having to do any working.

****

$\beta_4y_4 + c_4y_5 = \gamma_4,$ *where* $\beta_4 = b_4 - \frac{a_4}{\beta_3}c_3,$ $\gamma_4 = d_4 - \frac{a_4}{\beta_3}\gamma_3$

Continuing in the same way right through the original set of equations leads to, at the end,

$$\beta_{m-1}y_{m-1} + c_{m-1}y_m = \gamma_{m-1} \quad (62.1), \qquad \beta_m y_m = \gamma_m \quad (62.2)$$

There is only one term on the L.H.S. of this last equation as (60.6) does not contain a term in y_{m+1}.

Back substitution now leads to the solution.

(62.2) gives $y_m = \gamma_m/\beta_m$

Then (62.1) gives $y_{m-1} = \dfrac{\gamma_{m-1} - c_{m-1}y_m}{\beta_{m-1}}$

and so on, until, finally $y_1 = \dfrac{\gamma_1 - c_1 y_2}{\beta_1}$.

A flow diagram for the solution of m equations in m unknowns where the coefficient matrix is tridiagonal is shown on page 339.

Miscellaneous Examples

In this frame a collection of miscellaneous examples is given for you to try. Answers are provided in FRAME 65, together with such working as is considered helpful.

1. Use a Taylor series method to obtain values of y to three decimal places at $x = -0 \cdot 1$, $0 \cdot 1$ and $0 \cdot 2$, if $y'' + (x + 1)y^2 = 0$ with $y(0) = 1$ and $y'(0) = 1$. Then use the following predictor-corrector pair of formulae to obtain the value of y at $x = 0 \cdot 3$.

   Predictor: $y_{n+1} = 2y_{n-1} - y_{n-3} + \dfrac{4h^2}{3}(y_n'' + y_{n-1}'' + y_{n-2}'')$

   Corrector: $y_{n+1} = 2y_n - y_{n-1} + \dfrac{h^2}{12}(y_{n+1}'' + 10y_n'' + y_{n-1}'')$

   Describe briefly how such forward integration formulae can be used to solve a boundary value problem. (C.E.I.)

2. Show that the boundary-value problem $\dfrac{d^2y}{dx^2} = \dfrac{0 \cdot 06912x(x - 1)}{I(x)}$, $y(0) = y(1) = 0$, can be approximated at the points $x_1, x_2, \ldots\ldots\ldots, x_r = x_0 + rh, \ldots\ldots\ldots$ by the algebraic equations

   $$y_{r-1} - 2y_r + y_{r+1} = \dfrac{0 \cdot 06912h^2 x_r(x_r - 1)}{I(x_r)} , \quad \text{for} \quad r = 1, 2, \ldots, \left(\dfrac{1}{h} - 1\right).$$

   Write down these equations given

x	1/6	1/3	1/2	2/3	5/6
$I(x)$	$0 \cdot 012\ 81$	$0 \cdot 013\ 27$	$0 \cdot 013\ 42$	$0 \cdot 013\ 27$	$0 \cdot 012\ 81$

   (contd.)

Flow diagram for FRAME 63.

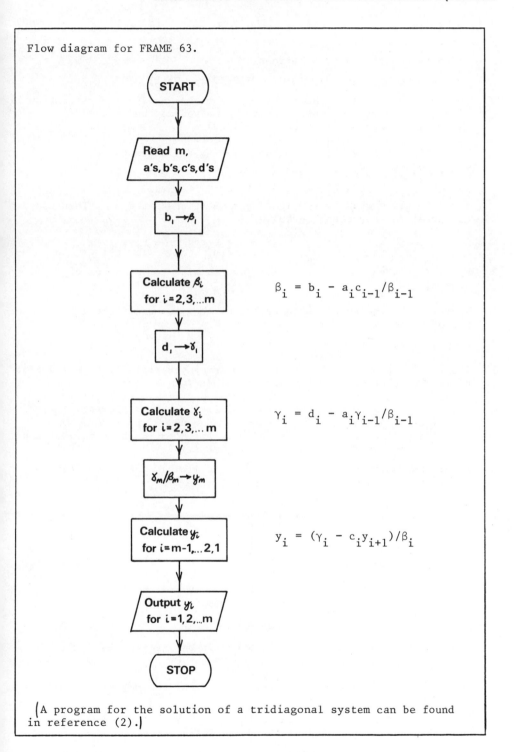

$$\beta_i = b_i - a_i c_{i-1}/\beta_{i-1}$$

$$\gamma_i = d_i - a_i \gamma_{i-1}/\beta_{i-1}$$

$$y_i = (\gamma_i - c_i y_{i+1})/\beta_i$$

(A program for the solution of a tridiagonal system can be found in reference (2).)

and taking h = 1/6.

Given the initial approximations

$(y_1, y_2, y_3, y_4, y_5)$ = (0·07, 0·12, 0·14, 0·12, 0·07)

obtain more accurate results by applying the Gauss-Seidel iteration. (The actual number of equations to be solved may be reduced by assuming symmetry of y about x = 1/2.)   (C.E.I.)

3. The motion of a certain non-linear spring satisfies the equation
$$\ddot{y} = -\frac{a}{4k^2}\left[(1 - k^2)y + 2k^2 y^3\right].$$

If a = 4 and k = 5, find, by forming a Taylor series, y when t = 0·2 given that y = 0 and $\dot{y}$ = 0·04 when t = 0.   Then use the predictor-corrector formulae

$$y_{n+1}^{p} = 2y_n - y_{n-1} + h^2 f_n, \qquad y_{n+1}^{c} = 2y_n - y_{n-1} + \frac{h^2}{12}(f_{n+1} + 10f_n + f_{n-1})$$

to find y when t = 0·4.   Give your results to a maximum of 6 decimal places.

4. What will be the Runge-Kutta equations that you would use to solve the set of equations

$$\dot{x} = F_1(t, x, u, v), \qquad \dot{u} = F_2(t, x, u, v), \qquad \dot{v} = F_3(t, x, u, v)$$

5. The Blasius problem relating to boundary layer fluid flow past a flat plate can be reduced to the dimensionless form $f(\eta)f''(\eta) + 2f'''(\eta) = 0$ with boundary conditions f(0) = f'(0) = 0, f'(∞) = 1.   Express this as a set of first-order differential equations.   Assuming f''(0) = 0·4, take one step $\Delta\eta = 1$ in the solution of the resulting initial value problem, using Runge-Kutta or any equivalent method.   (C.E.I.)

Note: For a third order equation, you will need three simultaneous first order equations.

6.

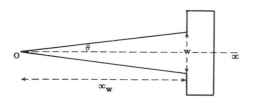

The temperature $T^{\circ}C$ at a point distant x m from O of a triangular fin of an air heater is given by the Bessel equation

$xy'' + y' - (2hx_w y \sec\theta)/kw = 0$ where $y = T - T_a$, $T_a = 20^{\circ}C$,

h = heat transfer coefficient between fin and air = 15 W/m² °K

k = heat conductivity of fin = 366 W/m °K

If x_w = 1 m, w = 0·2 m, the equation reduces to $xy'' + y' - 0·4119y = 0$.

Given that $T = 80^{\circ}C$ and $\frac{dT}{dx} = 12$ when x = 0·3, find T when x = 0·32, using one step of the Runge-Kutta process.

7. Describe how a second order boundary-value problem (second order
    ordinary differential equation) may be solved over a given interval
    using the Euler method with trapezoidal correction.

    For the boundary value problem

    $$yy'' + y'^2 + 1 = 0, \quad y(0) = 1, \quad y(1) = 2,$$

    solve for y consistent to two decimal places at $x = 0 \cdot 5$ and $x = 1 \cdot 0$
    using the above method and a trial value $y'(0) = 1$.   From the result
    estimate a new value for $y'(0)$.   (C.E.I.)

    (Note: The Euler method with trapezoidal correction is that described
    in this programme under the title "The Modified Euler Method".

    This question was set at an examination and the whole process
    described in FRAMES 46 - 51 would have been too lengthy for this
    purpose.   The candidates were therefore only asked to go so far with
    the solution.)

8. The equation $\dfrac{d^2y}{dt^2} + c_2 \dfrac{dy}{dt} - c_1 \dfrac{y}{(1 + y^2)^3} = 0$ occurs in the study of

    bubble motion.   Assuming $c_1 = c_2 = 1$ and, when $t = 0$,

    $y = 0 \cdot 000\ 01$ and $\dfrac{dy}{dt} = 0$,   use Runge-Kutta to find y when $t = 0 \cdot 2$,

    $0 \cdot 4$, $0 \cdot 6$ and modified Hamming to find y when $t = 0 \cdot 8$, $1 \cdot 0$.

9. It is required to solve the equation $y'' - xy' + \dfrac{1}{y^2} = 0$, subject to

    the conditions that $y = 2$ when $x = 0$ and $y = 4$ when $x = 1$.
    By splitting the x-interval $0 - 1$ into two steps, find, to two decimal
    places, the best initial value for y'.

Answers to Miscellaneous Examples

1. $y = 1 + x - \dfrac{x^2}{2} - \dfrac{x^3}{2} - \dfrac{x^4}{6} + \dfrac{x^5}{10} + \ldots\ldots\ldots$

    $0 \cdot 895$, $1 \cdot 094$, $1 \cdot 176$: $1 \cdot 240$ (predicted value), $1 \cdot 241$ (corrected value)

2. $y_0 - 2y_1 + y_2 = -0 \cdot 020\ 82,$ $y_1 - 2y_2 + y_3 = -0 \cdot 032\ 15,$
    $y_2 - 2y_3 + y_4 = -0 \cdot 035\ 77,$ $y_3 - 2y_4 + y_5 = -0 \cdot 032\ 15,$
    $y_4 - 2y_5 + y_6 = -0 \cdot 020\ 82.$

    $0 \cdot 071$, $0 \cdot 121$, $0 \cdot 139$, $0 \cdot 121$, $0 \cdot 071$.

3. $y = 0 \cdot 04t + 0 \cdot 0064t^3 + 0 \cdot 000\ 300\ 8t^5$
    When $t = 0 \cdot 2$, $y = 0 \cdot 008\ 051$.   When $t = 0 \cdot 4$, $y = 0 \cdot 016\ 413$.

4. $k_1 = hF_1(t_0, x_0, u_0, v_0),$ $m_1 = hF_2(t_0, x_0, u_0, v_0)$
    $q_1 = hF_3(t_0, x_0, u_0, v_0).$

    $k_2 = hF_1(t_0 + \tfrac{1}{2}h, x_0 + \tfrac{1}{2}k_1, u_0 + \tfrac{1}{2}m_1, v_0 + \tfrac{1}{2}q_1)$
    $m_2 = hF_2(t_0 + \tfrac{1}{2}h, x_0 + \tfrac{1}{2}k_1, u_0 + \tfrac{1}{2}m_1, v_0 + \tfrac{1}{2}q_1)$
    $q_2 = hF_3(t_0 + \tfrac{1}{2}h, x_0 + \tfrac{1}{2}k_1, u_0 + \tfrac{1}{2}m_1, v_0 + \tfrac{1}{2}q_1)$

$$k_3 = hF_1(t_0 + \tfrac{1}{2}h, \; x_0 + \tfrac{1}{2}k_2, \; u_0 + \tfrac{1}{2}m_2, \; v_0 + \tfrac{1}{2}q_2)$$
$$m_3 = hF_2(t_0 + \tfrac{1}{2}h, \; x_0 + \tfrac{1}{2}k_2, \; u_0 + \tfrac{1}{2}m_2, \; v_0 + \tfrac{1}{2}q_2)$$
$$q_3 = hF_3(t_0 + \tfrac{1}{2}h, \; x_0 + \tfrac{1}{2}k_2, \; u_0 + \tfrac{1}{2}m_2, \; v_0 + \tfrac{1}{2}q_2)$$

$$k_4 = hF_1(t_0 + h, \; x_0 + k_3, \; u_0 + m_3, \; v_0 + q_3)$$
$$m_4 = hF_2(t_0 + h, \; x_0 + k_3, \; u_0 + m_3, \; v_0 + q_3)$$
$$q_4 = hF_3(t_0 + h, \; x_0 + k_3, \; u_0 + m_3, \; v_0 + q_3)$$

$$x_1 = x_0 + (k_1 + 2k_2 + 2k_3 + k_4)/6$$
$$u_1 = u_0 + (m_1 + 2m_2 + 2m_3 + m_4)/6$$
$$v_1 = v_0 + (q_1 + 2q_2 + 2q_3 + q_4)/6$$

5. A suitable set is:

$$\begin{cases} u' = -\tfrac{1}{2}fu \\ y' = u \\ f' = y \end{cases}$$

Writing these as $u' = F_1(\eta, \; u, \; y, \; f)$
$$y' = F_2(\eta, \; u, \; y, \; f)$$
$$f' = F_3(\eta, \; u, \; y, \; f)$$

the Runge–Kutta equations will be those of question No. 4 with t
replaced by η,   x by u,   u by y   and   v by f.

$f(1) = 0 \cdot 2$.

6. A suitable pair of simultaneous equations is

$$y' = u$$
$$u' = \left(\frac{2hx_w \; y \sec \theta}{kw} - u \right) \Big/ x$$

$80 \cdot 247 \, ^{\circ}C$

7. A suitable pair of simultaneous equations is $y' = u$, $u' = -(u^2 + 1)/y$.
$1 \cdot 32$, $1 \cdot 37$.
3 (The value 1 has produced about one third of the rise required.)

8. A suitable pair of simultaneous equations is

$$y' = u$$
$$u' = c_1 \frac{y}{(1 + y^2)^3} - c_2 u$$

1019,   1071,   1153,   1262,   1397,   all times 10^{-8}.

9. $0 \cdot 85$.

Partial Differential Equations

Introduction

Partial differential equations frequently occur in practice.   Some examples are:

The displacement, y, of a tightly stretched string which is undergoing small transverse vibrations is given by $\frac{\partial^2 y}{\partial t^2} = c^2 \frac{\partial^2 y}{\partial x^2}$, t being the time, x the distance from one end and c a constant.

The displacement of a uniform beam undergoing transverse vibrations is similarly given by $EI \frac{\partial^4 y}{\partial x^4} + m \frac{\partial^2 y}{\partial t^2} = 0$.

The two-dimensional steady state heat flow equation is $\frac{\partial^2 \theta}{\partial x^2} + \frac{\partial^2 \theta}{\partial y^2} = 0$.

This equation gives the temperature θ at the point whose coordinates are (x, y) in a flat plate which is insulated so that no heat flows into or out of the plate, it being assumed that sufficient time has elapsed so that no further change takes place with time.   If the temperature is varying with time as well, then the equation becomes

$$\frac{\partial \theta}{\partial t} = c^2 \left(\frac{\partial^2 \theta}{\partial x^2} + \frac{\partial^2 \theta}{\partial y^2} \right)$$

We shall not consider the derivation of these equations here, but if you are interested, you will find the derivation of certain partial differential equations given in the second volume of our "Mathematics for Engineers and Scientists".

Taking, for example, the first equation given in the last frame, it is very easy to see that y = f(x + ct) and y = g(x - ct) are analytical solutions, where f and g represent arbitrary functions.   Furthermore y = f(x + ct) + g(x - ct) is also a solution.   Verify that this is so by differentiation and substitution.

****

2A

$\frac{\partial^2 y}{\partial t^2} = c^2 \{ f''\,(x + ct) + g''\,(x - ct) \}, \; \frac{\partial^2 y}{\partial x^2} = f''\,(x + ct) + g''\,(x - ct)$

Hence result.

Although it may seem that if an analytical solution like this exists all our troubles should be over, this is by no means the case.   The difficulty arises when we try to fit initial conditions, the form given in the last frame being, as it stands, rather unadaptable.   That is why special methods such as separation of variables are used.   But even these special methods do not always help, and in such cases a numerical technique is tried.   In this programme we shall show you just one method of dealing with simple examples of such problems.

PARTIAL DIFFERENTIAL EQUATIONS

Laplace's Equation $\quad \dfrac{\partial^2\theta}{\partial x^2} + \dfrac{\partial^2\theta}{\partial y^2} = 0$

Suppose the faces of a square plate are insulated so that no heat can flow into or out of them. Let one edge of the plate be maintained at 100°C and the other three at 0°C, these temperatures being accurate to the nearest tenth of a degree. After a sufficient time has elapsed, the temperature at any point in the plate will be time independent. Now this problem can be solved by the method of separation of variables but instead it will be used to afford an easy illustration of the numerical technique.

A start is made by dividing the square up into a number of smaller squares. Unless there is some reason for not doing so, the edges would quite likely

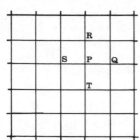

be divided first into two parts, then into four, next into eight and so on. A grid is thus produced and the method involves calculating the temperatures at the corners of the small squares. The diagram shows just a portion of the grid. Let the edge of each small square be h. Then, taking the coordinates of the grid point P as (x, y), the coordinates of Q, R, S and T are (x + h, y), (x, y + h), (x - h, y) and (x, y - h) respectively.

A process is now adopted similar to that used in FRAME 52 of the previous programme. There $\dfrac{d^2y}{dx^2}$ at the point (x_n, y_n) was replaced by $(y_{n+1} - 2y_n + y_{n-1})/h^2$. Can you write down the corresponding formulae for $\dfrac{\partial^2\theta}{\partial x^2}$ and $\dfrac{\partial^2\theta}{\partial y^2}$ at P? Denote the value of θ at P by $\theta(x, y)$, etc.

****

6A

$\left\{\theta(x + h, y) - 2\theta(x, y) + \theta(x - h, y)\right\}/h^2$
$\left\{\theta(x, y + h) - 2\theta(x, y) + \theta(x, y - h)\right\}/h^2$

Substituting these into the equation $\dfrac{\partial^2\theta}{\partial x^2} + \dfrac{\partial^2\theta}{\partial y^2} = 0$ gives, approximately,

$$\theta(x + h, y) + \theta(x - h, y) + \theta(x, y + h) + \theta(x, y - h) - 4\theta(x, y) = 0$$
$$(7.1)$$

i.e. $\theta(x, y) = \dfrac{1}{4}\left\{\theta(x + h, y) + \theta(x, y + h) + \theta(x - h, y) + \theta(x, y - h)\right\}$
$$(7.2)$$

This states that the value at any point P is approximately the average of the four values at Q, R, S and T.

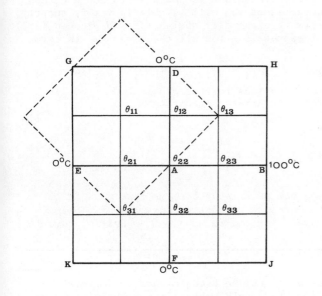

Now suppose the original square is divided first into 4, and then into 16, smaller squares. Let the temperatures at the corners of the small squares be as shown.

The next step is to find initial estimates of the nine temperatures indicated. A start is made by finding an estimate of θ_{22}. This is done by putting P, Q, R, S and T at A, B, D, E and F. Then an initial estimate of the value of θ_{22} is the average of the values at B, D, E and F, i.e.

$\frac{1}{4}(100 + 0 + 0 + 0) = 25$.

FRAME 9

The next value estimated is θ_{11}. As θ_{12} and θ_{21} are not yet known, (7.2) cannot be used directly. Instead, an oblique square (shown dashed) is drawn in and the value of θ_{11} taken as the average of the four values at the mid-points of the sides of this square, i.e. the average of the values at A, D, G and E. This gives an initial estimate of θ_{11} as

$\frac{1}{4}(25 + 0 + 0 + 0) \simeq 6 \cdot 2$.

What, to 1 decimal place, will be initial estimates of θ_{13}, θ_{31} and θ_{33}? Assume that the temperature at both H and J is 50°C.
****

9A

$\theta_{13} = 43 \cdot 8, \qquad \theta_{31} = 6 \cdot 2, \qquad \theta_{33} = 43 \cdot 8$

FRAME 10

Estimates of θ_{12}, θ_{21}, θ_{23} and θ_{32} can now be obtained using (7.2) directly. What will be the values of these estimates?
****

10A

$\theta_{12} = 18 \cdot 8, \qquad \theta_{21} = 9 \cdot 4, \qquad \theta_{23} = 53 \cdot 2, \qquad \theta_{32} = 18 \cdot 8$

PARTIAL DIFFERENTIAL EQUATIONS

We now have first estimates of all the nine temperatures at the grid points. A Gauss-Seidel iterative technique is now used to improve these values. This is possible because, as you will be able to verify shortly, the relevant coefficient matrix is diagonally dominant. To proceed (7.2) is used to write down general expressions for all the nine θ's in terms of their neighbours. Thus

$$\theta_{11} = (\theta_{12} + 0 + 0 + \theta_{21})/4, \qquad \theta_{12} = (\theta_{13} + 0 + \theta_{11} + \theta_{22})/4$$

What will be the corresponding equations for the other θ's?

****

$$\theta_{13} = (100 + 0 + \theta_{12} + \theta_{23})/4 \qquad \theta_{21} = (\theta_{22} + \theta_{11} + 0 + \theta_{31})/4$$
$$\theta_{22} = (\theta_{23} + \theta_{12} + \theta_{21} + \theta_{32})/4 \qquad \theta_{23} = (100 + \theta_{13} + \theta_{22} + \theta_{33})/4$$
$$\theta_{31} = (\theta_{32} + \theta_{21} + 0 + 0)/4 \qquad \theta_{32} = (\theta_{33} + \theta_{22} + \theta_{31} + 0)/4$$
$$\theta_{33} = (100 + \theta_{23} + \theta_{32} + 0)/4$$

In each case the four terms on the right have been written down in an anti-clockwise order from the point at which the temperature is being found. Doing this constitutes a systematic approach to writing down the equations.

Normally these nine equations would now be used in turn to improve the values of θ. In this particular example, there is obviously symmetry about EB and so $\theta_{31} = \theta_{11}$, $\theta_{32} = \theta_{12}$ and $\theta_{33} = \theta_{13}$. Making use of this symmetry reduces the nine equations to

$$\theta_{11} = (\theta_{12} + \theta_{21})/4 \quad (12.1), \qquad \theta_{12} = (\theta_{13} + \theta_{11} + \theta_{22})/4 \quad (12.2)$$
$$\theta_{13} = (100 + \theta_{12} + \theta_{23})/4 \qquad \theta_{21} = (\theta_{22} + 2\theta_{11})/4$$
$$\theta_{22} = (\theta_{23} + 2\theta_{12} + \theta_{21})/4 \qquad \theta_{23} = (100 + 2\theta_{13} + \theta_{22})/4$$

Using the initial estimates of θ_{12} and θ_{21} in (12.1) gives a second estimate of θ_{11} to be $(18 \cdot 8 + 9 \cdot 4)/4 = 7 \cdot 0$. Using this value, together with $\theta_{13} = 43 \cdot 8$, $\theta_{22} = 25$, (12.2) gives a second estimate of θ_{12} as $(43 \cdot 8 + 7 \cdot 0 + 25)/4 = 19 \cdot 0$.

What will be second estimates for the other four θ's? In each case use the latest values available for the calculations.

****

$$\theta_{13} = (100 + 19 \cdot 0 + 53 \cdot 2)/4 = 43 \cdot 0, \qquad \theta_{21} = (25 + 2 \times 7 \cdot 0)/4 = 9 \cdot 8$$
$$\theta_{22} = 25 \cdot 2 \qquad \theta_{23} = 52 \cdot 8$$

The process is now repeated as many times as is necessary until no further changes in the estimates take place. The working (when being done manually) is best set out in a table:

346

θ_{11}	θ_{12}	θ_{13}	θ_{21}	θ_{22}	θ_{23}
6·2	18·8	43·8	9·4	25·0	53·2
7·0	19·0	43·1	9·8	25·2	52·8
7·2	18·9	43·0	9·9	25·1	52·8
7·2	18·8	42·9	9·9	25·1	52·7
7·2	18·8	42·9	9·9	25·0	52·7
7·2	18·8	42·9	9·8	25·0	52·7
7·2	18·8	42·9	9·8		

These values have been given to one decimal place as it was stated that
the original temperatures quoted were correct to a tenth of a degree.
We stop the calculation when we reach the stage that the last six values
calculated are equal to, on a one to one basis, the previous six.   This
indicates that no further change will take place in the values if further
calculations are performed.   When using a computer, one would probably
instruct it, however, to test for convergence at the end of each row of
calculations.   In that case, the output would include the extra readings
25·0 and 52·7 at the end of the uncompleted row.

The original square could now be divided into smaller squares by dividing
each of the present small squares into four.   There would then be 64
small squares.   If this is done, two things will happen.   The first of
these is that the temperature will be obtained at 49 interior points
instead of only 9.   The second is that one would expect greater accuracy
in the result as, in general in numerical work, the smaller the value of
h, the more accurate the result.   As nothing fresh is involved in
reducing the size of the small squares, we shall not pursue the consequent
calculation.

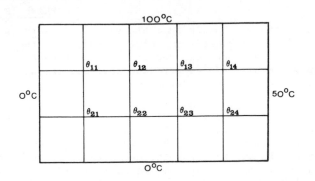

The diagram depicts a
similar problem to .that
just worked but this time
it is a rectangular area
that has been divided into
a number of small squares.
What will be the equations
similar to those in FRAME
11 and 11A for this
set-up?

****

16A

$\theta_{11} = (\theta_{12} + 100 + \theta_{21})/4$ $\theta_{12} = (\theta_{13} + 100 + \theta_{11} + \theta_{22})/4$
$\theta_{13} = (\theta_{14} + 100 + \theta_{12} + \theta_{23})/4$ $\theta_{14} = (150 + \theta_{13} + \theta_{24})/4$
$\theta_{21} = (\theta_{22} + \theta_{11})/4$ $\theta_{22} = (\theta_{23} + \theta_{12} + \theta_{21})/4$
$\theta_{23} = (\theta_{24} + \theta_{13} + \theta_{22})/4$ $\theta_{24} = (50 + \theta_{14} + \theta_{23})/4$

In the previous example it was possible to obtain reasonably accurate starting values for the iterations and, if this can be done, it is better to do so.   Here it isn't quite so simple as, for a start, the mid-point of the rectangle is not at a grid point.   However, one can write down by common sense a set of values and start the iterations from this set.

Now it should be obvious that the values of the θ's in the top row will be greater than those in the bottom.   Similarly one will expect those on the right of the diagram to be greater than those on the left.   Thus, a possible set of figures from which to start might be taken as

$$\theta_{11} = 50, \quad \theta_{12} = 60, \quad \theta_{13} = 70, \quad \theta_{14} = 80$$
$$\theta_{21} = 25, \quad \theta_{22} = 30, \quad \theta_{23} = 35, \quad \theta_{24} = 40$$

Using these figures and the equations in 16A, find the values of the θ's, working to the nearest integer.

****

θ_{11}	θ_{12}	θ_{13}	θ_{14}	θ_{21}	θ_{22}	θ_{23}	θ_{24}
50	60	70	80	25	30	35	40
46	62	69	65	19	29	34	37
45	61	65	63	18	28	32	36
45	60	64	62	18	28	32	36
44	59	63	62	18	27	32	36
44	58	63	62	18	27	32	36
44	58						

A flow diagram for the solution of Laplace's equation for a rectangular grid with n intervals across and m down is shown on page 349.

An alternative method of doing the arithmetic in examples of this sort is to enlarge the diagram and insert the successive values found on to the diagram.   This gives a more automatic process requiring less thought when you are doing an example manually but is not adaptable for computer use.   The diagram, shown on page 350, illustrates the idea for the example in FRAME 16.

The figures have been calculated in exactly the same order as in 17A, but instead of having to look up previous values in this table for insertion into the relevant equation, each value is found by taking the average of the latest values to the right, above, to the left, and below.   Thus, for example, the double starred 64 is the average of the four single starred numbers, i.e., 63, 100, 60 and 32.

The equation $\dfrac{\partial^2\theta}{\partial x^2} + \dfrac{\partial^2\theta}{\partial y^2} = 0$ is known as Laplace's equation in two dimensions.   The other equations that will be dealt with in this programme are $\dfrac{\partial^2 y}{\partial t^2} = c^2 \dfrac{\partial^2 y}{\partial x^2}$ and $\dfrac{\partial\theta}{\partial t} = h^2 \dfrac{\partial^2\theta}{\partial x^2}$.

Flow diagram for FRAME 18.

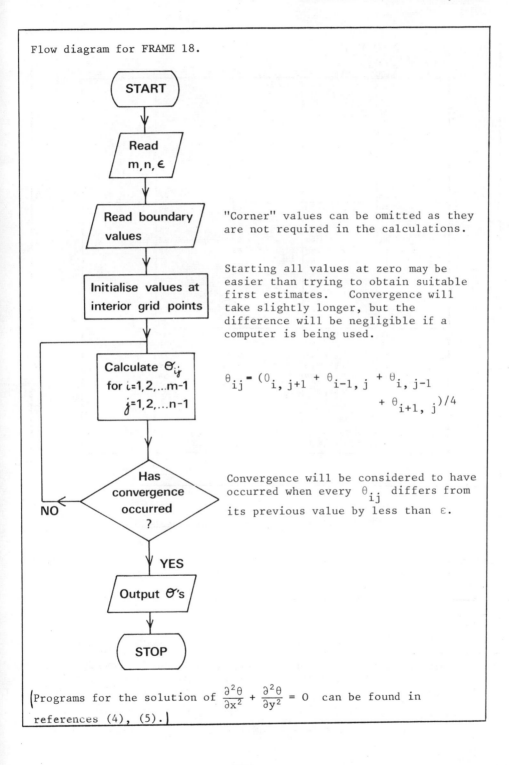

START

Read
m, n, ε

Read boundary values

"Corner" values can be omitted as they are not required in the calculations.

Initialise values at interior grid points

Starting all values at zero may be easier than trying to obtain suitable first estimates. Convergence will take slightly longer, but the difference will be negligible if a computer is being used.

Calculate θ_{ij} for $i = 1, 2, \ldots m-1$ $j = 1, 2, \ldots n-1$

$$\theta_{ij} = (\theta_{i,\,j+1} + \theta_{i-1,\,j} + \theta_{i,\,j-1} + \theta_{i+1,\,j})/4$$

Has convergence occurred ?

NO

Convergence will be considered to have occurred when every θ_{ij} differs from its previous value by less than ε.

YES

Output θ's

STOP

(Programs for the solution of $\dfrac{\partial^2 \theta}{\partial x^2} + \dfrac{\partial^2 \theta}{\partial y^2} = 0$ can be found in references (4), (5).)

349

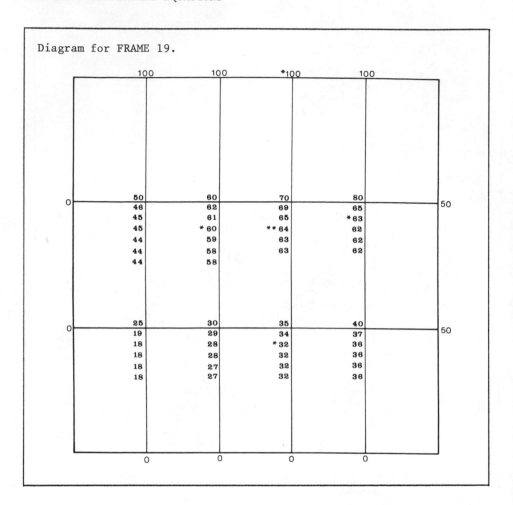

Diagram for FRAME 19.

FRAME 20   (continued)

Basically, the method we shall use for each of these is similar to that for Laplace's equation.   There are, however, certain points to notice about each of these other equations.

FRAME 21

The Equation $\dfrac{\partial^2 y}{\partial t^2} = c^2 \dfrac{\partial^2 y}{\partial x^2}$

When solving Laplace's equation, h was used for the increments in both x and y.   This is reasonable, as x and y are both distances, and the equal increments occurred as the x, y diagram was divided into a number of small squares.   Now $\dfrac{\partial^2 y}{\partial t^2} = c^2 \dfrac{\partial^2 y}{\partial x^2}$ is the equation of a vibrating string and here x represents a distance, while t represents time.   There is thus no particular reason for taking equal increments for these two variables.   In this case let the increments in x and t be h and k respectively.

Now, for this differential equation, what will be the equation corresponding to (7.1)?
****

21A

$$\frac{y(x,\ t+k)\ -\ 2y(x,\ t)\ +\ y(x,\ t-k)}{k^2}\ =\ c^2\ \frac{y(x+h,\ t)\ -\ 2y(x,\ t)\ +\ y(x-h,\ t)}{h^2}$$

$$(21A.1)$$

FRAME 22

This equation takes on a particularly simple form if a certain relation is chosen between h and k.   Remember that x and t are independent variables so that we can choose h and k to suit ourselves.

Can you suggest the best relation between h and k in order that (21A.1) becomes as simple as possible?   What will be this simplified form?
****

22A

If $\frac{1}{k^2} = \frac{c^2}{h^2}$,   i.e.   if $h = ck$,   (21A.1) becomes

$$y(x,\ t+k)\ +\ y(x,\ t-k)\ =\ y(x+h,\ t)\ +\ y(x-h,\ t) \qquad (22A.1)$$

or an alternative version of this in which either h or k does not appear. (One assumes here that h and k are both positive.)

FRAME 23

Thus, $y(x,\ t+k) = y(x-h,\ t) + y(x+h,\ t) - y(x,\ t-k)$ (23.1) and this is the form in which we shall use this equation.

To illustrate, let us take the equation $\frac{\partial^2 y}{\partial t^2} = 4 \frac{\partial^2 y}{\partial x^2}$ and find a solution which satisfies the conditions

i) $y_{x=0} = 0$ 　　　　　　　　ii) $y_{x=2} = 0$

iii) $y_{t=0} = \frac{1}{100} \sin \frac{1}{2}\pi x$ 　　iv) $\left(\frac{\partial y}{\partial t}\right)_{t=0} = 0$

for $0 \leqslant x \leqslant 2$.

Now it has already been stated that the differential equation is that of a vibrating string.   What information about the motion and how it is started do the given conditions tell you?
****

23A

The points $x = 0$ and $x = 2$ are permanently at rest with zero displacement.   The motion is started by bending the string into one half of a sine wave and releasing it from rest in that position.

PARTIAL DIFFERENTIAL EQUATIONS

FRAME 24

Now suppose increments of x to be 0·2.   What value would you take for increments of t?
****

24A

$0·1$, from $h = ck$ with $h = 0·2$ and $c = 2$

FRAME 25

The values of x are limited to lie between 0 and 2 inclusive.   The values of t will increase indefinitely.   This means that, as far as time is concerned, only the beginning of the table will be shown.   It can then be extended as necessary.

Some information can immediately be inserted into the table, as follows:

t \ x	0	0·2	0·4	0·6	0·8	1·0	1·2	1·4	1·6	1·8	2·0
0	0	3090	5878	8090	9511	10 000	9511	8090	5878	3090	0
0·1	0										0
0·2	0										0
0·3	0										0
0·4	0										0
⋮											

All values in the body of the table are to be multiplied by 10^{-6}.   The values in the row t = 0 are those of the function $\frac{1}{100} \sin \frac{1}{2}\pi x$ for $0(0·2)2$.

You will notice that there is symmetry about x = 1 and this continues throughout the table.

What initial information hasn't so far been included in the table?
****

25A

$$\left(\frac{\partial y}{\partial t}\right)_{t=0} = 0$$

FRAME 26

Now, given a portion of the table represented by the dots below

equation (23.1) gives a formula for the squared dot in terms of the circled ones, as, for the position indicated, the dots picked out are, respectively,

352

$$y(x, t - k)$$

$$y(x - h, t) \qquad\qquad\qquad y(x + h, t)$$

$$y(x , t + k)$$

So, when a number of lines have been filled in, the next one is found by moving the squared dot along the line currently being worked out.   But can you see how difficulty is going to occur if we try to start the table with this procedure?

****

26A

In working out the entries for line t = 0·1, the uppermost dot in each case is off the table.

FRAME 27

To find the entries in this awkward row, we have to incorporate the so far unused initial condition, i.e. $\left(\dfrac{\partial y}{\partial t}\right)_{t=0}$ = 0.   This is done in the following way:

In our actual problem, motion is started in a particular way at time t = 0.   But, instead of starting the motion at this time, let us assume that it is taking place prior to t = 0 in such a way that, at t = 0, the string is doing exactly what the problem says it should be.

When t = 0,   (22A.1) becomes

$$y(x, k) + y(x, -k) = y(x + h, 0) + y(x - h, 0) \qquad\qquad (27.1)$$

and, although we do not know the value of y(x, -k), it is assumed to be that which agrees with our "invention".

FRAME 28

Now, an approximate expression for $\dfrac{\partial y(x, t)}{\partial t}$ is $\dfrac{y(x, t+k) - y(x, t-k)}{2k}$

and so $y(x, t - k) = y(x, t + k) - 2k \dfrac{\partial y(x, t)}{\partial t}$ which, when t = 0, gives

$y(x, -k) = y(x, k) - 2k \dfrac{\partial y(x, 0)}{\partial t}$.

This result is now incorporated into (27.1).   What do you get for y(x, k) when this is done?

****

28A

$$2y(x, k) - 2k \frac{\partial y(x, 0)}{\partial t} = y(x + h, 0) + y(x - h, 0)$$

$$\therefore \quad y(x, k) = \frac{1}{2}\{y(x - h, 0) + y(x + h, 0)\} + k \frac{\partial y(x, 0)}{\partial t} \qquad (28A.1)$$

FRAME 29

This equation now enables us to fill in the awkward row.   In the particular example being worked out, $\dfrac{\partial y(x, 0)}{\partial t} = 0$ and so (28A.1) here becomes $y(x, k) = \frac{1}{2}\{y(x - h, 0) + y(x + h, 0)\}$.

Thus, when x = 0·2, the entry in row t = 0·1, i.e., y(0·2, 0·1), is $\frac{1}{2}$(0 + 5878) and that when x = 0·4 is $\frac{1}{2}$(3090 + 8090).

Complete the row of entries for t = 0·1.
****

29A

| 0 | 2939 | 5590 | 7694 | 9045 | 9511 | 9045 | 7694 | 5590 | 2939 | 0 |

FRAME 30

After this, (22A.1) is used directly to give the entries, in order, in the following rows.   Thus, when t = 0·2, x = 0·2, the entry is given by 0 + 5590 − 3090 = 2500 and when t = 0·2, x = 0·4, the corresponding entry is 2939 + 7694 − 5878 = 4755.

Now complete the row of entries for t = 0·2.
****

30A

| 0 | 2500 | 4755 | 6545 | 7694 | 8090 | 7694 | 6545 | 4755 | 2500 | 0 |

FRAME 31

The following rows can now be filled in and the table, as far as t = 1·0, becomes

t \ x	0	0·2	0·4	0·6	0·8	1·0
0	0	3090	5878	8090	9511	10 000
0·1	0	2939	5590	7694	9045	9511
0·2	0	2500	4755	6545	7694	8090
0·3	0	1816	3455	4755	5590	5877
0·4	0	955	1816	2500	2938	3090
0·5	0	0	0	−1	0	−1
0·6	0	−955	−1817	−2500	−2940	−3090
0·7	0	−1817	−3455	−4756	−5590	−5879
0·8	0	−2500	−4756	−6545	−7695	−8090
0·9	0	−2939	−5590	−7695	−9045	−9511
1·0	0	−3090	−5878	−8090	−9511	−10 000

Don't forget that each entry is to be multiplied by 10^{-6}.   Also, due to the symmetry mentioned in FRAME 25, only that part of the table between x = 0 and x = 1 has been shown.

FRAME 32

The problem just solved can easily be dealt with analytically by separating the variables and so the numerical technique is not really necessary.   The analytical solution is $y = \frac{1}{100} \sin \frac{1}{2}\pi x \cos \pi t$ and, if you wish, you can insert into this equation any pair of values of x and t given in the table above and so check how accurate the numerical solution is.

The following problem is one in which you would find yourself in trouble if you attempted to solve it by the separation of variables technique: A stretched string is at rest in the position shown. When oscillating, the motion satisfies the equation $\frac{\partial^2 y}{\partial t^2} = 16 \frac{\partial^2 y}{\partial x^2}$. Motion is imparted to the string by giving a forced

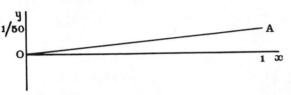

oscillation $y = \frac{1}{50} \cos 2\pi t$ to the end A, the other end remaining fixed. The problem is to find the displacement of other points on the string at later times.

If you are familiar with the separation of variables method, try it this way and see what happens. Otherwise proceed directly to the next frame.

****

Separating the variables leads to

$$y = (A \cos kx + B \sin kx)(A_1 \cos 4kt + B_1 \sin 4kt)$$

As $y = 0$ *when* $x = 0$, $A = 0$. *As* $\frac{\partial y}{\partial t} = 0$ *when* $t = 0$, $B_1 = 0$.

$$\therefore \quad y = C \sin kx \cos 4kt$$

When $x = 1$, $y = \frac{1}{50} \cos 2\pi t$, $\therefore$ $k = \frac{1}{2}\pi$ *and so* $C = \frac{1}{50}$.

$$y = \frac{1}{50} \sin \tfrac{1}{2}\pi x \cos 2\pi t$$

When $t = 0$, $y = \frac{1}{50} x$ *and this condition cannot be satisified.*

If x is taken at intervals of $0 \cdot 1$, at what intervals will t be taken?

****

$0 \cdot 025$

Construct a table similar to that in FRAME 25, filling in the top row and the first and last columns. Take values of t from 0 to $0 \cdot 5$ and work to 5 decimal places.

****

t \ x	0	0·1	0·2	0·3	0·4	0·5	0·6	0·7	0·8	0·9	1·0
0	0	200	400	600	800	1000	1200	1400	1600	1800	2000
0·025	0										1975
0·050	0										1902
0·075	0										1782
0·100	0										1618
0·125	0										1414
0·150	0										1176
0·175	0										908
0·200	0										618
0·225	0										313
0·250	0										0
0·275	0										−313
0·300	0										−618
0·325	0										−908
0·350	0										−1176
0·375	0										−1414
0·400	0										−1618
0·425	0										−1782
0·450	0										−1902
0·475	0										−1975
0·500	0										−2000

All entries in the main body of the table to be multiplied by 10^{-5}. *The last column was obtained by using the function* $\frac{1}{50} \cos 2\pi t$.

Now complete the table, working on the same lines as before.
****

The table is shown on page 357.

If you examine the figures in the table you have just completed, you will notice one or two interesting points.   Firstly, it is some little time before the effect of the motion imparted to the end A reaches along to the other end of the string.   Secondly, the displacement of some points of the string does, at certain times, exceed the maximum displacement of the end A.   Thirdly, the point of maximum displacement varies with time, thus giving rise to a wave effect.

Table for FRAME 36A.

t \ x	0	0·1	0·2	0·3	0·4	0·5	0·6	0·7	0·8	0·9	1·0
0	0	200	400	600	800	1000	1200	1400	1600	1800	2000
0·025	0	200	400	600	800	1000	1200	1400	1600	1800	1975
0·050	0	200	400	600	800	1000	1200	1400	1600	1775	1902
0·075	0	200	400	600	800	1000	1200	1400	1575	1702	1782
0·100	0	200	400	600	800	1000	1200	1375	1502	1582	1618
0·125	0	200	400	600	800	1000	1175	1302	1382	1418	1414
0·150	0	200	400	600	800	975	1102	1182	1218	1214	1176
0·175	0	200	400	600	775	902	982	1018	1014	976	908
0·200	0	200	400	575	702	782	818	814	776	708	618
0·225	0	200	375	502	582	618	614	576	508	418	313
0·250	0	175	302	382	418	414	376	308	218	113	0
0·275	0	102	182	218	214	176	108	18	-87	-200	-313
0·300	0	7	18	14	-24	-92	-182	-287	-400	-513	-618
0·325	0	-84	-161	-224	-292	-382	-487	-600	-713	-818	-908
0·350	0	-168	-326	-467	-582	-687	-800	-913	-1018	-1108	-1176
0·375	0	-242	-474	-684	-862	-1000	-1113	-1218	-1308	-1376	-1414
0·400	0	-306	-600	-869	-1102	-1288	-1418	-1508	-1576	-1614	-1618
0·425	0	-358	-701	-1018	-1295	-1520	-1683	-1776	-1814	-1818	-1782
0·450	0	-395	-776	-1127	-1436	-1690	-1878	-1989	-2018	-1982	-1902
0·475	0	-418	-821	-1194	-1522	-1794	-1996	-2120	-2157	-2102	-1975
0·500	0	-426	-836	-1216	-1552	-1828	-2036	-2164	-2204	-2150	-2000

FRAME 38

A flow diagram for the solution of $\dfrac{\partial^2 y}{\partial t^2} = c^2 \dfrac{\partial^2 y}{\partial x^2}$ with $\dfrac{\partial y}{\partial t} = 0$ at $t = 0$, y being fixed at the two ends for all values of t, as in FRAME 23, is shown on page 358.

FRAME 39

The Equation $\dfrac{\partial \theta}{\partial t} = c^2 \dfrac{\partial^2 \theta}{\partial x^2}$

This is the equation governing heat flow in a rod, the sides of which are insulated. θ, the temperature, is a function of distance and time and once again increments of h and k will be taken in x and t respectively.

For the numerical solution of this equation, the same type of expression as before is used for the second derivative and $\{\theta(x, t + k) - \theta(x, t)\}/k$ for the first. What will be the equation corresponding to (7.1) in this case?

****

Flow diagram for FRAME 38.

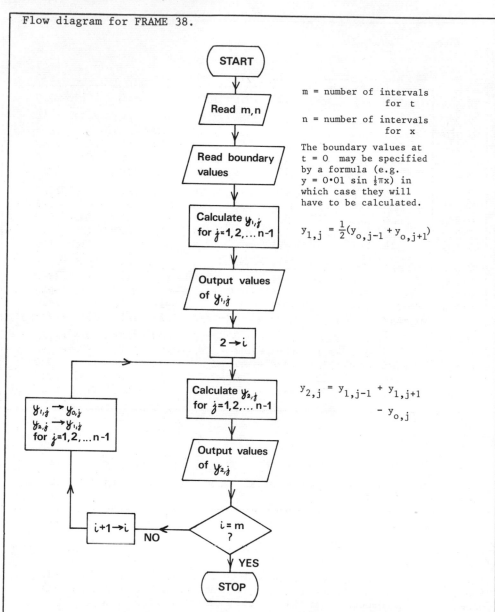

m = number of intervals for t

n = number of intervals for x

The boundary values at $t = 0$ may be specified by a formula (e.g. $y = 0\cdot01 \sin \frac{1}{2}\pi x$) in which case they will have to be calculated.

$$y_{1,j} = \frac{1}{2}(y_{0,j-1} + y_{0,j+1})$$

$$y_{2,j} = y_{1,j-1} + y_{1,j+1} - y_{0,j}$$

This flow diagram shows how it is not necessary to use double subscripts in programming, if each row of values is printed out as it is calculated. Only three rows of values need be kept in store at any one time.   In FORTRAN these could be denoted by the variables YO(J), Y1(J) and Y2(J).

A slight modification to this diagram will accomodate the type of situation in FRAME 33, where an end is not fixed.

(Programs for the solution of the equation $\dfrac{\partial^2 y}{\partial t^2} = c^2 \dfrac{\partial^2 y}{\partial x^2}$ can be found in references (2), (4), (5).)

$$\frac{\theta(x,\ t+k) - \theta(x,\ t)}{k} = c^2 \frac{\theta(x+h,\ t) - 2\theta(x,\ t) + \theta(x-h,\ t)}{h^2}$$

Upon rearrangement, this can be written as

$$\theta(x,\ t+k) = \frac{kc^2}{h^2}\theta(x-h,\ t) + (1 - \frac{2kc^2}{h^2})\theta(x,\ t) + \frac{kc^2}{h^2}\theta(x+h,\ t)$$

or $\quad \theta(x,\ t+k) = \alpha\theta(x-h,\ t) + (1-2\alpha)\theta(x,\ t) + \alpha\theta(x+h,\ t) \quad (40.1)$

if $\frac{kc^2}{h^2}$ is denoted by α.

Now it can be shown theoretically that α, which will be positive, should be less than or equal to $\frac{1}{2}$, otherwise large errors can build up in the solution. We shall not go into this theory here, but if you are interested, you will find it treated in Chapter 8 of "Numerical Solution of Differential Equations" by W.E. Milne (Wiley). Furthermore it can be shown that, from the point of view of accuracy, the best value to take for α is 1/6. When this is done, (40.1) becomes

$$\theta(x,\ t+k) = \frac{1}{6}\left[\theta(x-h,\ t) + 4\theta(x,\ t) + \theta(x+h,\ t)\right] \quad (40.2)$$

where h and k are connected by the relation $\frac{kc^2}{h^2} = \frac{1}{6}$.

To see how this works, suppose an insulated rod of length 1 m is heated by maintaining one end at 100°C and the other at 0°C. Sufficient time is allowed to elapse so that steady state conditions prevail. The temperature of the hot end is then suddenly reduced to zero. Assuming that the value of c^2 is 1/60, the temperature will be found at subsequent times at 10 cm intervals along the length of the rod.

In order to use equation (40.2), what will be the required value for k? (Remember all lengths should be expressed in metres.)
****

41A

$$k = \frac{\alpha h^2}{c^2} = \frac{1}{6} \times \frac{0 \cdot 01}{1/60} = 0 \cdot 1$$

When steady state conditions prevail, the temperature at 10 cm intervals along the rod will be 0(10)100. (The temperature will increase uniformly from one end to the other.) It will be assumed that the drop in temperature at the hot end occurs at time t = 0. So that immediately before t = 0, θ at this end is 100, while immediately after, it is zero. This change does however occupy a finite time and so it would appear reasonable to take the temperature of this end as being 50°C at t = 0 and zero immediately after. Our initial table of values will thus be:

t \ x	0	0·1	0·2	0·3	0·4	0·5	0·6	0·7	0·8	0·9	1
0	0	10	20	30	40	50	60	70	80	90	50
0·1	0										0
0·2	0										0

We can now start filling in the values in row t = 0·1.   When x = 0·1,
using (40.2) gives   θ = (0 + 4 × 10 + 20)/6 = 10.   Similarly, when
x = 0·2,   θ = (10 + 4 × 20 + 30)/6 = 20.   What will be the remainder of
the values to be filled in in this row?   (Work to 1 decimal place where
necessary.)   **

30,   40,   50,   60,   70,   80,   81·7.

Continuing in the same way for subsequent rows, the temperature during the
first two and a half seconds is as given below:

t \ x	0	0·1	0·2	0·3	0·4	0·5	0·6	0·7	0·8	0·9	1
0	0	10	20	30	40	50	60	70	80	90	50
0·1	0	10	20	30	40	50	60	70	80	81·7	0
0·2	0	10	20	30	40	50	60	70	78·6	67·8	0
0·3	0	10	20	30	40	50	60	69·8	75·4	58·3	0
0·4	0	10	20	30	40	50	60	69·1	71·6	51·4	0
0·5	0	10	20	30	40	50	59·8	68·0	67·8	46·2	0
0·6	0	10	20	30	40	50	59·5	66·6	64·2	42·1	0
0·7	0	10	20	30	40	49·9	59·1	65·0	60·9	38·8	0
0·8	0	10	20	30	40	49·8	58·6	63·3	57·9	36·0	0
0·9	0	10	20	30	40	49·6	57·9	61·6	55·2	33·6	0
1·0	0	10	20	30	39·9	49·4	57·1	59·9	52·7	31·6	0
1·1	0	10	20	30	39·8	49·1	56·3	58·2	50·4	29·8	0
1·2	0	10	20	30	39·7	48·8	55·4	56·6	48·3	28·3	0
1·3	0	10	20	30	39·6	48·4	54·5	55·0	46·4	26·9	0
1·4	0	10	20	29·9	39·5	48·0	53·6	53·5	44·6	25·7	0
1·5	0	10	20	29·8	39·3	47·5	52·6	52·0	42·9	24·6	0
1·6	0	10	20	29·8	39·1	47·0	51·6	50·6	41·4	23·6	0
1·7	0	10	20	29·7	38·9	46·4	50·7	49·2	40·0	22·6	0
1·8	0	10	20	29·6	38·6	45·9	49·7	47·9	38·6	21·7	0
1·9	0	10	19·9	29·5	38·3	45·3	48·8	46·6	37·3	20·9	0
2·0	0	10	19·8	29·4	38·0	44·7	47·8	45·4	36·1	20·2	0
2·1	0	10	19·8	29·2	37·7	44·1	46·9	44.2	35·0	19·5	0
2·2	0	10	19·7	29·0	37·4	43·5	46·0	43·1	34·0	18·8	0
2·3	0	10	19·6	28·8	37·0	42·9	45·1	42·1	33·0	18·2	0
2·4	0	9·9	19·5	28·6	36·6	42·3	44·2	41·1	32·0	17·6	0
2·5	0	9·8	19·4	28·4	36·2	41·7	43·4	40·1	31·1	17·1	0

If you examine the figures in the table on the previous page, you will
see that they show very clearly exactly the effect that you would expect.

FRAME 45

A flow diagram for the solution of $\dfrac{\partial y}{\partial t} = a \dfrac{\partial^2 y}{\partial x^2}$ for a one-dimensional

flow system with constant end conditions is shown on page 362.   x and t
are such that $0 \leqslant x \leqslant L, \quad 0 \leqslant t \leqslant T.$

FRAME 46

A slight modification to the procedure in the last problem is illustrated
if the conditions are changed as follows:

Let the end   x = 0   be insulated and suppose heat is lost from the rod by
fluid flowing across the other end.   As it does so, the fluid will, of
course, become heated.   Let the initial temperature of the rod be $1000^{\circ}C$
and the temperature of the fluid flowing be $50^{\circ}C$ before contact with the
rod.

The difference here from the last problem is that the end being cooled
does not have its temperature suddenly reduced to a final value but that
the change in its temperature takes place gradually.   To incorporate this

gradual change, it is necessary to introduce the temperature gradient $\dfrac{\partial \theta}{\partial x}$
at this end.

FRAME 47

Now it can be shown that, under these circumstances, the rate of loss of
heat from a surface is given by each of two different expressions:

i) $HA(\theta - \theta_c)$ where H is the heat transfer coefficient, A the area of
     the surface, θ the temperature of the surface and θ_c the temperature
     of the fluid approaching the surface,

ii) $-KA \dfrac{\partial \theta}{\partial x}$ where K is the thermal conductivity of the rod.

Equating these two expressions, $\dfrac{\partial \theta}{\partial x} = - \dfrac{H}{K} (\theta - \theta_c)$ (47.1)

for x = L,   L being the length of the rod, which is here 1 m.

FRAME 48

To make use of this equation, an artifice somewhat similar to that in
FRAME 27 is adopted.   We imagine that the rod is increased in length by h
(0·1 m here) and that the temperature at x = 1·1 varies in such a way
that the temperature at the actual end is always what it should be under
the conditions of the actual problem.   The temperature at the actual end
is now calculated using (40.2).   Using L and h rather than 1 and 0·1
(in order to get a more general result), (40.2) gives, for x = L

$$\theta(L, t + k) = \frac{1}{6}\left[\theta(L - h, t) + 4\theta(L, t) + \theta(L + h, t)\right] \qquad (48.1)$$

and, at each stage, the unknown quantity on the R.H.S. is $\theta(L + h, t)$.

Flow diagram for FRAME 45.

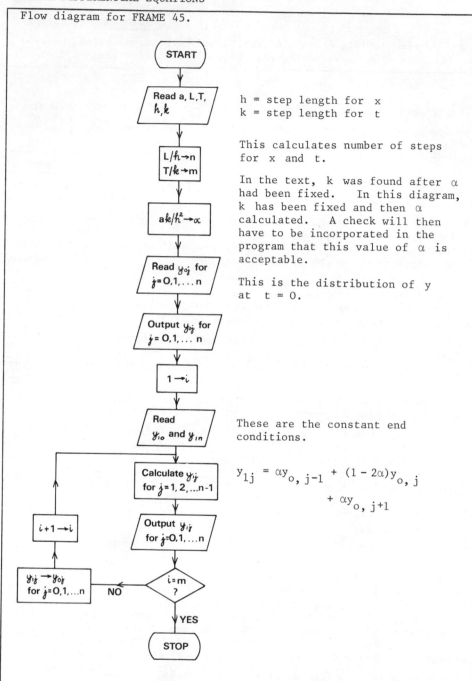

h = step length for x
k = step length for t

This calculates number of steps for x and t.

In the text, k was found after α had been fixed. In this diagram, k has been fixed and then α calculated. A check will then have to be incorporated in the program that this value of α is acceptable.

This is the distribution of y at t = 0.

These are the constant end conditions.

$$y_{1j} = \alpha y_{o,\,j-1} + (1 - 2\alpha)y_{o,\,j} + \alpha y_{o,\,j+1}$$

Note that, in programming, double subscripts are not necessary. Only two rows of y's need to be in store at a time. In FORTRAN these could be denoted by Y0(J) and Y1(J).

This unknown quantity is found by making use of (47.1). $\frac{\partial \theta}{\partial x}$, for x = L

and time t, can be approximated to by $\frac{\theta(L + h, t) - \theta(L - h, t)}{2h}$, and

so (47.1) becomes, for x = L,

$$\frac{\theta(L + h, t) - \theta(L - h, t)}{2h} = -\frac{H}{K}\left[\theta(L, t) - \theta_c\right] \qquad (49.1)$$

Incorporate this equation into (48.1) so that the result is an equation for $\theta(L, t + k)$ which does not involve $\theta(L + h, t)$.

****

49A

$$\theta(L, t + k) = \frac{1}{6}\left[\theta(L - h, t) - \frac{2hH}{K}\{\theta(L, t) - \theta_c\} + 4\theta(L, t) + \theta(L - h, t)\right]$$

$$= \frac{1}{6}\left[2\theta(L - h, t) + (4 - \frac{2hH}{K})\theta(L, t) + \frac{2hH}{K}\theta_c\right] \qquad (49A.1)$$

Now our value of h is 0·1, and so also is that of k (from 41A) and taking, as reasonably typical values for H and K, 3000 W/m² °C and 200 W/m °C respectively, (49A.1) becomes, with L = 1 and θ_c = 50,

$$\theta(1, t + 0·1) = \frac{1}{6}\left[2\theta(0·9, t) + \theta(1, t) + 150\right] \qquad (50.1)$$

(40.2) is now used for values of x = 0·1(0·1)0·9 and this last equation for x = 1.

We are still left with the problem of finding successive values of θ when x = 0. This is done by a process similar to that just adopted for x = L. The result, however, is somewhat simpler since at the end x = 0,

$\frac{\partial \theta}{\partial x} = 0$ as the rod is insulated there. By suitably modifying (48.1) and

(49.1) can you work out the formula, similar to (50.1), for $\theta(0, t + 0·1)$?

****

51A

(48.1) *will be replaced by*

$$\theta(0, t + k) = \frac{1}{6}\left[\theta(-h, t) + 4\theta(0, t) + \theta(h, t)\right] \qquad (51A.1)$$

(49.1) *will be replaced by*

$$\frac{\theta(h, t) - \theta(-h, t)}{2h} = 0, \quad \text{giving} \quad \theta(-h, t) = \theta(h, t)$$

(51A.1) *then becomes* $\theta(0, t + k) = \frac{1}{6}\left[4\theta(0, t) + 2\theta(h, t)\right]$

$$\text{i.e.} \quad \theta(0, t + 0·1) = \frac{1}{3}\left[2\theta(0, t) + \theta(0·1, t)\right]$$

PARTIAL DIFFERENTIAL EQUATIONS

The three equations

$$\theta(0,\ t + 0\cdot1) = \frac{1}{3}\left[2\theta(0,\ t) + \theta(0\cdot1,\ t)\right]$$

$$\theta(x,\ t + 0\cdot1) = \frac{1}{6}\left[\theta(x - 0\cdot1,\ t) + 4\theta(x,\ t) + \theta(x + 0\cdot1,\ t)\right],$$

$$x = 0\cdot1(0\cdot1)0\cdot9$$

$$\theta(1,\ t + 0\cdot1) = \frac{1}{6}\left[2\theta(0\cdot9,\ t) + \theta(1,\ t) + 150\right]$$

are now used for each successive stage in time.

Starting with $\theta(x,\ 0) = 1000$ for all x, work out the table corresponding to that in FRAME 44 for this new problem. Continue the solution as far as $t = 1\cdot0$, giving the values of θ to the nearest integer.

****

t \ x	0	0·1	0·2	0·3	0·4	0·5	0·6	0·7	0·8	0·9	1·0
0	1000	1000	1000	1000	1000	1000	1000	1000	1000	1000	1000
0·1	1000	1000	1000	1000	1000	1000	1000	1000	1000	1000	525
0·2	1000	1000	1000	1000	1000	1000	1000	1000	1000	921	446
0·3	1000	1000	1000	1000	1000	1000	1000	1000	987	855	406
0·4	1000	1000	1000	1000	1000	1000	1000	998	967	802	378
0·5	1000	1000	1000	1000	1000	1000	1000	993	945	759	355
0·6	1000	1000	1000	1000	1000	1000	999	986	922	723	337
0·7	1000	1000	1000	1000	1000	1000	997	978	900	692	322
0·8	1000	1000	1000	1000	1000	1000	994	968	878	665	309
0·9	1000	1000	1000	1000	1000	999	991	957	858	641	298
1·0	1000	1000	1000	1000	1000	998	987	946	838	620	288

A flow diagram for calculating the temperature distribution, at successive times, in a one-dimensional heat flow system with one end $(x = 0)$ insulated and a heat gain (or loss) at the other end $(x = L)$, is shown on page 365.

The differential equations are:

$$\frac{\partial\theta}{\partial t} = a\frac{\partial^2\theta}{\partial x^2},\qquad \frac{\partial\theta}{\partial x} = 0\ \text{ at }\ x = 0,\qquad \frac{\partial\theta}{\partial x} = -\frac{H}{K}(\theta - \theta_c)\ \text{ at }\ x = L.$$

Errors in the Methods

It is not proposed to go into a detailed analysis of the errors in the solutions of the problems treated in this programme. These will be due to two causes. The first arises due to the replacement of the partial derivatives by expressions such as $\left[y(x+h,\ t) - 2y(x,\ t) + y(x-h,\ t)\right]/h^2$, which are themselves only approximations. The effect of this is to replace the partial differential equation by what is known as a DIFFERENCE EQUATION.

Flow diagram for FRAME 53.

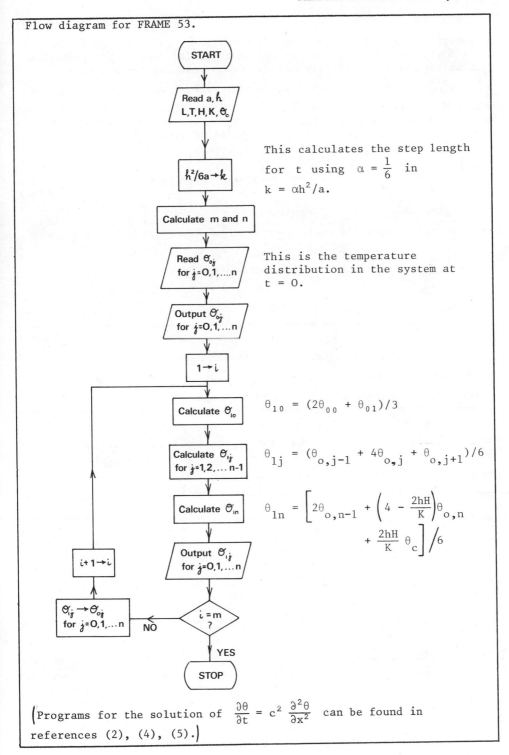

START

Read a, h
L,T,H,K, θ_c

$h^2/6a \rightarrow k$

This calculates the step length
for t using $\alpha = \dfrac{1}{6}$ in
$k = \alpha h^2/a$.

Calculate m and n

Read θ_{oj}
for j=0,1,....n

This is the temperature
distribution in the system at
t = 0.

Output θ_{oj}
for j=0,1,...n

$1 \rightarrow i$

Calculate θ_{io}

$$\theta_{10} = (2\theta_{00} + \theta_{01})/3$$

Calculate θ_{ij}
for j=1,2,... n-1

$$\theta_{1j} = (\theta_{o,j-1} + 4\theta_{o,j} + \theta_{o,j+1})/6$$

Calculate θ_{in}

$$\theta_{1n} = \left[2\theta_{o,n-1} + \left(4 - \frac{2hH}{K}\right)\theta_{o,n} + \frac{2hH}{K}\theta_c\right]\Big/6$$

Output θ_{ij}
for j=0,1,...n

$i+1 \rightarrow i$

$\theta_{ij} \rightarrow \theta_{oj}$
for j=0,1,...n   NO

$i = m$?

YES

STOP

$\Big($Programs for the solution of $\dfrac{\partial \theta}{\partial t} = c^2 \dfrac{\partial^2 \theta}{\partial x^2}$ can be found in
references (2), (4), (5).$\Big)$

In general, one would not expect the solution of the difference equation to have the same solution as the original partial differential equation that it replaces.   Unusually though, in the case of the equation $\frac{\partial^2 y}{\partial t^2} = c^2 \frac{\partial^2 y}{\partial x^2}$, the two solutions are the same if h and k satisfy the relation h = ck, and you will remember that this was the relation used. So that, as well as simplifying the difference equation, this is also the best value to take from the accuracy point of view.

Verify that   y = f(x + ct) + g(x − ct)   satisfies
y(x, t+k) = y(x − h, t) + y(x + h, t) − y(x, t − k)   if h = ck.
****

54A

On substitution in the difference equation

$L.H.S. = f\{x + c(t + k)\} + g\{x - c(t + k)\}$

$R.H.S. = f(x - h + ct) + g(x - h - ct) + f(x + h + ct) + g(x + h - ct)$
$$- f\{x + c(t - k)\} - g\{x - c(t - k)\}$$
$$= f(x - ck + ct) + g(x - ck - ct) + f(x + ck + ct) + g(x + ck - ct)$$
$$- f(x + ct - ck) - g(x - ct + ck) = L.H.S.$$

FRAME 55

Now you have already verified in 2A that this solution satisfies $\frac{\partial^2 y}{\partial t^2} = c^2 \frac{\partial^2 y}{\partial x^2}$ and so both difference equation and partial differential equation have the same solution under the condition h = ck.

FRAME 56

The other source of error is that due to round−off in the arithmetic. Even this type of error can lead to serious trouble in the case of the equation $\frac{\partial \theta}{\partial t} = c^2 \frac{\partial^2 \theta}{\partial x^2}$ if α is taken greater that $\frac{1}{2}$.

FRAME 57

In this programme, we have concentrated on showing you a basic method for dealing with certain partial differential equations.   There are other methods and also refinements to the basic method that has been considered, but we shall not go into these here.   Neither shall we go into the modifications that have to be made to deal with other partial differential equations.   You should, however, now have the background to enable you to pursue further your study of the subject if you find this to be necessary.

FRAME 58

Miscellaneous Examples

In this frame a collection of miscellaneous examples is given for you to try.   Answers are supplied in FRAME 59, together with such working as is considered helpful.

1. A thin square plate has its sides maintained at temperatures -10°, 50°, 100° and 75°C as shown.

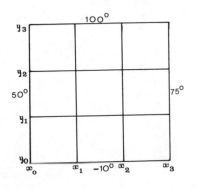

Assuming that the steady-state temperature distribution $u(x, y)$ satisfies Laplace's equation

$$\frac{\partial^2 u}{\partial x^2} + \frac{\partial^2 u}{\partial y^2} = 0, \quad \text{show that this can}$$

be approximated by the difference equation

$$u_{i+1,j} + u_{i-1,j} + u_{i,j+1}$$
$$+ u_{i,j-1} - 4u_{i,j} = 0$$

over any uniform mesh spanning the plate.

Hence determine approximate values for u at the four internal nodal points of the given uniform mesh.   (C.E.I.)

2. A square plate is bounded by the lines $x = \pm 2$; $y = \pm 2$. The temperature θ of the plate obeys Laplace's equation $\dfrac{\partial^2 \theta}{\partial x^2} + \dfrac{\partial^2 \theta}{\partial y^2} = 0$. The boundary temperatures are:-

on the line $y = 2$, $\theta = -80$
on the lines $x = \pm 2$, $\theta = -40y$
on the line $y = -2$, $\theta = 240 + 80x$ for $-2 \leqslant x \leqslant 0$
                             $\theta = 240 - 80x$ for $0 \leqslant x \leqslant 2$.

Find the temperatures at the nine points (x_i, y_j) where $i = -1, 0, 1$, $j = -1, 0, 1$.   (L.U.)

3. If $f(x, y)$ satisfies $\dfrac{\partial^2 f}{\partial x^2} + \dfrac{\partial^2 f}{\partial y^2} = 0$ when f has the value 100 on AB, CD and EF, and is zero on PQ and RS, find the approximate value of f at O.

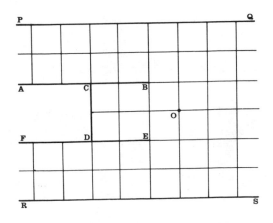

(The lines PQ and RS extend to infinity in both directions;   CA and DF extend to infinity on the left.)   (L.U.)

4. If the zero on the R.H.S. of Laplace's equation is replaced by $f(x, y)$, $\dfrac{\partial^2 V}{\partial x^2} + \dfrac{\partial^2 V}{\partial y^2} = f(x, y)$ results and this is known as Poisson's equation.

What will be the formula corresponding to (7.2) for this equation?

5.

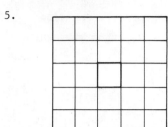

The Poisson equation $\dfrac{\partial^2 \phi}{\partial x^2} + \dfrac{\partial^2 \phi}{\partial y^2} + 2 = 0$ occurs in certain torsion problems. Solve this equation for ϕ at the mesh points in the hollow bar shown. It is divided into unit squares and the value of ϕ at all points on both the outer and inner boundaries is zero.

6. Within the region OABCDE, ϕ satisfies the equation
$$\frac{\partial^2 \phi}{\partial x^2} + \frac{\partial^2 \phi}{\partial y^2} = 4x(5 - y).$$

Determine, to the nearest integer, the values of ϕ at the lattice points of the unit lattice shown, given the values shown for ϕ at the lattice points on the boundary. (L.U.)

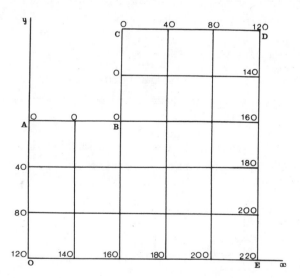

7. Find the solution of $\dfrac{\partial^2 y}{\partial t^2} = 25 \dfrac{\partial^2 y}{\partial x^2}$, subject to the conditions:

$y(0, t) = 0$, $y(1 \cdot 0, t) = 0$, $\dfrac{\partial y(x, 0)}{\partial t} = 0$, $y(x, 0)$ given by the table

x	0	0·1	0·2	0·3	0·4	0·5	0·6	0·7	0·8	0·9	1·0
y	0	0·01	0·02	0·03	0·04	0	−0·04	−0·03	−0·02	−0·01	0

For the solution, choose a suitable value of k and proceed
sufficiently far to get a complete picture of the motion.

8. Find the solution of $\dfrac{\partial^2 y}{\partial t^2} = 100 \dfrac{\partial^2 y}{\partial x^2}$ subject to the conditions:

$y(0, t) = 0$, $y(2, t) = 0$, $y(x, 0)$ and $\dfrac{\partial y(x, 0)}{\partial t}$ as given by the
table

x	0	0·25	0·50	0·75	1	1·25	1·50	1·75	2
y	0	0·0025	0·0050	0·0075	0·0100	0·0075	0·0050	0·0025	0
$\dfrac{\partial y}{\partial t}$	0	−0·012	−0·024	−0·036	−0·048	−0·060	−0·040	−0·020	0

Continue the solution sufficiently far time-wise for one oscillation
to be evident.

9. Find the solution of the heat conduction equation $\dfrac{\partial \theta}{\partial t} = c^2 \dfrac{\partial^2 \theta}{\partial x^2}$ for a
rod, with $c^2 = \dfrac{1}{96}$, taking the length as 1 and dividing it into
eight intervals.   At time $t = 0$, $\theta = 100^{\circ}$ throughout but both ends
are then suddenly reduced to 0°.   Assuming that the units are
compatible, proceed with the solution as far as $t = 3$.

10. A metre rod is lagged so that the flow of heat along it satisfies the
equation $\dfrac{\partial \theta}{\partial t} = c^2 \dfrac{\partial^2 \theta}{\partial x^2}$,   c being $\dfrac{1}{\sqrt{96}}$.   At time $t = 0$, $\theta = 20^{\circ}C$
throughout the rod.   Fluid at $400^{\circ}C$ is then made to flow past the end
at $x = 0$ and fluid at $0^{\circ}C$ to flow past the other end.   How long is
it before the temperature of the centre of the rod exceeds $35^{\circ}C$?   The
values of H and K are $3000 \text{ W/m}^2 {}^{\circ}C$ and $200 \text{ W/m} {}^{\circ}C$ respectively
and h is to be taken as 0·125 m.

Answers to Miscellaneous Examples

1. 64·4   70·6
    36·9   43·1

2. −32·2   −28·6   −32·2
    20·0   30·0   20·0
    82·2   108·6   82·2

3. 59.   You will have found it necessary to include more points in your
    calculation as the iterations proceeded.

4. $V(x, y) = \dfrac{1}{4}\{V(x + h, y) + V(x, y + h) + V(x - h, y) + V(x, y - h)$
$$- h^2 f(x, y)\} \qquad\qquad (59.1)$$

FRAME 59   (continued)

5.

From symmetry, there will be only two distinct values of ϕ, ϕ_A and ϕ_B, to find.

Applying (59.1) gives

$$\phi_A = \frac{1}{4}(2\phi_B + 2), \qquad \phi_B = \frac{1}{4}(\phi_A + \phi_B + 2)$$

from which $\phi_A = \phi_B = 1.$

6.

		39	83
		43	90
39	43	66	106
83	90	106	137

7.   k = 0·02

t \ x	0	0·1	0·2	0·3	0·4	0·5
0	0	1	2	3	4	0
0·02	0	1	2	3	1·5	0
0·04	0	1	2	0·5	-1	0
0·06	0	1	-0·5	-2	-1	0
0·08	0	-1·5	-3	-2	-1	0
0·10	0	-4	-3	-2	-1	0
0·12	0	-1·5	-3	-2	-1	0
0·14	0	1	-0·5	-2	-1	0
0·16	0	1	2	0·5	-1	0
0·18	0	1	2	3	1·5	0
0·20	0	1	2	3	4	0

From x = 0·5 to x = 1, for all values of t, f(1 − x, t) = −f(x, t).

All entries in the body of the table to be multiplied by 0·01.

After t = 0·20, the values repeat so that, for all values of x, f(x, t + 0·20) = f(x, t).

8. k = 0·025

t \ x	0	0·25	0·50	0·75	1·00	1·25	1·50	1·75	2
0	0	25	50	75	100	75	50	25	0
0·025	0	22	44	66	63	60	40	20	0
0·050	0	19	38	32	26	28	30	15	0
0·075	0	16	7	-2	-3	-4	3	10	0
0·100	0	-12	-24	-28	-32	-28	-24	-12	0
0·125	0	-40	-47	-54	-53	-52	-43	-34	0
0·150	0	-35	-70	-72	-74	-68	-62	-31	0
0·175	0	-30	-60	-90	-87	-84	-56	-28	0
0·200	0	-25	-50	-75	-100	-75	-50	-25	0
0·225	0	-20	-40	-60	-63	-66	-44	-22	0
0·250	0	-15	-30	-28	-26	-32	-38	-19	0
0·275	0	-10	-3	4	3	2	-7	-16	0
0·300	0	12	24	28	32	28	24	12	0
0·325	0	34	43	52	53	54	47	40	0
0·350	0	31	62	68	74	72	70	35	0
0·375	0	28	56	84	87	90	60	30	0
0·400	0	25	50	75	100	75	50	25	0

9. $k = \dfrac{1}{6} \times \dfrac{1}{64} \times 96 = 0\cdot25$

t \ x	0	0·125	0·250	0·375	0·500	0·625	0·750	0·875	1
0	50	100	100	100	100				
0·25	0	92	100	100	100				
0·50	0	78	99	100	100				
0·75	0	68	96	100	100				
1·00	0	61	92	99	100				
1·25	0	56	88	98	100				
1·50	0	52	84	97	99				
1·75	0	49	81	95	98				
2·00	0	46	78	93	97				
2·25	0	44	75	91	96				
2·50	0	42	72	89	94				
2·75	0	40	70	87	92				
3·00	0	38	68	85	90				

From symmetry,
$\theta(1 - x, t) = \theta(x, t).$

10. $k = \dfrac{1}{6} \times \dfrac{1}{64} \times 96 = 0\cdot25$

The equations corresponding to those in FRAME 52 are

$$\theta(0, t + k) = \frac{1}{6}\Big\{2\theta(0\cdot125, t) + \frac{1}{4}\,\theta(0, t) + 1500\Big\};$$

$$\theta(x, t + k) = \frac{1}{6}\Big\{\theta(x - 0\cdot125, t) + 4\theta(x, t) + \theta(x + 0\cdot125, t)\Big\},$$

$$x = 0\cdot125(0\cdot125)0\cdot875; \qquad \theta(1, t + k) = \frac{1}{6}\Big\{2\theta(0\cdot875, t) + \frac{1}{4}\theta(1, t)\Big\}$$

3·5 s

1. During the interval of free fall immediately after a parachutist jumped from a plane, his velocity, v m/s at time t s was given by $\frac{dv}{dt} = 9 \cdot 81 - 0 \cdot 0068v^{3/2}$ with $v = 0$ when $t = 0$.

   Working to 5 decimal places, use the Runge-Kutta method to find his velocity after $0 \cdot 4$ s, taking two step lengths.

2. Use a Taylor series to find y when $x = 0 \cdot 1$, $0 \cdot 2$, $0 \cdot 3$ for the equation $\frac{dy}{dx} = \frac{1}{6} x^3 + y^2$ given that $y = 0 \cdot 5$ when $x = 0$. Then obtain y when $x = 0 \cdot 4$ by Milne's method and when $x = 0 \cdot 5$ by Milne's modified method. Give your result to 4 decimal places.

3. When solving beam problems, the formula for the radius of curvature,

   i.e., $\dfrac{\left\{1 + \left(\dfrac{dy}{dx}\right)^2\right\}^{3/2}}{\dfrac{d^2y}{dx^2}}$, is often taken approximately as $1 \Big/ \dfrac{d^2y}{dx^2}$, on

   the assumption that $\left(\dfrac{dy}{dx}\right)^2$ is small in comparison with 1. The differential equation for the beam is then taken as $EI \dfrac{d^2y}{dx^2} = M$. If this approximation cannot be made then the differential equation for

   the beam becomes $EI \dfrac{d^2y}{dx^2} = M\left\{1 + \left(\dfrac{dy}{dx}\right)^2\right\}^{3/2}$

   In a particular instance $\dfrac{M}{EI} = 10x^2$, so that $\dfrac{d^2y}{dx^2} = 10x^2\left\{1 + \left(\dfrac{dy}{dx}\right)^2\right\}^{3/2}$

   Find the slope and deflection of the beam when $x = 0 \cdot 2$ and $0 \cdot 4$, taking $0 \cdot 2$ as the step length. Assume that the slope and deflection are both zero when $x = 0$.

4. A rectangular plate 20 cm $\times$ 25 cm is divided into squares of edge 5 cm. One long edge is maintained at $100°C$ and the other three edges at $20°C$. Working to the nearest integer, find the steady state temperatures at the grid corners.

5.

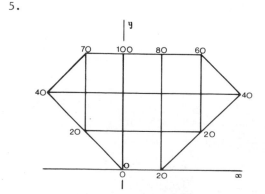

Working to the nearest integer, find θ at the internal points indicated, given the values shown in the diagram. θ satisfies the equation

$$\frac{\partial^2\theta}{\partial x^2} + \frac{\partial^2\theta}{\partial y^2} = 0$$

Note: The method used here is the same as for a rectangular plate. The outer edges need not all be parallel to the axes provided however that they pass through a series of nodal points at which θ is known.

6. Re-work question 5 if θ now satisfies the equation

   $\dfrac{\partial^2\theta}{\partial x^2} + \dfrac{\partial^2\theta}{\partial y^2} = 32(x^2 - y)$, the sides of the squares each being $\frac{1}{2}$ unit.

7. A rod, $0\cdot5\,$m long, is heated, one end being at 0°C and the other at 1000°C. Sufficient time is allowed to elapse for the temperature throughout the rod to become steady. The cold end is then insulated and the hot end simultaneously cooled by allowing liquid initially at 20°C to flow past it. The temperature of the rod satisfies the equation $\dfrac{\partial\theta}{\partial t} = \dfrac{1}{96}\dfrac{\partial^2\theta}{\partial x^2}$, x being measured from the cold end and t from the commencement of the liquid flow.

   Taking, as in the text, $\dfrac{kc^2}{h^2} = \dfrac{1}{6}$, $H = 3000\ \text{W/m}^2\,^{\circ}\text{C}$, $K = 200\ \text{W/m}\,^{\circ}\text{C}$, find the temperature distribution along the rod at $t = 0\cdot2\,$s if h is taken as $0\cdot05\,$m. Work throughout to the nearest integer.

8. The ends of a stretched string are fixed at two points, $(0, 0)$ and $(1, 0)$. The string is then made to take up the shape $y = x/25$ for $0 \leqslant x \leqslant 0\cdot5$, $y = (1 - x)/25$ for $0\cdot5 \leqslant x \leqslant 1$. When in this position, each point on the string is given a velocity towards its equilibrium position equal in magnitude to its displacement. Measuring time from this instant and taking $h = 0\cdot1$, follow through the calculation for y until one half of a cycle has been completed. The differential equation for the particular motion is $\dfrac{\partial^2 y}{\partial t^2} = 16\dfrac{\partial^2 y}{\partial x^2}$.

9. Given the starting values

x	0	$0\cdot2$	$0\cdot4$	$0\cdot6$
y	0	$0\cdot0002$	$0\cdot0032$	$0\cdot0162$

   for the equation $\dfrac{dy}{dx} = \dfrac{1}{2}x^3 + y^2$, use Hamming's method to find y when $x = 0\cdot8$ and his modified method to find y when $x = 1\cdot0$.

10. By using the shooting method with $h = 0\cdot5$ find, to 2 decimal places, the value of y' when $x = 0$ that it is necessary to take for the equation $y'' - xy' + x^2 y = 0\cdot5x$ to pass through the points $(0, 0)$ and $(1, 0)$.

11. Using the simultaneous algebraic equations method, with $h = 0\cdot25$ find, to 3 decimal places, the values of y when $x = 0\cdot25, 0\cdot5, 0\cdot75$ for the equation and conditions given in question 10.

12. By means of a Taylor series, find the value of y when $x = 0\cdot5$ for the equation $y'' + y^2 = 0$ given that $y = 0$ and $y' = 1$ when $x = 0$. Then use a predictor-corrector method to find y when $x = 1\cdot0$ and $1\cdot5$.

13. Given the table

θ (degrees)	70	69	68	67
ω (rad/s)	$2\cdot083$	$2\cdot181$	$2\cdot275$	$2\cdot365$
$(g\sin\theta)/\omega\ell$	$5\cdot810$	$5\cdot513$	$5\cdot249$	$5\cdot013$

and that $g/\ell = 12 \cdot 88 \text{ s}^{-2}$, solve the pendulum equation
$$\frac{d\omega}{d\theta} + \frac{g}{\ell}\,\frac{\sin\theta}{\omega} = 0 \quad \text{for } \omega \text{ at } \theta = 65^{\circ} \text{ correct to three decimal places,}$$
using a high-order finite-difference method.

Hint: The Adams-Bashforth formulae for integrating the equation
$\dfrac{d\omega}{d\theta} = f(\theta, \omega)$ over the interval $\Delta\theta(\theta_r, \theta_{r+1})$ are

$$\omega^P_{r+1} = \omega_r + \Delta\theta(1 + \tfrac{1}{2}\nabla + \tfrac{5}{12}\nabla^2 + \tfrac{3}{8}\nabla^3 + \ldots\ldots)f_r$$

$$\omega^c_{r+1} = \omega_r + \Delta\theta(1 - \tfrac{1}{2}\nabla - \tfrac{1}{12}\nabla^2 - \tfrac{1}{24}\nabla^3 \ldots\ldots)f^P_{r+1} \qquad \text{(C.E.I.)}$$

ANSWERS

1. $3 \cdot 9156$ m/s.

2. $0 \cdot 5263$, $0 \cdot 5555$, $0 \cdot 5898$, $0 \cdot 6262$, $0 \cdot 6697$.

3. At $x = 0 \cdot 2$, slope $= 0 \cdot 0267$, deflection $= 0 \cdot 0013$
   At $x = 0 \cdot 4$, slope $= 0 \cdot 191\,90$, deflection $= 0 \cdot 020\,16$.

4. If the top edge of the rectangle is that kept at 100°, the temperatures are:

56	67	67	56
37	45	45	37
27	30	30	27

5. | 46 | 56 | 52 | 43 |
   |----|----|----|----|
   | | 26 | 30 | |

6. | 49 | 60 | 56 | 44 |
   |----|----|----|----|
   | | 29 | 32 | |

7. $k = 0 \cdot 04$

   $$\theta(0, t + k) = \tfrac{1}{3}\{2\theta(0, t) + \theta(0 \cdot 05, t)\}$$

   $$\theta(x, t + k) = \tfrac{1}{6}\{\theta(x - 0 \cdot 05, t) + 4\theta(x, t) + \theta(x + 0 \cdot 05, t)\},$$

   $x = 0 \cdot 05(0 \cdot 05)0 \cdot 45$,

   $$\theta(0 \cdot 5, t + k) = \tfrac{1}{6}\{2\theta(0 \cdot 45, t) + 2 \cdot 5\theta(0 \cdot 5, t) + 30\}$$

98	129	206	300	400	500	600	696	762	718	466

8. $k = 0 \cdot 025$

   $$y = (x, k) = \tfrac{1}{2}\{y(x - h, 0) + y(x + h, 0) - 2ky(x, 0)\}, \quad \text{replacing}$$
   $\dfrac{\partial y(x, 0)}{\partial t}$ in (28A.1) page 353 by $-y(x, 0)$.

$$y(x, t + k) = y(x - h, t) + y(x + h, t) - y(x, t - k), \quad t = nk, \quad n > 1$$

t \ x	0	0·1	0·2	0·3	0·4	0·5	0·6	0·7	0·8	0·9	1·0
0	0	40	80	120	160	200	160	120	80	40	0
0·025	0	39	78	117	156	155	156	117	78	39	0
0·050	0	38	76	114	112	112	112	114	76	38	0
0·075	0	37	74	71	70	69	70	71	74	37	0
0·100	0	36	32	30	28	28	28	30	32	36	0
0·125	0	-5	-8	-11	-12	-13	-12	-11	-8	-5	0
0·150	0	-44	-48	-50	-52	-52	-52	-50	-48	-44	0
0·175	0	-43	-86	-89	-90	-91	-90	-89	-86	-43	0
0·200	0	-42	-84	-126	-128	-128	-128	-126	-84	-42	0
0·225	0	-41	-82	-123	-164	-165	-164	-123	-82	-41	0
0·250	0	-40	-80	-120	-160	-200	-160	-120	-80	-40	0

Each reading in the body of the table to be multiplied by 10^{-4}

9. 0·0514, 0·1268.

10. -0·10

11. -0·020, -0·033, -0·030.

12. 0·4949, 0·9211, 1·1384.

13. 2·532.

REFERENCES

Specimen programs in FORTRAN can be found in

1.   Engineering Mathematics, A.C. Bajpai, L.R. Mustoe and D. Walker, John Wiley 1974

2.   Applied Numerical Analysis, C.F. Gerald, Addison-Wesley 1970

3.   Computer Applications of Numerical Methods, S.S. Kuo, Addison-Wesley 1972

4.   Applied Numerical Methods for Digital Computation with FORTRAN, M.L. James, G.M. Smith and J.C. Wolford, International Textbook Company, 1967

5.   Numerical Methods in FORTRAN, J.M. McCormick and M.G. Salvadori, Prentice-Hall 1964

6.   Application of Computers to Engineering Analysis, J.R. Wolberg, McGraw-Hill 1971

7.   Numerical Methods with FORTRAN IV Case Studies, W.S. Dorn and D.D. McCracken, John Wiley 1972

8.   Introduction to Numerical Methods and FORTRAN Programming, T.R. McCalla, John Wiley 1967

Specimen programs in ALGOL can be found in

9.   Advanced Mathematics for Technical Students Part Three, H.V. Lowry, H.A. Hayden and K.E. Pitman, Longman 1971

****

Fortran and Algol
A.C. Bajpai, H.W. Pakes, R.J. Clarke, J.M. Doubleday, T.J. Stevens, Wiley 1972.

Mathematics for Engineers and Scientists, Volume 1, A.C. Bajpai, I.M. Calus, J.A. Fairley, Wiley 1972.

Mathematics for Engineers and Scientists, Volume 2, A.C. Bajpai, I.M. Calus, J.A. Fairley, D. Walker, Wiley 1972.

INDEX

The two figures following each entry are respectively the page number and frame number.

DATE DUE

~~MAR 25 1996~~		
8/1/96		